Intermediate A
A Graphing Ap

Graphing Calculator Resource Manual

DEMANA/WAITS/CLEMENS/GREENE

Intermediate Algebra
A Graphing Approach

Graphing Calculator Resource Manual

Ray Barton
David Lawrence
Joan McCarter

ADDISON-WESLEY PUBLISHING COMPANY, INC.
Reading, Massachusetts • Menlo Park, California
New York • Don Mills, Ontario • Wokingham, England • Amsterdam
Bonn • Sydney • Singapore • Tokyo • Madrid • San Juan • Milan • Paris

ISBN 0-201-59055-7

1 2 3 4 5 6 7 8 9 10 CRW 97969594

CONTENTS

Intermediate Algebra
A Graphing Approach

Graphing Calculator Resource Manual

Preliminary Chapter Getting Started Using the TI-81

Section 1 The Keyboard

The keys on the TI-81 are organized by color and position. See Fig. 1. The grey keys on the lower part of the keyboard are used to enter numbers. The dark blue keys on the lower right-hand side of the keyboard are used to perform arithmetic. The dark blue keys with arrows in the upper right-hand corner of the keyboard are used to move the cursor on the screen; they are called the cursor-movement keys. The black keys on the top row just under the screen are used to draw graphs of equations. The other black keys are used for scientific functions and menus.

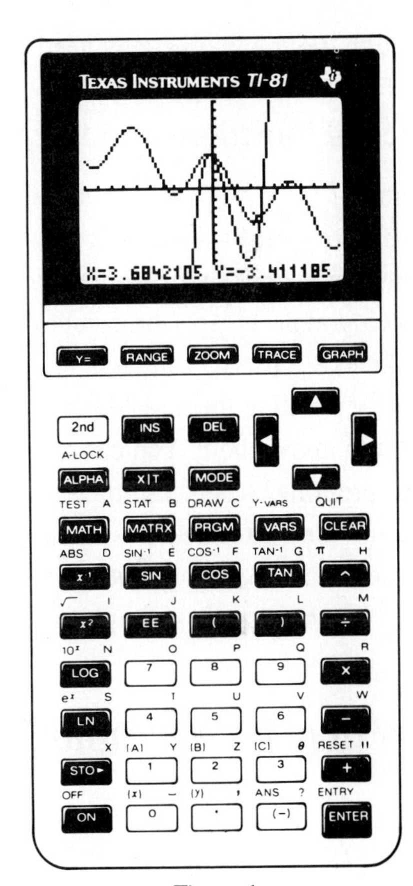

Figure 1

The [ON] and [OFF] Keys and the Automatic Off

The black key in the lower left-hand corner of the keyboard is used to turn the calculator on. Notice the key is labeled 'ON' and the word 'OFF' appears in light blue lettering above the key. Press [ON] to turn the calculator on. If you press [2nd] (the light blue key in the upper left-hand corner of the keyboard) and then press [ON], you will turn the calculator off. Pressing [2nd] lets you use the feature printed in light blue above the key. If you leave the TI-81 on for about five minutes without pressing any keys, it will turn itself off. This automatic off feature saves the batteries.

The Cursor and Adjusting the Contrast

After you turn the calculator on, you should see a blinking rectangle displayed on the screen. This rectangle is called the cursor. The cursor indicates where the first number or letter you enter will be placed on the screen. If the screen seems too light, you can darken the display by pressing [2nd] and then holding [▲] until the contrast looks right. If the screen seems too dark, you can lighten the display by pressing [2nd] and holding [▼] until you have the contrast you want. ([▲] and [▼] are dark blue keys in the upper right-hand corner of the keyboard.) If you are not able to make the display dark enough, your batteries may need to be replaced.

Second Functions and the Alpha Key

Many of the keys have light blue labels above them. These blue labels are called second functions. You can access the second function above a key by first pressing [2nd] and then pressing the key. When you press [2nd] the cursor changes to " ■.' While the cursor looks like " ■.' you are able to access the second function above a key by pressing the key. Many of the keys also have a grey letter or symbol above them. You can access the grey letter above a key by first pressing [ALPHA] and then pressing the key. When you press [ALPHA] the cursor changes to ' ■ '. We say the cursor is in the Alpha mode. The role of [2nd] and [ALPHA] is similar to the shift key on a typewriter or computer keyboard.

Section 2 Keyboarding Instructions

When you see instructions to press a certain key, the instructions will include a box containing the label on the key. If the key is a number key, however, the label will not be enclosed in a box. For example, if you follow the directions to turn on the calculator and press

[CLEAR] **2** [+] **4** [ENTER],

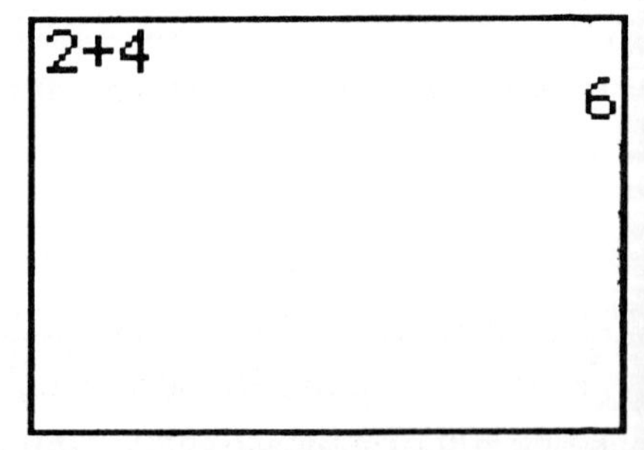

Figure 2

your screen should look like Fig. 2. When you are instructed to access a second function, the directions will include [2nd] followed by a box containing the second function label above the key rather than the label on the key. For example, notice the second function of [x^2] is $\sqrt{\ }$ ([x^2] is midway down the first column). The directions to calculate the square root of two would be to press

[CLEAR] [2nd] [$\sqrt{\ }$] **2** [ENTER].

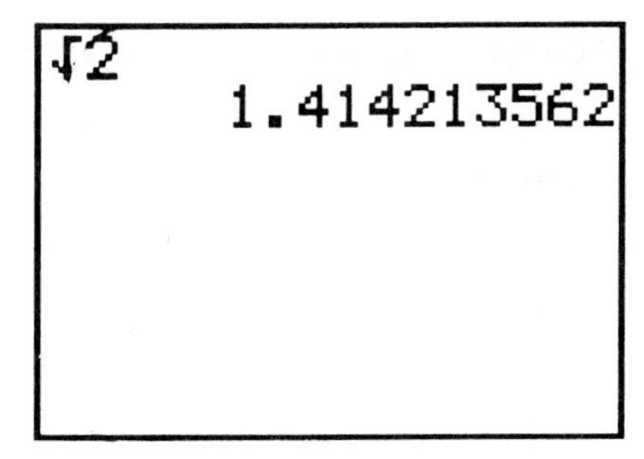

Figure 3

See Fig. 3. The reference to the [$\sqrt{\ }$] key in these directions means to press the key with x^2 on the key and the $\sqrt{\ }$ above the key. The instructions to access a letter will include [ALPHA] followed by a box containing the letter above the key.

Resetting Your Grapher

Figure 4

If you wish to reset the TI-81 to its factory settings and erase all previously entered data and programs, turn on the calculator, press [2nd], and then press [+]. (The [+] key is a dark blue key in the lower right-hand corner of the keyboard. Notice the word 'RESET' in light blue lettering above the [+] key.) After you press [2nd] and [+] you should see the **RESET** menu on the screen. The RESET menu should look like Fig. 4 except the numbers on the menu may be different on your calculator if data or programs have previously been entered. Now press **2** to select **Reset,** the second item in the RESET menu. The display should show the message **Mem cleared.** Press [CLEAR] to clear the screen. ([CLEAR] is a black key on the fourth row, fifth column of the keyboard.) You may need to adjust the contrast after you reset your TI-81.

Section 3 The Home Screen, CLEAR, Menus, and QUIT

When you first turn on the TI-81, you see the **Home screen**. The calculations you performed in Keyboarding Instructions, Section 2, were done on the Home screen. The [CLEAR] key you used before each calculation clears the Home screen of previous entries. If you make a mistake in a calculation, you can start over by pressing [CLEAR]. You can access **Menu screens** from the Home screen. If you reset your TI-81 earlier, you used the RESET menu. You can return from a Menu screen to the Home screen by pressing

[2nd] [QUIT].

(QUIT is the second function of CLEAR.) Sometimes you are automatically returned to the Home screen when you make a selection from the Menu. Press [MATH] to see the **MATH** menu. [MATH] is on the first column, fourth row of the keyboard. See Fig. 5. There are seven features visible on this menu, and the arrow by the seventh feature indicates there are more features available. To see the eighth feature, press and hold [▼]. Press

when you are ready to return to the Home screen.

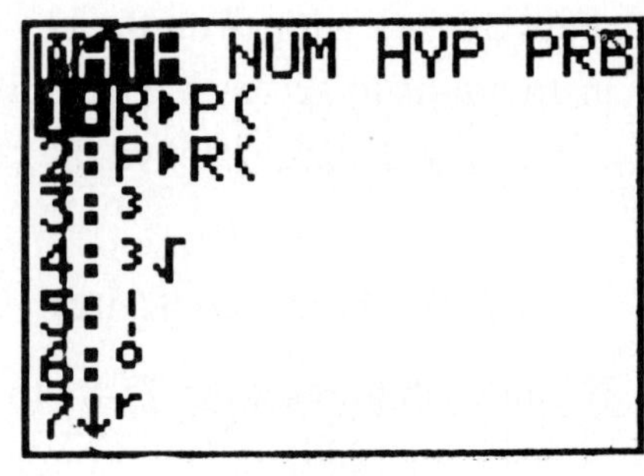

Figure 5

You can calculate 7^3 by using a feature in the MATH menu. From the Home screen press

[CLEAR] **7** [MATH] [3:³] [ENTER].

The result should be 343. See Fig. 6. The notation [3:³] in the keystroke instructions means to press 3 after you are in the MATH menu. Pressing **3** selects the third option in the menu. Did you notice that you were automatically returned to the Home screen after you selected the third option from the MATH menu? Selecting this option also caused the exponent of three to be printed on the Home screen.

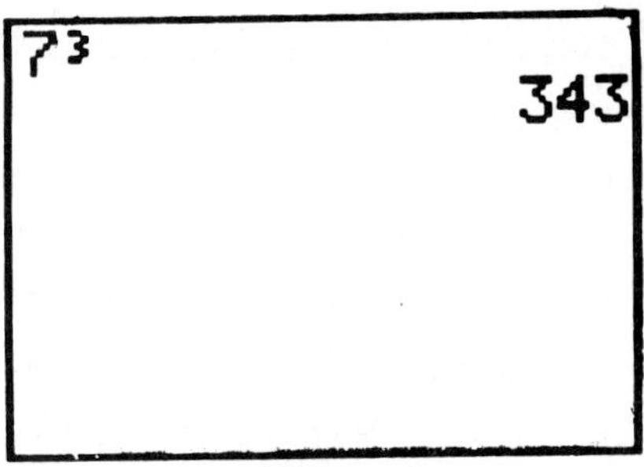

Figure 6

The MODE Menu

From the Home screen press [MODE]. Your screen should look like Fig. 7. Each of the left-hand entries should be highlighted. We will discuss the meaning of these entries in later sections. If any of the left-hand entries are not highlighted on your calculator, use the cursor-movement keys to move the cursor to the unhighlighted item and then press [ENTER] to highlight that item. Press [CLEAR] to return to the Home screen.

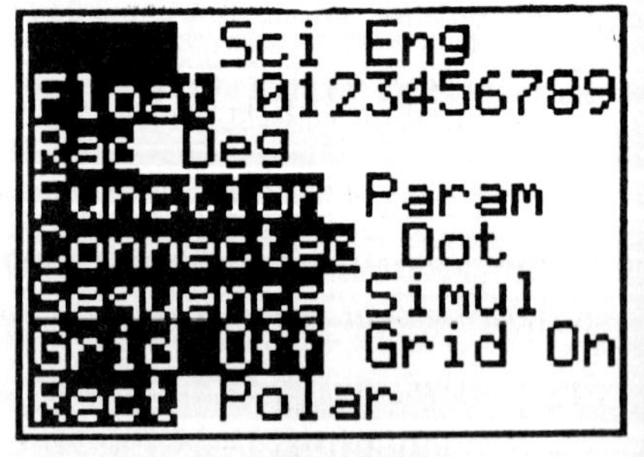

Figure 7

Section 4 The Basic Operations Keys:

The four dark blue keys

on the lower right-hand side of the keyboard are used for addition, subtraction, multiplication, and division. Press the following keys to perform the calculation 19 minus 27 plus 23:

[CLEAR] **19** [−] **27** [+] **23** [ENTER].

See Fig. 8. Notice you don't see the result of the calculation until you press [ENTER]. Then you see the expression on the left-hand side of the Home screen and the result of the calculation on the right-hand side of the next line. Press the following keys to multiply 18 by 145:

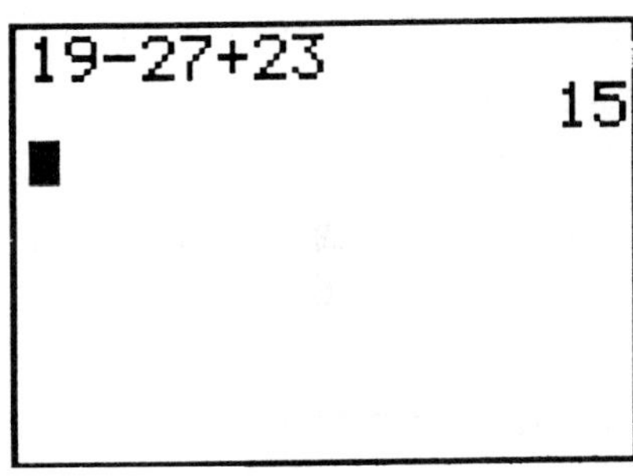

Figure 8

[CLEAR] **18** [X] **145** [ENTER].

The result is 2610. See Fig. 9. Notice that pressing [X] produces the '*' symbol on the Home screen. Both '*' and '×' mean multiplication. Now press

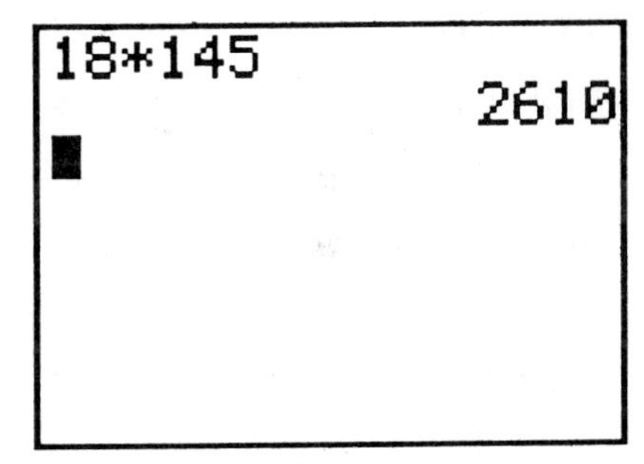

Figure 9

8 [÷] **156** [ENTER]

to divide 8 by 156. See Fig. 10. Notice that pressing [÷] produces the '/' symbol on the Home screen. Both '÷' and '/' mean division. Also notice the previous calculation '18*145' remains on the screen. This is because you did not clear the screen by pressing [CLEAR] before entering the division problem.

The Subtraction Key [−] versus the Opposite Key [(−)]

There are two keys that appear to be minus signs on the TI-81. One is the dark blue subtraction key [−] mentioned earlier. The other is the grey key [(−)] on the bottom row. This grey key is the "opposite of" or negative key. For example, –3 is denoted on the Home screen

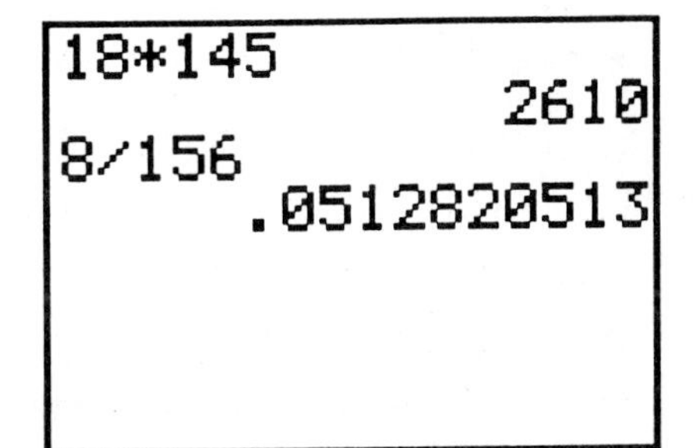

Figure 10

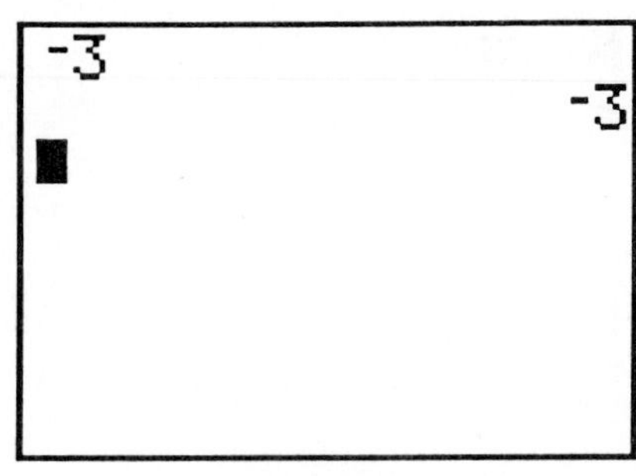

Figure 11

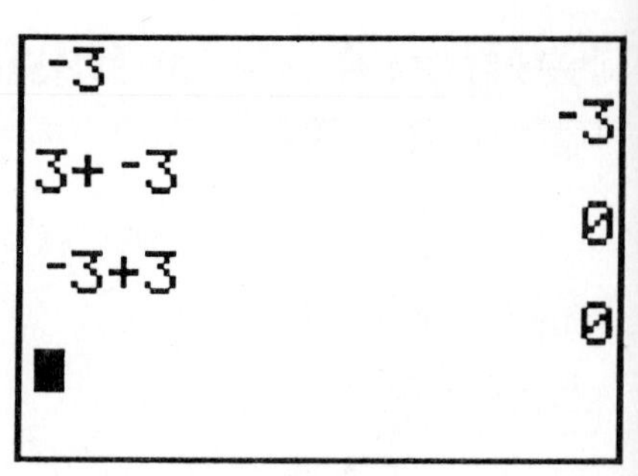

Figure 12

by pressing

[CLEAR] [(–)] **3** [ENTER].

See Fig. 11. When the number –3 is added to 3, the result is zero. To see this on your calculator, press

3 [+] [(–)] **3** [ENTER] or [(–)] **3** [+] **3** [ENTER].

See Fig. 12. The negative key is different from the subtraction key. You can see the difference if you press [CLEAR] [(–)] [–]. The negative key screen symbol is smaller in length and higher than the subtraction symbol. See Fig. 13. Press [CLEAR] to clear the Home screen for the next example.

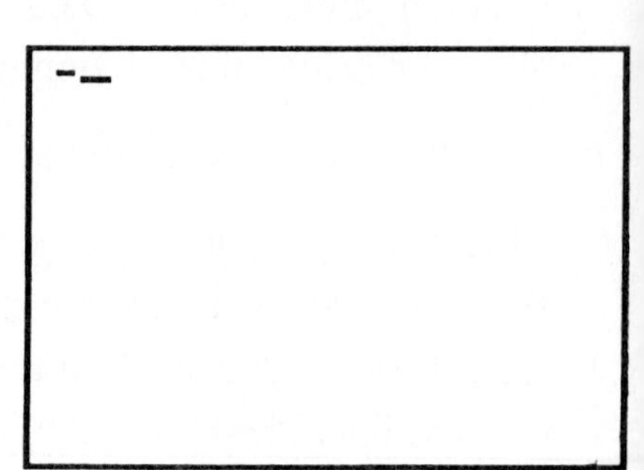

Figure 13

Section 5 Error Messages

If you want to see the result of subtracting 3 from 7 on the calculator, you must press

7 [–] **3** [ENTER].

If you press

7 [(–)] **3** [ENTER],

the TI-81 will give you a syntax error message. See Figs. 14 and 15. If you type **1** from the Error Message screen, the TI-81 will return you to the expression on the Home screen. The cursor blinking on the negative symbol tells you this is the mistake. You should have

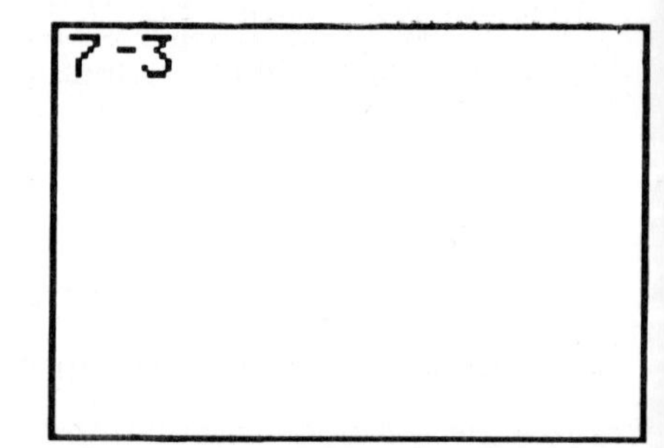

Figure 14

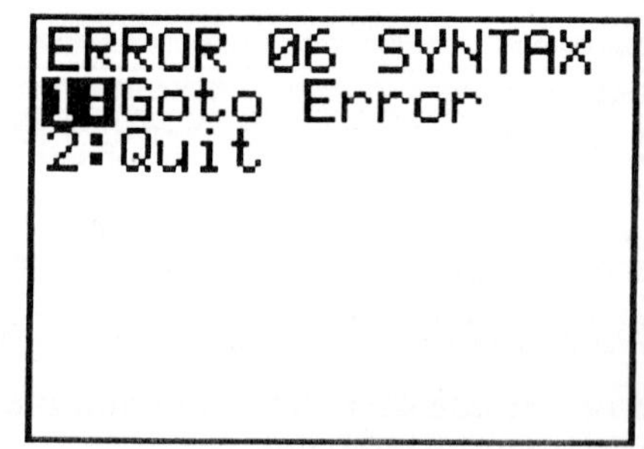

Figure 15

used the subtract key instead of the negative key. If you want to calculate 2(–3), press

[CLEAR] **2** [×] [(–)] **3** [ENTER] or **2** [(] [(–)] **3** [)] [ENTER].

If you press

2 [(–)] **3** [ENTER],

the TI-81 will return a syntax error message. Press **2** from the Error Message screen to return to the Home screen. Why do you see a MATH error when you press

2 [÷] **0** ?

Section 6 How to Input with the [x^2] and [x^{-1}] Keys

You can calculate squares with [x^2]. This key is in the first column. For example, the square of 5 can be calculated by pressing

[CLEAR] **5** [x^2] [ENTER].

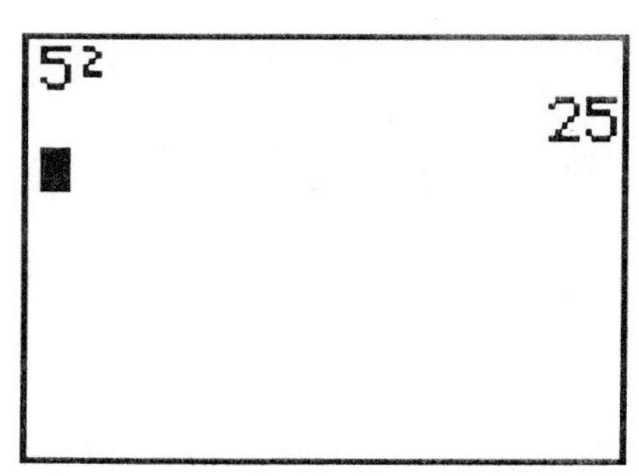

Figure 16

See Fig. 16. Notice when you pressed [x^2], the ² symbol was printed on the Home screen next to the 5, but the answer did not appear until you pressed [ENTER]. You can calculate reciprocals with [x^{-1}]. This key is also in the first column. The reciprocal of 4 can be determined by pressing

4 [x^{-1}] [ENTER].

See Fig. 17.

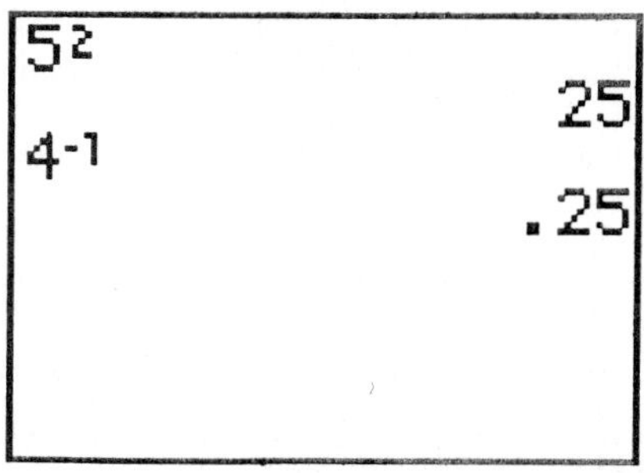

Figure 17

Section 7 Insert and Delete

You can edit expressions with the insert key [INS] and the delete key [DEL]. These keys are in the second row. Press

36 [+] **10**,

but don't press [ENTER].

The cursor should be at the end of the expression. If you would like to change the expression to 316 + 10 by inserting a 1 between the 3 and the 6, then press the left cursor-movement key until the cursor is blinking on the '6' in '36.' Now press [INS]. Notice the cursor changes from a blinking rectangle to a blinking underline. Press **1**. The expression changes to '316 + 10.' Press [ENTER] to evaluate this new expression. The result is 326. See Fig. 18.

Now we will use the [DEL] to delete a digit. Press

123 [−] **10**

but don't press [ENTER]. The cursor should be blinking at the end of this new expression. Press the left cursor-movement key until the cursor is blinking on the '2' in '123.' Now press [DEL]. The expression has been changed to '13 – 10.' Press [ENTER] to evaluate this new expression. See Fig. 18.

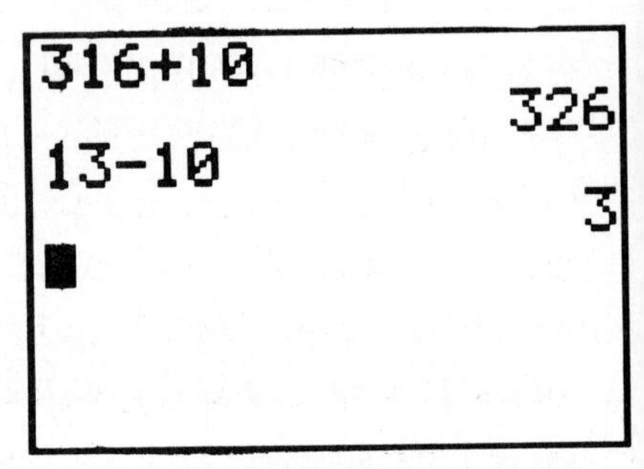

Figure 18

Chapter 1 Numerical Mathematics and the Graphing Calculator

Section 1.1 Real Numbers and the Graphing Calculator

Many of the computational skills you will need in this chapter are explained in the Preliminary Chapter. Be sure to read that chapter before beginning Chapter 1.

Changing Decimals to Fractions

The TI-81 does not have the built-in ability to change decimals to fractions.

Decimal Representation of Fractions

Example 1: Convert the following fractions to decimals: $^{11}/_{25}$, $^{143}/_{999}$, $^{1}/_{17}$.

Press

[CLEAR] **11** [÷] **25** [ENTER] **143** [÷] **999** [ENTER] **1** [÷] **17** [ENTER].

See Fig. 1.1. The decimal representation of $^{11}/_{25}$ is exact because it is a terminating decimal. The TI-81 decimal representations of $^{143}/_{999}$ and $^{1}/_{17}$ are approximations because the correct representations of these fractions are repeating decimals. Although the screen shows an approximation to $^{143}/_{999}$, you can probably tell what the repeating pattern is from looking at your calculator result. Although $^{1}/_{17}$ is a repeating decimal, you cannot see the repeating block in the first 10 digits. The TI-81 can only show 10 decimals in a number. If the decimal repeats or terminates after the first 10 digits, you may not be able to determine the exact form from the TI-81 representation.

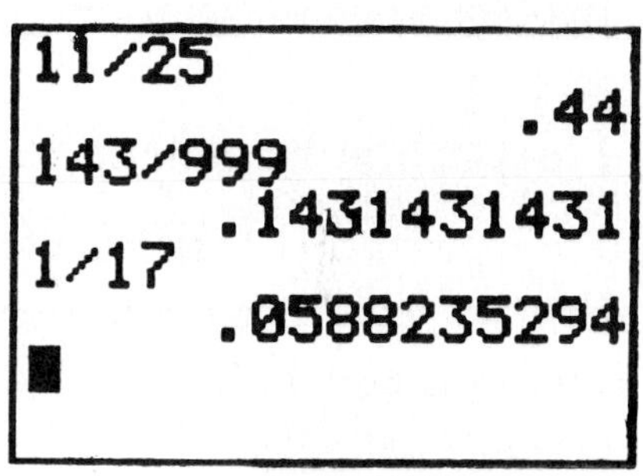

Figure 1.1

Floating Decimal Display versus Fixed-Point Display

When your calculator displayed the first 10 digits of the decimal representation of $^{143}/_{999}$ in the previous example, it was in Floating Decimal mode. If you only want to see the first two digits, you can switch to Fixed Decimal mode. Do this by pressing

[MODE] [▼] [▶] [▶] [▶] [ENTER].

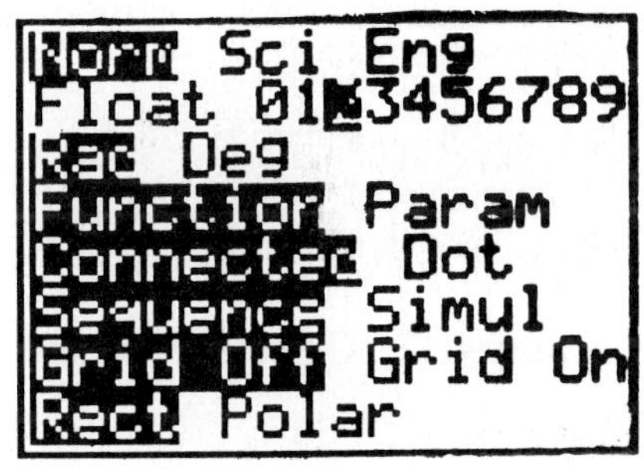

Figure 1.2

See Fig. 1.2. Press [CLEAR] to exit the Mode screen. Now press

[CLEAR] **143** [÷] **999** [ENTER].

See Fig. 1.3. You should see the first two decimals in the answer: .14. Press [MODE] and change the number of decimals in Fixed mode to four by using the cursor-movement keys to highlight the '4' on the second line of the Mode screen, then press [ENTER]. Press [CLEAR] to return to the Home screen and again divide 143 by 999. See Fig.

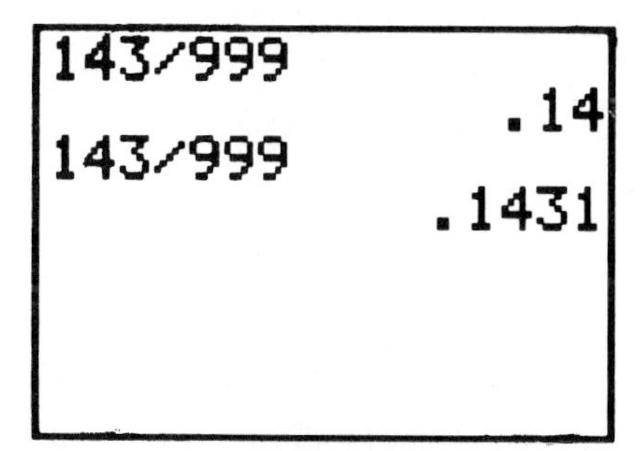

Figure 1.3

1.3. You should see the first four digits of the decimal representation: .1431. Now switch back to Floating Decimal mode by pressing [MODE], highlighting 'Float,' and pressing [ENTER] and [CLEAR].

The Square Root Key and Use of [2nd] Keys

The use of these keys was described in the Preliminary Chapter. The decimal representation of $\sqrt{2}$ shown in Fig. 3 is an approximation because $\sqrt{2}$ is a nonterminating, nonrepeating decimal; however, it is a very accurate approximation.

Powers and Exponentiation

The Preliminary Chapter described how to square and cube numbers. Now we will see how to raise numbers to other powers. You can calculate powers with the power key on the fifth row, fifth column of the keyboard, [^].

Example 2: Compute 3^5.

To see 3^5 press

[CLEAR] **3** [^] **5** [ENTER].

See Fig. 1.4.

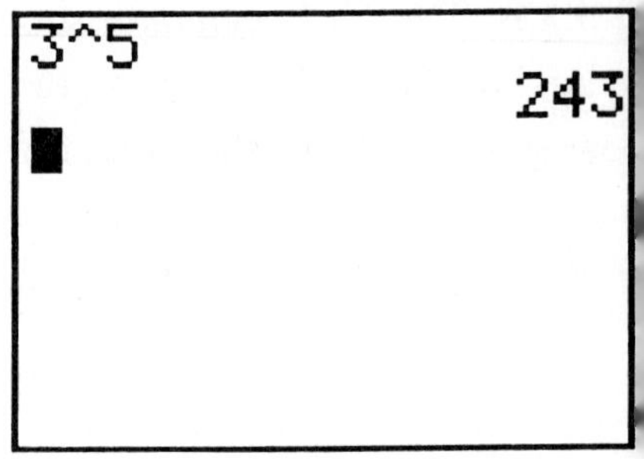

Figure 1.4

Example 3: Calculate -3^2 and $(-3)^2$.

Press

[CLEAR] [(-)] **3** [^] **2** [ENTER]

to see -3^2, and press

[(] [(-)] **3** [)] [^] **2** [ENTER]

to see $(-3)^2$.

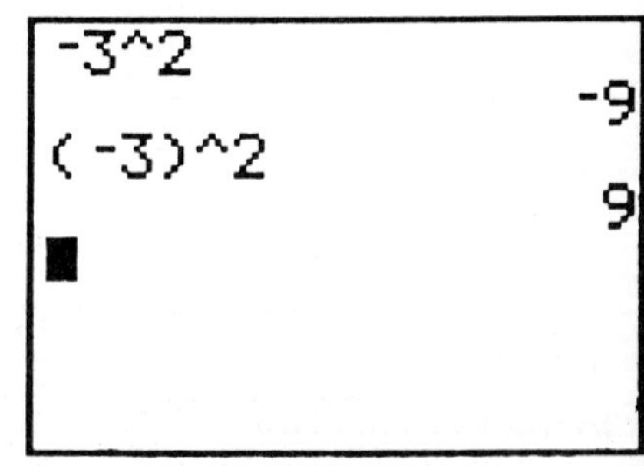

Figure 1.5

See Fig. 1.5. Notice the results of the two calculations are different. In the first calculation, the TI-81 squared three first and then negated. In the second calculation, the parentheses forced the TI-81 to negate three first and then square negative three. You can square numbers with the [x^2] or the [^] keys.

Example 4: Find $(2/5)^3$.

You must use parentheses to raise a fraction to a power. Press

[CLEAR] [(] **2** [÷] **5** [)] [^] **3** [ENTER].

The answer is .064.

Order of Operations

Example 5: Calculate $2 + 3 \times 4^2 - 6 \div 2$.

Press

[CLEAR] **2** [+] **3** [×] **4** [^] **2** [−] **6** [÷] **2** [ENTER].

See Fig. 1.6. The TI-81 does powers first, which reduces the expression to

$2 + 3 \times 16 - 6 \div 2$.

Next the calculator does multiplication and division, which simplify the expression further to $2 + 48 - 3$. Last it does addition and subtraction to yield a result of 47.

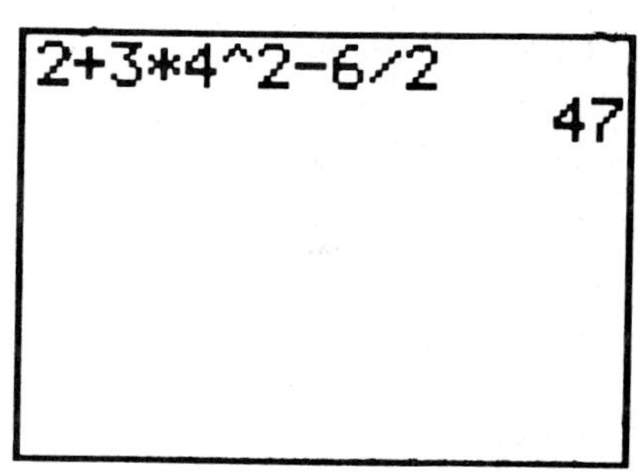

Figure 1.6

Using the TEST Menu

To see how the TI-81 interprets $3 < 5$, we will need to use the TEST menu. TEST is the second function of MATH. Before we evaluate the expression $3 < 5$, let's look at the TEST menu by pressing

[2nd] [TEST].

See Fig. 1.7. This menu contains equality and inequality symbols. We will indicate which symbol to select in our keystroke instructions by writing the menu number followed by the symbol. Press [CLEAR] to return to the Home screen and [CLEAR] again to clear the Home

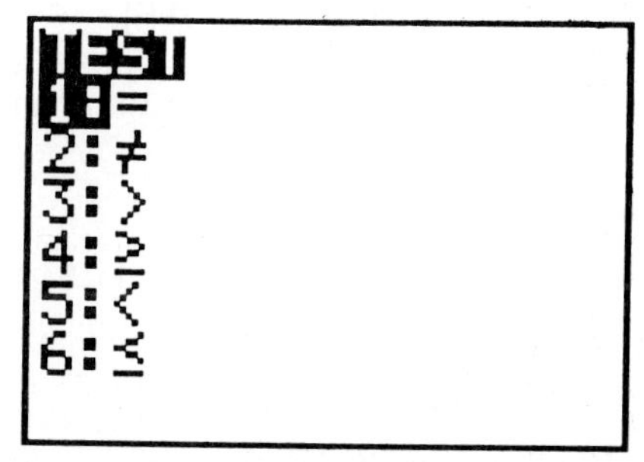

Figure 1.7

screen. Now evaluate the expression 3 < 5 by pressing

3 [2nd] [TEST] [5:<] **5** [ENTER].

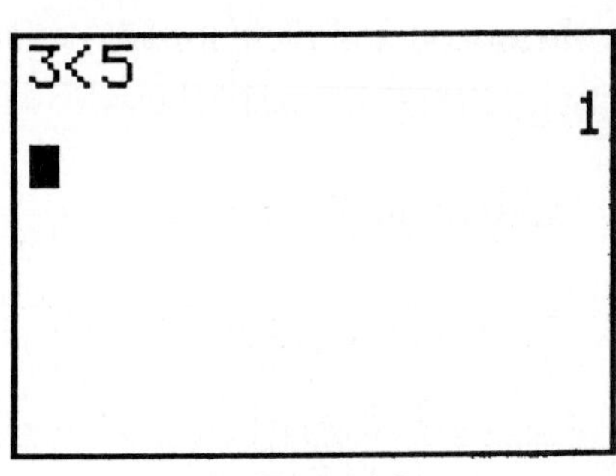

Figure 1.8

See Fig. 1.8. The TI-81 returns the value 1 because the expression is true. To see the result of 3 > 5 press

3 [2nd] [TEST] [3:>] **5** [ENTER].

The calculator shows a result of 0 because this expression is false. You should also see a result of 0 when you press

[(-)] **1** [2nd] [TEST] [4:≥] [(-)] **0** [.] **5** [ENTER]

because –1 is not greater than or equal to –0.5.

Section 1.2 Properties of Real Numbers and the Basic Rules of Operations

Opposites and the Subtraction Key

These keys were discussed in the Preliminary Chapter.

The Absolute Value Key

Absolute value is the second function of [x^{-1}].

Example 1: Calculate | 7|, | –7|, and –| –7|.

Press

[CLEAR] [2nd] [ABS] **7** [ENTER]

to see the absolute value of 7, and then press

[2nd] [ABS] [(-)] **7** [ENTER]

to see the absolute value of –7. Next press

[(-)] [2nd] [ABS] [(-)] **7** [ENTER]

to see –|–7|. See Fig. 1.9.

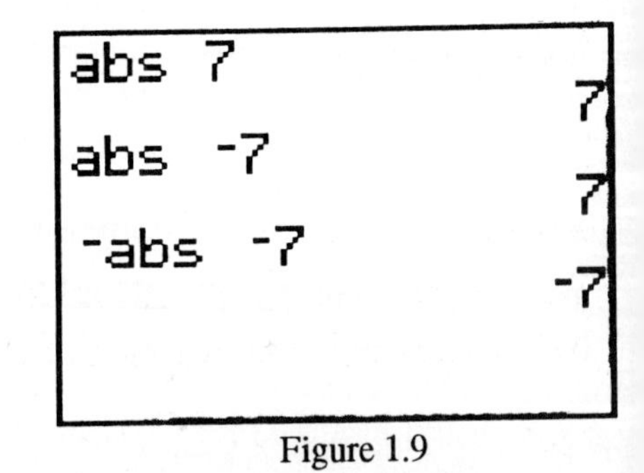

Figure 1.9

Example 2: Calculate |5.2 – 17.8|.

Press

CLEAR 2nd ABS (5 . 2 – 17 . 8) ENTER.

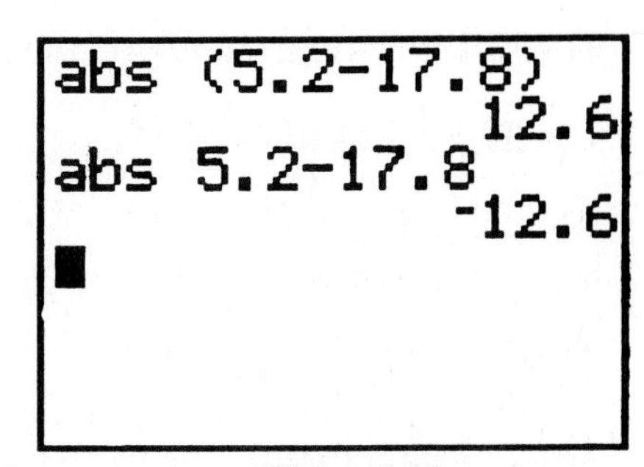

Figure 1.10

The result is 12.6. If you repeat the same keystroke sequence but leave out the parentheses, you will obtain an incorrect result. See Fig. 1.10. With the parentheses, the TI-81 subtracts 17.8 from 5.2 first and then finds the absolute value of the difference. Without the parentheses, the TI-81 takes the absolute value of 5.2 and then subtracts 17.8.

Section 1.3 Algebraic Expressions and Problem Situations

Using a Grapher to Evaluate an Expression by Using STO▶ and REPLAY (RECALL)

You can assign values to variables on the TI-81 and then use those variables in calculations. TI-81 variables are the letters of the alphabet found above many of the keys. You assign values to variables with STO▶. This key is just above the ON key. When you press STO▶ the cursor changes to the Alpha mode, which means you do not have to press ALPHA to access the letters above the keys. When the TI-81 is in the Alpha mode and you press a key, the letter above the key will be printed.

Example 1: Evaluate $x^3 + 17$ at $x = 25$, 39, and 52.

Press

CLEAR **25** STO▶ X ENTER.

The reference to X in these instructions means to press the key with the 'X' above it. This key happens to be STO▶. Now press

ALPHA X ^ **3** + **17** ENTER.

The screen should show you that

$x^3 + 17 = 15{,}642$

when x is 25. See Fig. 1.11.

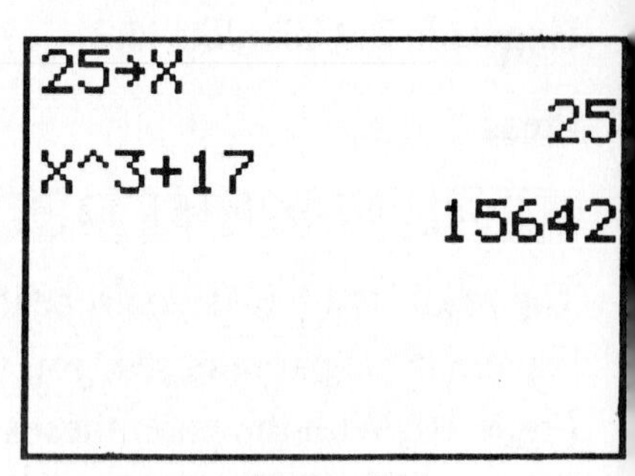

Figure 1.11

We will evaluate $x^3 + 17$ at $x = 39$ and 52 by a different approach. Instead of storing 39 in X and then evaluating the expression, we will enter 39 directly into the expression. Press

[CLEAR] **39** [^] **3** [+] **17** [ENTER].

The answer is 59,336. Now we will use the **REPLAY** feature to edit the expression and replace 39 with 52. You can REPLAY an expression by using [ENTRY], which is the second function of [ENTER]. Recall the expression by pressing

[2nd] [ENTRY].

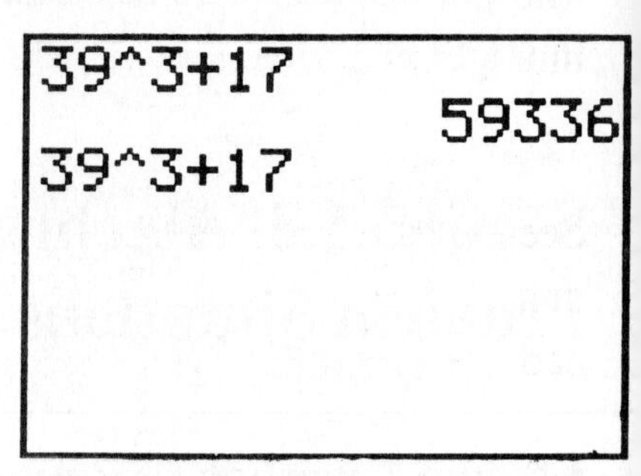

Figure 1.12

The previous expression is copied on the next line and the cursor appears at the end of the line. See Fig. 1.12. Now press the left cursor-movement key until the cursor is on the 3 in 39. Replace 39 with 52 by simply typing 52. Finish the calculation by pressing [ENTER]. See Fig. 1.13.

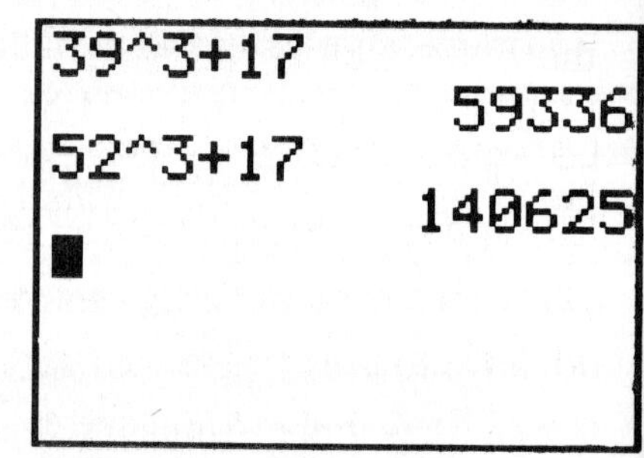

Figure 1.13

Example 2: Evaluate $4T - \dfrac{3S^3}{5R}$ at $R = 13.5$, $S = -23.7$, and $T = 3.28$.

Assign values to the variables by pressing

[CLEAR] **13** [.] **5** [STO▶] [R] [ENTER] [(-)] **23** [.] **7** [STO▶] [S] [ENTER] **3** [.] **28** [STO▶] [T] [ENTER].

Now evaluate the expression by pressing

4 [ALPHA] [T] [−] **3** [ALPHA] [S] [MATH] [3:³] [÷] [(] **5** [ALPHA] [R] [)] [ENTER].

See Fig. 1.14.

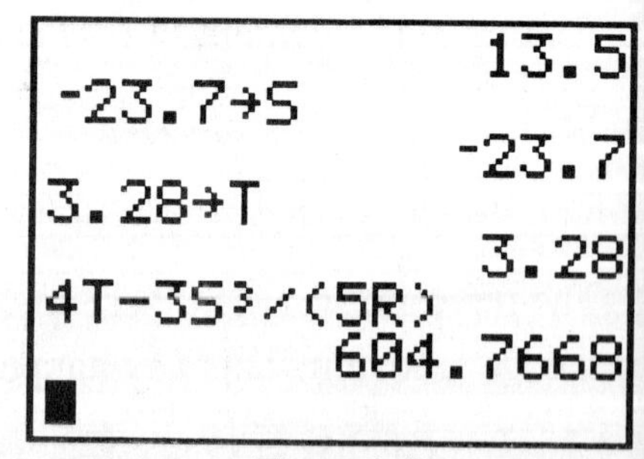

Figure 1.14

Concatenation

The TI-81 does not have a concatenation key.

Implied Multiplication

Press

[CLEAR] **3** [STO▶] [X] [ENTER] **2** [X|T] [ENTER].

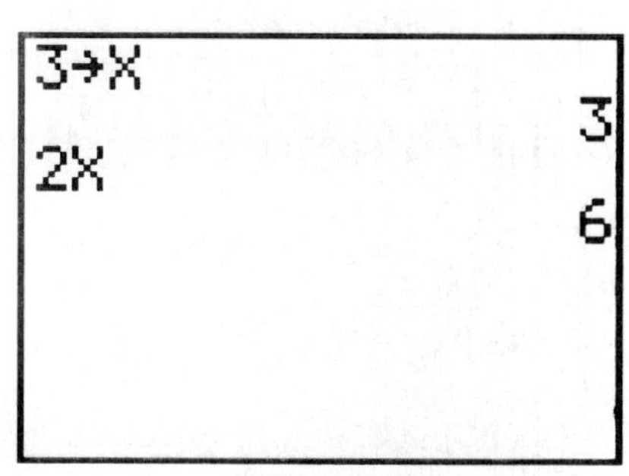

Figure 1.15

See Fig. 1.15. Using the [X|T] key is a shortcut for pressing [ALPHA] [X]. The TI-81 multiplies the 2 and the value of X even though there is no multiplication sign. This is called implied multiplication because their position next to one another implies multiplication. The expression 2 * X is explicit multiplication because there is a multiplication sign between the two factors. Now press

12 [÷] **2** [X|T] [ENTER].

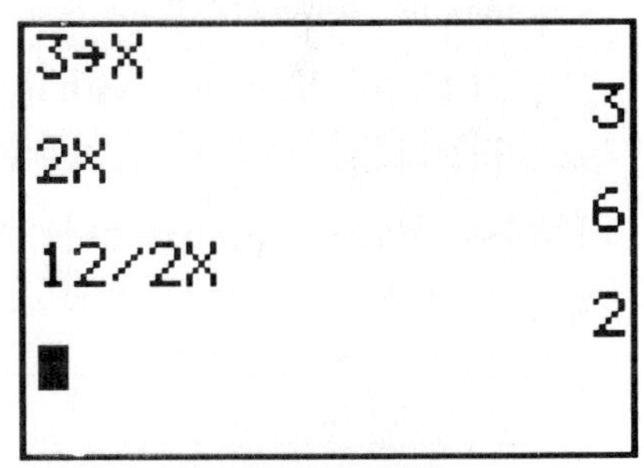

Figure 1.16

See Fig. 1.16. The result is 2 because the implied multiplication is done before the division. The TI-81 carries out an implied multiplication before an explicit multiplication or division.

Example 3: Evaluate $\frac{5}{9}(41-32)$.

You will need to use parentheses to enter this expression correctly in the TI-81. Press

[CLEAR] [(] **5** [÷] **9** [)] [(] **41** [−] **32** [)] [ENTER].

The answer is 5. The TI-81 performs the calculations inside both pairs of parentheses first and then multiplies the results. Now see what happens if you leave off the first set of parentheses. Press

5 [÷] **9** [(] **41** [−] **32** [)] [ENTER].

The result is different. See Fig. 1.17. This is because the TI-81 performs the implied multiplication between the 9 and the value in the parentheses before it does the division. The expression 5/9(41–32) is equivalent to 5/[9(41–32)]. Now redo Example 2 and omit the pair of parentheses around the 5R. The result should be the same as with the parentheses because the implied multiplication between 5 and R is done before the division.

```
(5/9)(41-32)
                5
5/9(41-32)
      .0617283951
```

Figure 1.17

Section 1.5 Numerical Representations of Problem Situations

Making Tables

It is not possible to make tables on the TI-81.

Finding Percents

Example 1: Find the wholesale price for a book that sells retail for $17.10 if the markup is 20%.

We combine a formula with the method of *guess and correct*. The retail price is equal to the wholesale price plus 20% of the wholesale price. If we let X be the wholesale price, then the retail price is equal to

$X + .20X$.

We can make an initial guess for X and then check its correctness with the calculator. Since the numbers involved represent money, we are only interested in two decimal places. Use the Mode screen to change from Floating Decimal format to Fixed Decimal format by highlighting the 2 on the second line of the Mode screen. (Don't forget to press [ENTER] after you move the cursor to '2'.) To check an initial guess of $15.00, press

[CLEAR] **15** [.] **00** [+] [.] **20** [×] **15** [.] **00** [ENTER].

Since the answer is $18.00, our initial guess is too high. To change the guess to $14.00, press

[2nd] [ENTRY]

to Replay the expression and use the cursor-movement keys to replace each '5' with a '4.' This is done by placing the cursor on each '5' and then pressing 4. After you have changed the expression, press [ENTER]. This guess is too low. Use the Replay feature to change the guess to \$14.50. Since the answer is \$17.40, we are still high. Now Replay and change the expression to guess \$14.25. This value works. See Fig. 1.18. Change to Floating Decimal mode when you are finished with this problem.

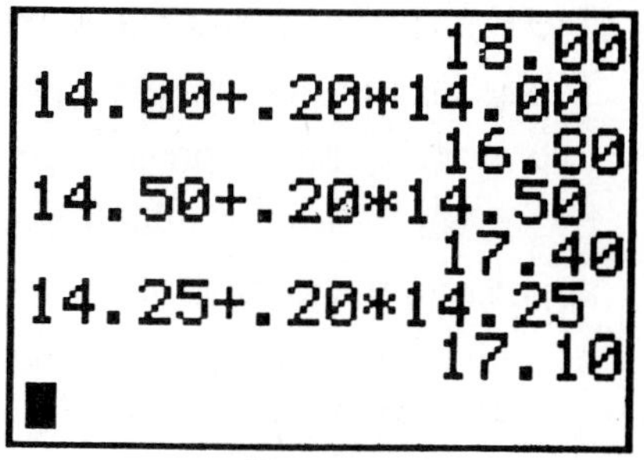

```
                 18.00
14.00+.20*14.00
                 16.80
14.50+.20*14.50
                 17.40
14.25+.20*14.25
                 17.10
█
```

Figure 1.18

Section 1.6 Algebraic Representations of Problem Situations

The Parametric Mode and the Simultaneous Mode

Example 1: Suppose that at the instant Juan is running 30 feet per second, he is 10 feet behind Carlos, who is running at 28 feet per second. How long will it take for Juan to catch Carlos? The runners in this problem can be simulated by the TI-81 with the **Parametric Graphing mode** and **Simultaneous Graphing mode**. Press

[MODE]

and highlight 'Param' and 'Simul.' Next, call up the **Range** screen by pressing

[RANGE].

The [RANGE] key is on the top row. See Fig. 1.19. This screen is used to set up the viewing window for the simulation. The values on this screen can be changed by using the cursor-movement keys to move the cursor to the line you want to change and then typing in the new value. Change the Range values to the following: *T*min = 0, *T*max = 5.5, *T*step = 0.01, *X*min = 0, *X*max = 150, *X*scl = 10, *Y*min = 0, *Y*max = 8, *Y*scl = 1. See Fig. 1.20. Now press [Y=] (also on the top row). We will enter the distance equations on this screen. Remember that

distance = rate × time.

Press

30 [X|T] [ENTER] **5** [ENTER] **28** [X|T] [+] **10** [ENTER] **2** [ENTER].

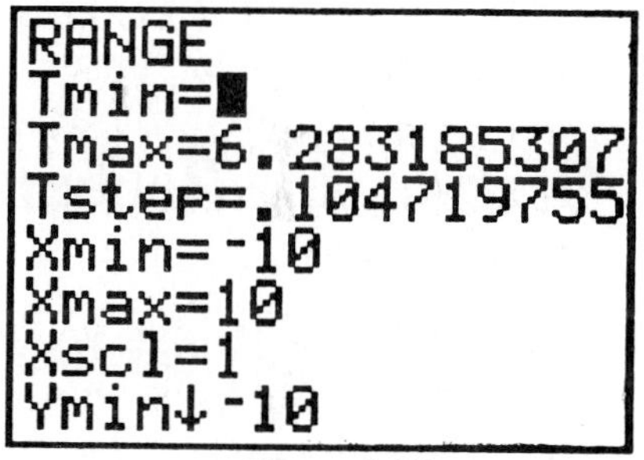

```
RANGE
Tmin=█
Tmax=6.283185307
Tstep=.104719755
Xmin=-10
Xmax=10
Xscl=1
Ymin↓-10
```

Figure 1.19

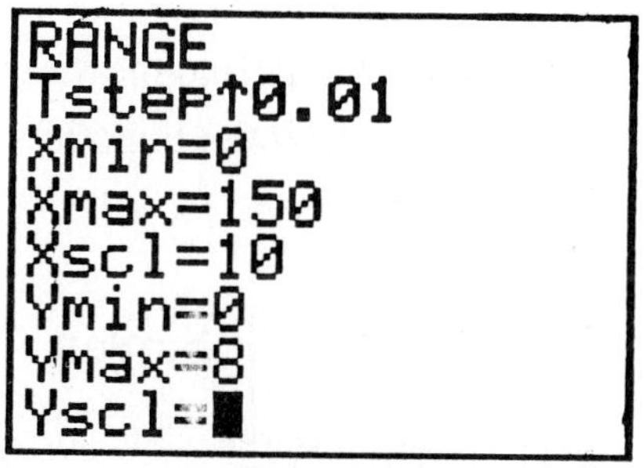

```
RANGE
Tstep↑0.01
Xmin=0
Xmax=150
Xscl=10
Ymin=0
Ymax=8
Yscl=█
```

Figure 1.20

See Fig. 1.21. In Parametric mode, [X|T] is a shortcut for [ALPHA] [T]. To see the simulation, press the [GRAPH] key (on the top row). The two moving lines represent the runners. You can trace their progress by pressing [TRACE], which is next to [GRAPH]. The Trace cursor appears in the middle of the top line. The value of T printed at the bottom of the screen is the elapsed time, and the value of X is the distance traveled by Juan. To see the distance traveled by Carlos at the same time, press [▼]. Press [▲] to move the cursor back to the top graph. Press [▶] repeatedly to see the progress of the runners each 0.01 second along the way. When $T = 5$ both runners have traveled a distance of 150 feet. See Fig. 1.22. Juan will catch Carlos after five seconds. Reselect the Function and Sequence modes for future examples.

```
:X1T=30T
:Y1T=5
:X2T=28T+10
:Y2T=2
:X3T=
:Y3T=
```

Figure 1.21

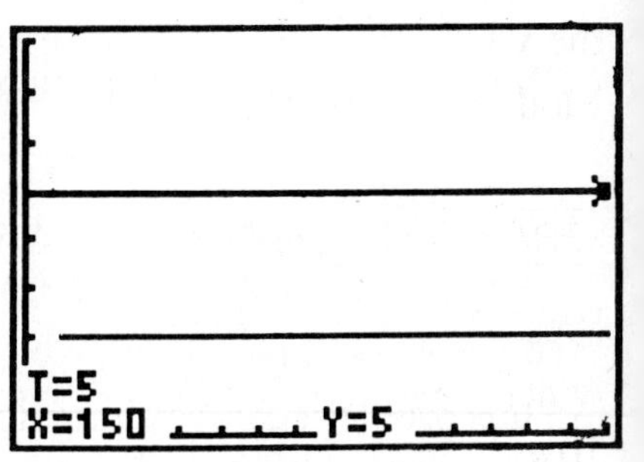

Figure 1.22

Chapter 2 Graphing Equations and Functions

Section 2.1 Rectangular Coordinate Plane

Displaying the Coordinate Plane

The TI-81 can use the screen to display a portion of the coordinate plane. This portion is called a **viewing window**. Graphs of functions can be displayed in the viewing window. The black keys on the top row are used to define the viewing window and create graphs. Press [MODE] and select the Function mode before continuing with the following examples.

The Range Edit Screen

The Range is used to determine which portion of the coordinate plane will be displayed in the viewing window. Press [RANGE] to see the **Range edit screen**. See Fig. 2.1. Your screen probably doesn't have the same numbers as those shown in Fig. 2.1. The variables on the Range edit screen are called Range variables. *X*min is the coordinate of the left side of the viewing window, and *X*max is the coordinate of the right side of the window. *X*scl is the distance between "tick marks" that are drawn on the *x*-axis. *Y*min is the coordinate of the bottom of the window. *Y*max is the coordinate of the top, and *Y*scl is the distance between "tick marks" on the *y*-axis. You can change the values of your Range variables by moving the cursor to the value you would like to change and typing in the new value. Modify your Range edit screen until it looks like Fig. 2.1.

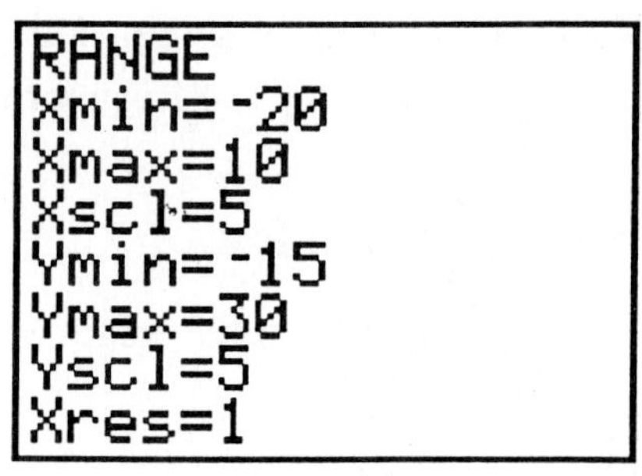

Figure 2.1

GRAPH, Screen Coordinates, and Pixels

Press GRAPH to see the viewing window. See Fig. 2.2. This viewing window represents points whose *x*-coordinates are between –20 and 10 and whose *y*-coordinates are between –15 and 30. The "tick marks" are each five units apart. We say that this is the window [–20, 10] by [–15, 30]. This notation means: *X*min = –20, *X*max = 10, *Y*min = –15, *Y*max = 30.

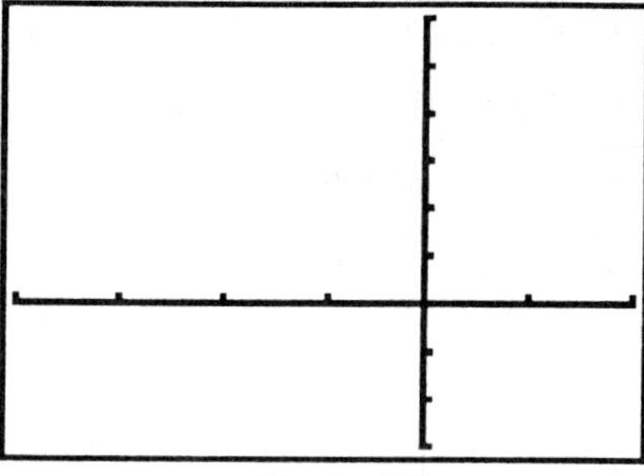
Figure 2.2

Now press ▶. A cursor in the shape of a '+' with a blinking center appears on the screen. This is called the **free-moving cursor**. The blinking rectangle at the center of the free-moving cursor is called a pixel. The *x*- and *y*-coordinates of this pixel are printed at the bottom of the screen. See Fig. 2.3. You can move the cursor around the viewing window with the cursor-movement keys. The coordinates printed at the bottom of the screen are approximations to the actual coordinates of the pixel at the center of the free-moving cursor. Move the cursor to the origin. The coordinates should be (0, 0) but the *x*-coordinate is slightly off. Press RANGE and change the Range values to [–10, 10] by [–10, 10] with *X*scl and *Y*scl values of 1, then press GRAPH and use the cursor-movement keys to move the free-moving cursor to the origin. Notice the cursor coordinates are slightly different from zero, (0.10526316, 0.15873016), even though

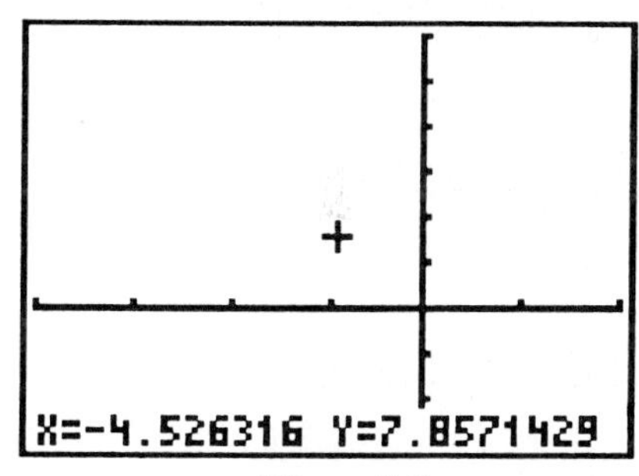

Figure 2.3

the cursor is at the origin. This is because the viewing window has changed and the TI-81 is approximating the coordinates of the origin.

Integer Windows

With the cursor at the origin of the [–10, 10] by [–10, 10] window, press

ZOOM [8:Integer] ENTER.

Now use the cursor-movement keys to move the free-moving cursor. The x- and y-coordinates of the cursor are always integers. This is because the viewing window has been changed. Press RANGE to see how the values of the Range variables have changed. They are [–48, 47] by [–32, 31]. This is called the **Integer window**. There are other windows where the cursor coordinates are integers, but since this window has the origin at the center, your textbook calls it the Integer window. You can find other Integer windows by moving the screen cursor away from the origin after you press

ZOOM [8:Integer]

but before you press ENTER. The range [0, 95] by [0, 63] is also an Integer window, as is [0, 190] by [0, 126]. In an Integer window, Xmax – Xmin is a multiple of 95 and Ymax – Ymin is a multiple of 63.

The Decimal or 0.1 Window and the 0.2 Window

Now change the Range values to Xmin = –4.8, Xmax = 4.7, Xscl = 1, Ymin = –3.2, Ymax = 3.1, Yscl = 1. Press

GRAPH

and use the cursor-movement keys. The x- and y-coordinates are always tenths. This is called the **Decimal window** or **0.1 window**. [–4.8, 4.7] by [–10, 10] is a window where the x-coordinates are always tenths, but not the y-coordinates. The **0.2 window** is [–9.6, 9.4] by [–6.4, 6.2]. Why is this called the 0.2 window?

Section 2.2 Equations in Two Variables and Their Graphs

Plotting Points with the Point-On (PT-On( ) and STAT Commands

Example 1: Plot the following solution pairs to $14x + 28y = 168$, with the points (–15, 13.5), (–10, 11), (0,6), (15,–1.5).

Use [RANGE] to create a [–20, 20] by [–15, 15] viewing window. Press

[2nd] [QUIT]

to move from the Range edit screen to the Home screen. Now we will use the **point-on** option of the **DRAW menu** to plot these points. Press

[2nd] [DRAW] [3:PT-On(].

The PT-On(command is copied to the Home screen. Now press

[(-)] **15** [ALPHA] [,] **13** [.] **5** [)].

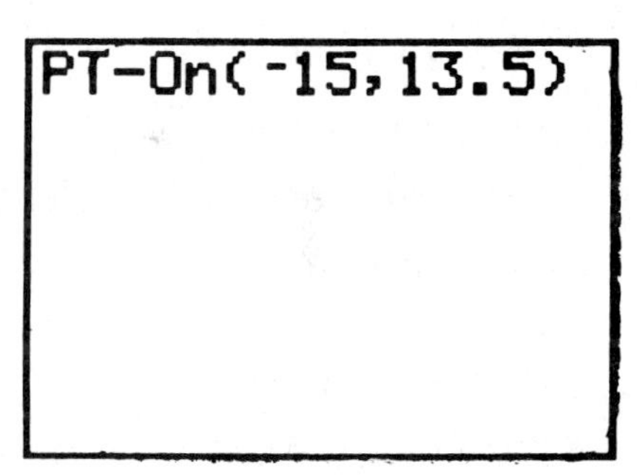

Figure 2.4

See Fig. 2.4. This command tells the TI-81 to plot the point (–15, 13.5). Press [ENTER] to see the point plotted on the viewing window. The point should be plotted in the upper left-hand corner of the viewing window. Now press [CLEAR] to return to the Home screen. Then press

[2nd] [DRAW] [3:PT-On(] [(-)] **10** [ALPHA] [,] **11** [)] [ENTER]

to plot the second point. To plot the third point press

[CLEAR] [2nd] [DRAW] [3:PT-On(] **0** [ALPHA] [,] **6** [)] [ENTER].

Plot the last point by pressing

[CLEAR] [2nd] [DRAW] [3:PT-On(] **15** [ALPHA] [,] [(-)] **1** [.] **5** [)] [ENTER].

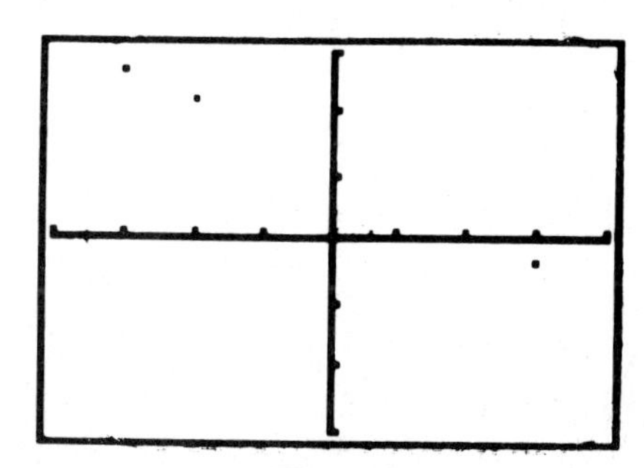
Figure 2.5

See Fig. 2.5. Press

[2nd] [DRAW] [1:ClrDraw] [ENTER]

to clear the viewing window.

Another way to plot the points is with the Statistical features of the TI-81. You should still have the same values for the Range variables that were just used in the first solution to Example 1. The Statistical menu is the second function of the key on the fourth row, second column. The first feature of the Statistical menu we will access is the Data Edit screen. To do this, press

[2nd] [STAT] [▶] [▶] [1:Edit].

See Fig. 2.6. If your Data Edit screen does not look like this, you can erase the existing numbers on the screen by pressing

[2nd] [STAT] [▶] [▶] [2:ClrStat] [ENTER].

Figure 2.6

This will clear the list of ordered pairs and return you to the Home screen so you will need to return to the Data Edit screen by the keystroke sequence described above. Now enter the ordered pairs by pressing

[(-)] **15** [ENTER] **13** [.] **5** [ENTER] [(-)] **10** [ENTER] **11** [ENTER] **0** [ENTER] **6** [ENTER] **15** [ENTER] [(-)] **1** [.] **5** .

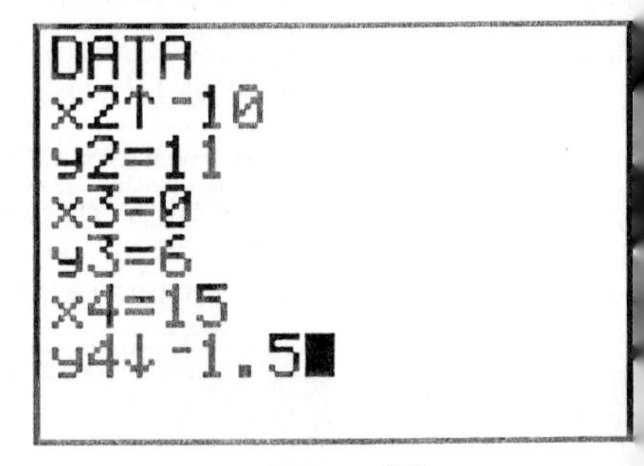

Figure 2.7

Your screen should look like Fig. 2.7. Now plot the points by pressing

[2nd] [STAT] [▶] [2:Scatter] [ENTER].

The points in this picture are the same ones you plotted previously with the PT-On(option. Press [CLEAR] to return to the Home screen.

Graphing Equations Using TRACE

Example 2: Graph $y = 2x - 8$.

The **Y=** key on the top row is used to enter the equation to be graphed. Press [Y=]. You can enter four different equations. You can erase an old equation by pressing [CLEAR]. We will enter an equation in Y_1. Press

2 [X|T] [−] **8**.

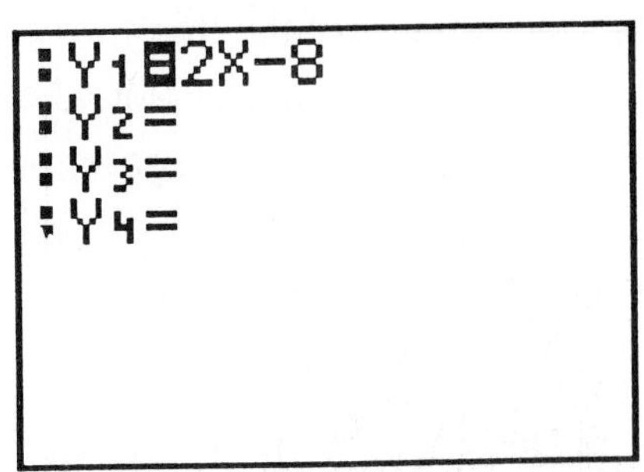

Figure 2.8

See Fig. 2.8. Now press [RANGE] and enter the [−48, 47] by [−32, 31] window. Use an *X*scl and *Y*scl of 10. Press [GRAPH] to see the graph of $y = 2x - 8$. Press [TRACE]. The Trace cursor appears on the graph. See Fig. 2.9. Notice the Trace cursor looks like a blinking 'X' with a box in the center. The coordinates at the bottom of the screen are the coordinates of the Trace cursor. Use the left and right cursor-movement keys to see the coordinates of points on the line. Each pair of coordinates you Trace on the screen should be a solution to the equation

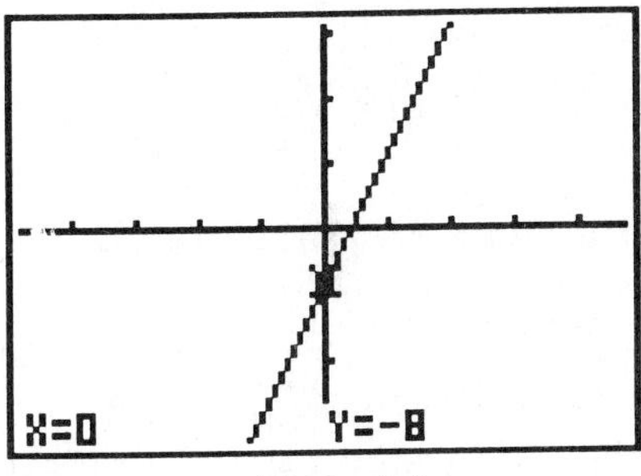

Figure 2.9

$y = 2x - 8.$

Example 3: Graph $y = \frac{3}{4}x - 6$.

Press [Y=] and press [CLEAR] to erase the previous equation. Press

[(] **3** [÷] **4** [)] [X|T] [−] **6**.

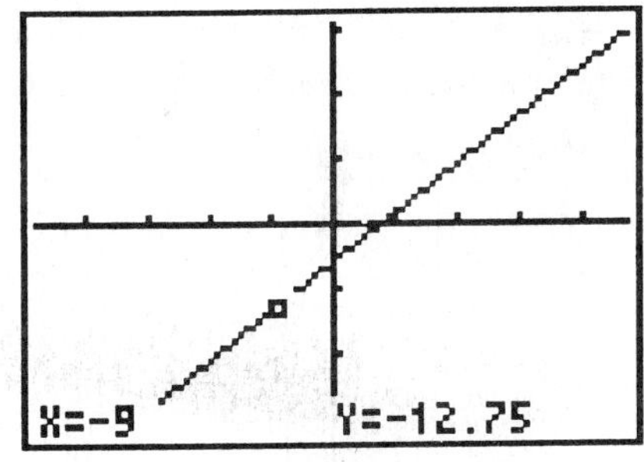

Figure 2.10

The parentheses are needed here because of implied multiplication (see Section 1.3). The viewing window from Example 2 will work here so there is no need to use the Range edit screen. Press [GRAPH] to see the graph. Trace to the point where $X = -9$ and $Y = -12.75$. See Fig. 2.10. This ordered pair is a solution to the equation

$y = (3/4)x - 6.$

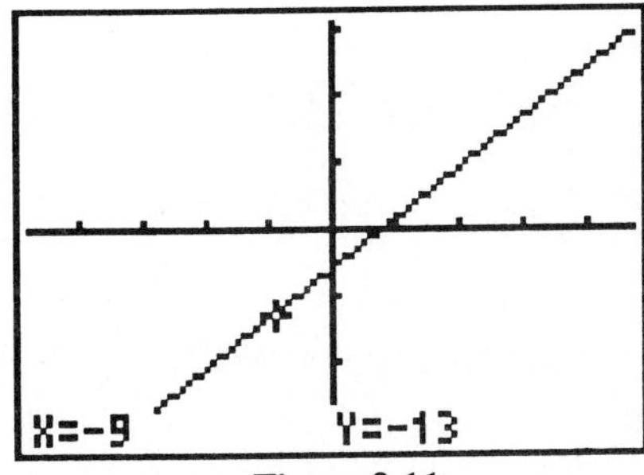

Figure 2.11

The *y*-coordinate of the Trace cursor is calculated by replacing 'X' from the equation in the Y= menu with the *x*-coordinate of the Trace cursor. Now we will move the free-moving cursor to the same point currently occupied by the Trace cursor. Press [GRAPH] to clear the Trace cursor from the viewing window and then use the cursor-movement keys to move the free-moving cursor until its *x*-coordinate is −9 and the cursor appears to be on the line. See Fig. 2.11. The

y-coordinate of the free-moving cursor is not the same as the y-coordinate of the Trace cursor when it was on this point. This ordered pair (–9, –13) is not a solution to the equation. This is because the y-coordinate of the free-moving cursor is not calculated from the equation in the Y= menu.

Section 2.3 Linear Equations and the Slope of a Line

Graphing Vertical Lines

Example 1: Graph $x = 3$.

The graph of $x = 3$ is a vertical line. This is not a function. When the TI-81 is in Function mode, it can only graph functions, so we need to change to Parametric mode (Param). After you change to Parametric mode, press [Y=]. This screen looks different than it did in Function mode. If there are equations on this screen, erase them by moving the cursor to each line and pressing [CLEAR]. Now move the cursor to the top line and press

3 [ENTER] [X|T].

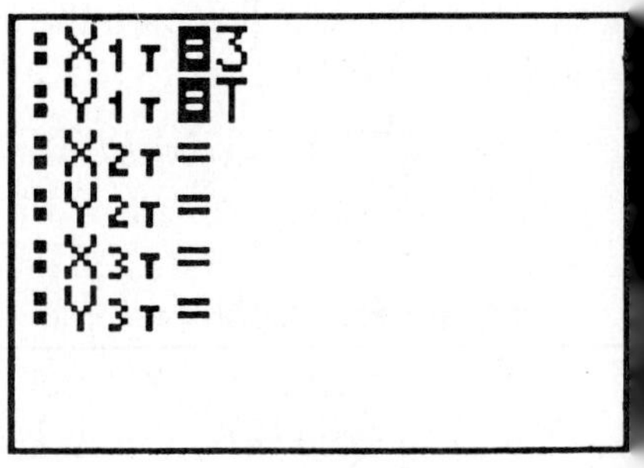

Figure 2.12

See Fig. 2.12. Now select the viewing window by pressing [RANGE] and entering the following values for the Range variables: *T*min = –5, *T*max = 5, *T*step = 0.1, *X*min = –5, *X*max = 5, *X*scl = 1, *Y*min = –5, *Y*max = 5, and *Y*scl = 1. Press [GRAPH]. See Fig. 2.13.

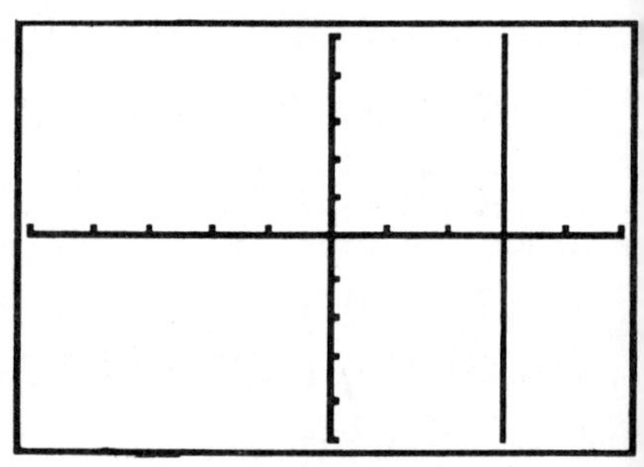
Figure 2.13

Section 2.4 The Slope-Intercept Form and Point-Slope Form of Linear Equations

Squaring Viewing Windows

Example 1: Graph $y = 0.5x + 2$ and $y = -2x - 3$ in a square viewing window.

Select Function mode and press [Y=]. Erase any existing equations by pressing [CLEAR] on each line that has an equation. You can use the cursor-movement keys to move the cursor to each line. With the cursor on the top line, enter the first equation in Y_1 by pressing

0 [.] **5** [X|T] [+] **2** [ENTER],

and then enter the second equation in Y_2 by pressing

[(-)] **2** [X|T] [−] **3**.

Press [RANGE] and select a [–10, 10] by [–10, 10] window, then press [GRAPH]. The lines should be perpendicular, but they don't look like they are. See Fig. 2.14. The window needs to be adjusted to make the lines look perpendicular. Press

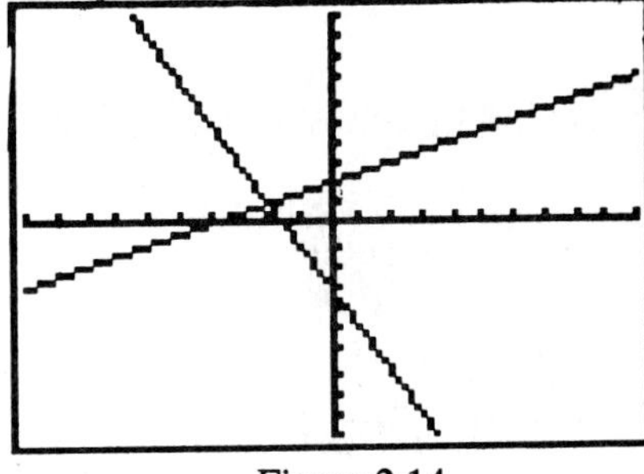

Figure 2.14

[ZOOM] [5:Square]

to make this adjustment. The Square option in the ZOOM menu adjusts the values of the Range variables so the slopes and proportions of the picture are correct. Press [RANGE] to see how these values have changed.

Section 2.5 Functions and Graphs of Functions

Functions on the TI-81 and the Standard Window

The method of graphing functions such as $y = x^2$, $y = \sqrt{x}$, and $y = x^{-1}$ is identical to the one you used in Section 2.2 to find the graph of a line. You enter the equation in the Y= menu, select an appropriate window, and graph.

Example 1: Graph $y = -x^2 - 3x$.

Press [Y=] and use [CLEAR] and the cursor-movement keys to erase any equations that may be in the menu. Move the cursor to 'Y_1' and press

[(-)] [X|T] [x^2] [−] 3 [X|T].

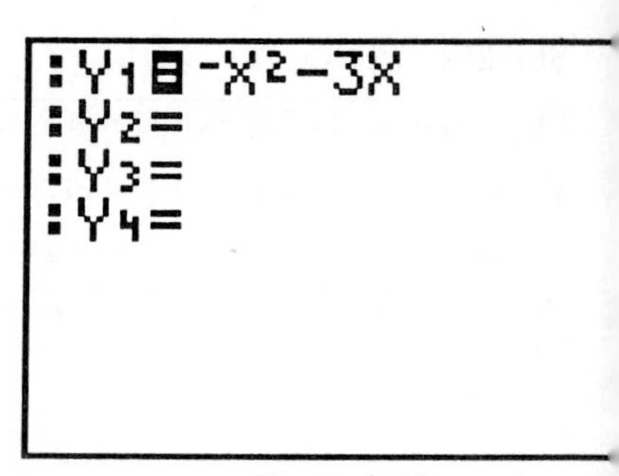

Figure 2.15

See Fig. 2.15. Most graphers use notation like '$Y_1 = -X^2 - 3X$' to represent the function $f(x) = -x^2 - 3x$. Now press [RANGE] and enter a [–5, 5] by [–10, 10] window, then press [GRAPH]. See Fig. 2.16. The graph is a parabola. Next we redraw the graph in the [–10, 10] by [–10, 10] window. This window is called the **Standard window**. There is a shortcut to enter the Standard window. Instead of using [RANGE], press

[ZOOM] [6:Standard].

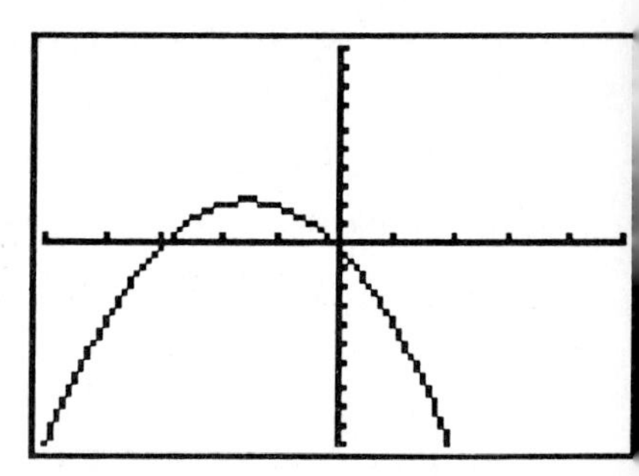
Figure 2.16

Notice the TI-81 redraws the graph in a new window as soon as you press '6.' See Fig. 2.17. Press [RANGE] to verify this is the Standard window.

Using [TRACE] to Find the Domain and Range of a Function and Scrolling

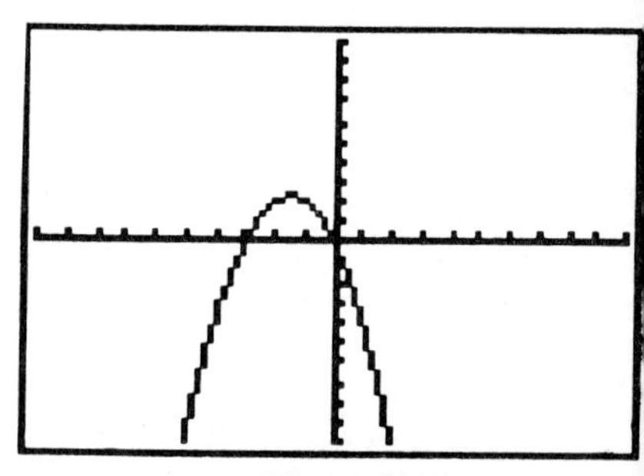
Figure 2.17

Example 2: Find the domain and range of $f(x) = \sqrt{(x-3)}$.

We will enter this equation in the Y= menu and graph in the Integer window. We are going to use a shortcut to enter the Integer window. Press [Y=] and erase existing functions from the menu. Press

[2nd] [√] [(] [X|T] [−] 3 [)].

Now press

[ZOOM] [8:Integer].

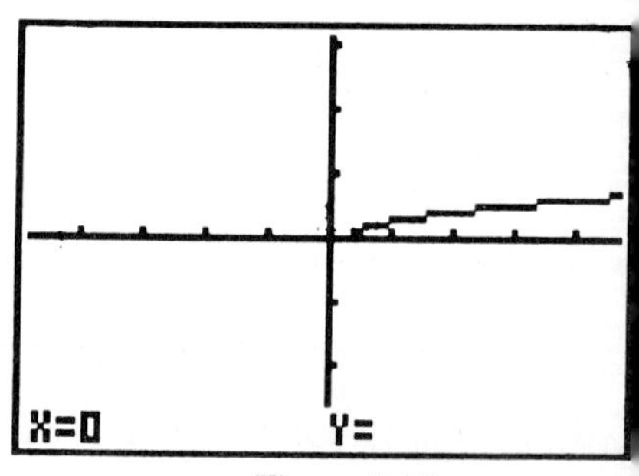

Figure 2.18

As soon as you press '8' you will see the graph drawn in the current window and the free-moving cursor blinking at the origin. Press [ENTER] to redraw the graph in the Integer window. This option in the ZOOM menu creates an Integer window centered on the free-moving cursor. Now press [TRACE]. See Fig. 2.18. The *X*-coordinate is 0 and the *Y*-coordinate is blank. This is because the function is not defined for $X = 0$. Press ▶ three times. Now $X = 3$ and $Y = 0$. Press and hold ▶ to scroll to the right on the graph. The results of TRACE indicate the domain is $[3, \infty)$ and the range is $[0, \infty)$.

Example 3: Find a complete graph of $f(x) = x^4 - 3x^2 + 5x - 6$ and estimate its range.

Press Y= and erase existing functions. Move the cursor to 'Y_1' and press

X|T ^ **4** − **3** X|T x^2 + **5** X|T − **6.**

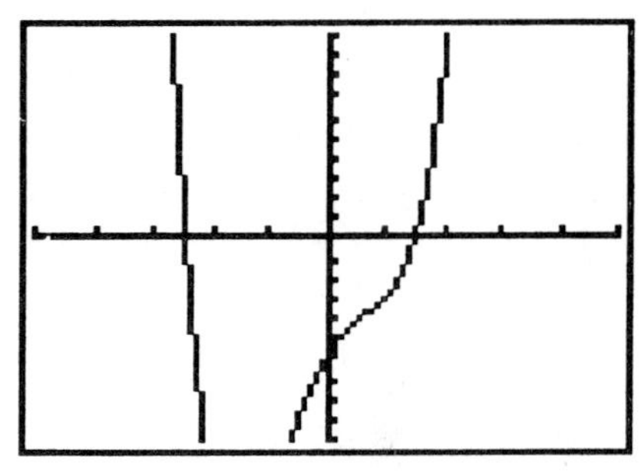

Figure 2.19

Press RANGE and enter a [–5, 5] by [–10, 10] window, then press GRAPH. See Fig. 2.19. This is not a complete graph because you can't see the bottom portion. Press

and repeatedly press ◀. Notice even when the Trace cursor moves off the screen, you can still see its coordinates. The smallest Y-coordinate that is printed as the Trace cursor moves to the left is about –15.19. If you continue to press ◀, the Trace cursor will reappear on the graph. The results of TRACE give us an idea of a better viewing window. Press RANGE and enter the [–5, 5] by [–20, 20] window with Xscl = 1 and Yscl = 2, then press GRAPH. See Fig. 2.20. This is a complete graph. The results of TRACE indicate the range is about $[-15.2, \infty)$.

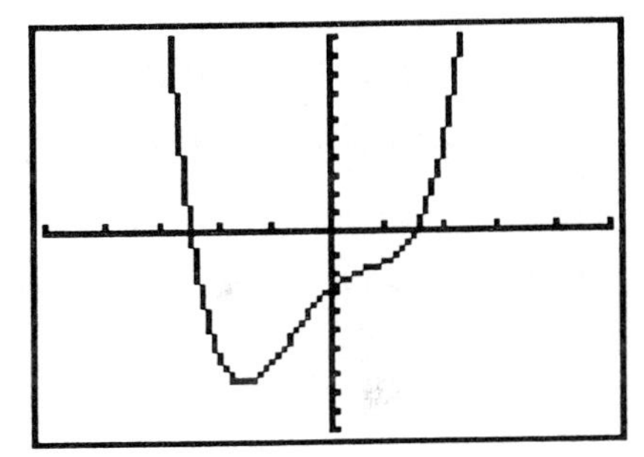

Figure 2.20

How to Graph a Relation that Is Not in Function Form

Example 4: Graph $y^2 = x$.

This relation is not a function and the TI-81 is in Function mode. However, the relation can be split into two functions by solving for y. If you do this you get

$y = -\sqrt{x}$ and $y = \sqrt{x}$.

Enter these functions in Y_1 and Y_2 by pressing

[Y=] [CLEAR] [(-)] [2nd] [√] [X|T] [ENTER] [CLEAR] [2nd] [√] [X|T],

and graph in the Standard window by pressing

[ZOOM] [6:Standard].

See Fig. 2.21.

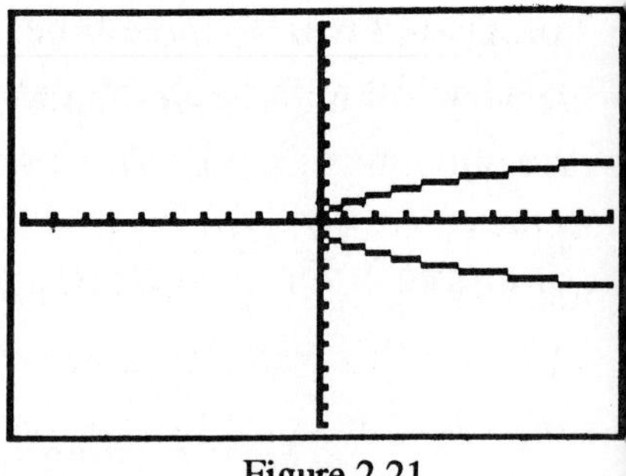

Figure 2.21

Chapter 3 Solving Equations and Systems of Equations

Section 3.1 Finding Graphical Solutions

Selecting and Deselecting Graphs and Sequential versus Simultaneous Graphing

When several functions have been entered in the Y= menu, you can select some of them for graphing and deselect others so that they won't be graphed. If you select several functions to be graphed in the same window, you can graph them sequentially (one at a time) or simultaneously.

Example 1: Graph $y = 2x + 5$ and $y = 21$.

Clear the Y= menu and enter $Y_1 = 2X + 5$ and $Y_2 = 21$. Notice the '=' signs next to both Y_1 and Y_2 are highlighted. This means both functions have been selected for graphing. Deselect Y_2 by using the cursor-movement keys to move the cursor onto the '=' next to Y_2 and then press [ENTER]. Move the cursor off the '=' to see it is no longer highlighted. This means Y_2 is deselected. See Fig. 3.1. Now graph in the Integer window. You should see the graph of Y_1 but not Y_2. Now reselect Y_2 by pressing

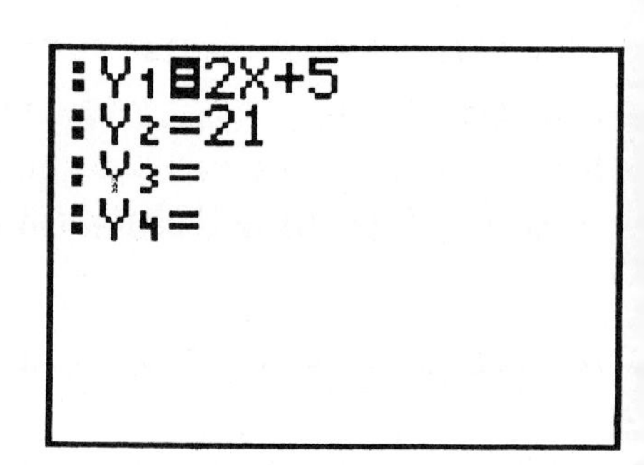

Figure 3.1

[Y=] [◄] [▼].

This should place the cursor on the '=' beside Y_2. Highlight it by pressing [ENTER]. Notice that if the '=' is highlighted, pressing [ENTER] unhighlights it. If the '=' is not highlighted, pressing [ENTER] highlights it. Press [GRAPH]. Since both functions are highlighted, you see the graph of $y = 2x + 5$, followed by the graph of $y = 21$. When the TI-81 is in **Sequential mode**, the functions are graphed one at a time. Now press [MODE] and select **Simultaneous mode** by highlighting 'Simul' on the Mode screen. Don't forget to press [ENTER] after you move the cursor to 'Simul.' Press [GRAPH]. Both graphs are drawn at the same time. See Fig. 3.2.

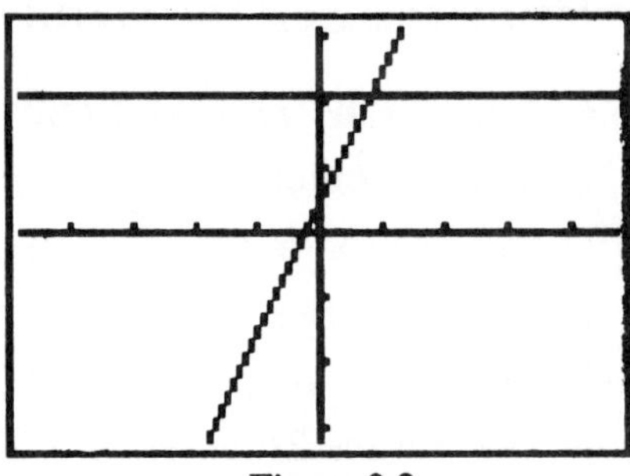

Figure 3.2

Solving Linear Equations Graphically

Example 2: Solve $17 - 1.5x = -10$.

Press [Y=] and use [CLEAR] and the cursor-movement keys to clear the Y= menu. Enter

$Y_1 = 17 - 1.5X$ and $Y_2 = -10$.

Graph in the Integer window. Now press [TRACE] and repeatedly press [▶] until the Trace cursor is on the intersection of the two lines. See Fig. 3.3. If you press [▼] while Tracing, the Trace cursor will move to the next graph on the screen. If you press [▼] when the Trace cursor is at the point of intersection, it won't move because the Trace cursor is on both graphs, but if you move the Trace cursor away from the point of intersection and then press [▼], the cursor will move from one graph to the other.

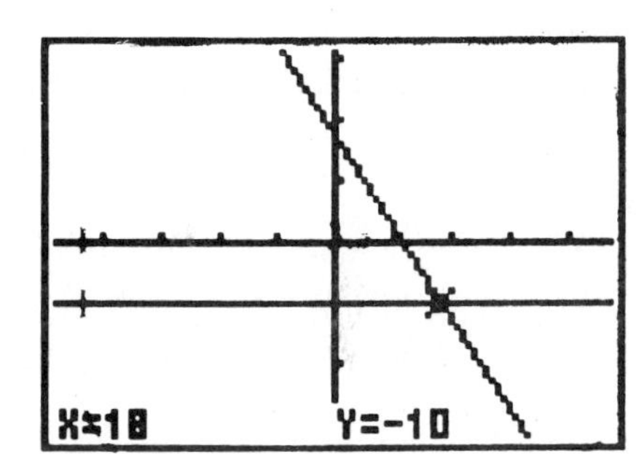

Figure 3.3

Section 3.2 Solving Equations Using Zoom In

Finding a Graphical Solution Using the 0.1 (Decimal) Window Settings

The 0.1 window settings were described in Section 2.1. They are $X\text{min} = -4.8$, $X\text{max} = 4.7$, $X\text{scl} = 1$, $Y\text{min} = -3.2$, $Y\text{max} = 3.1$, and $Y\text{scl} = 1$.

Using the ZOOM Menu on the Grapher

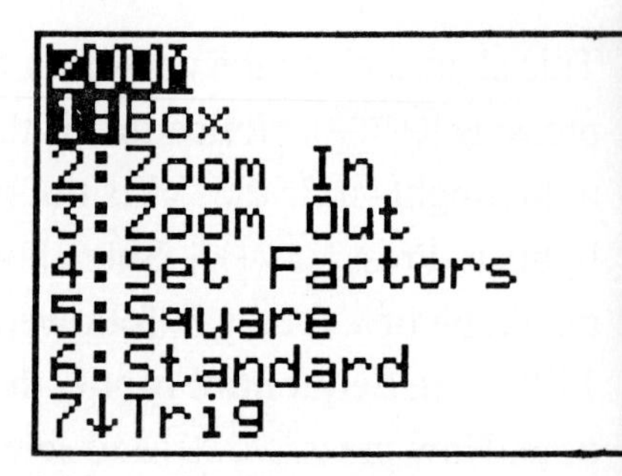

Figure 3.4

Options on the **ZOOM menu** can be used to quickly magnify a portion of the screen. This is helpful when you need to find an accurate approximation to the solution of an equation. Press [ZOOM] to see the Zoom menu. See Fig. 3.4. In this section we will discuss option [**4:Set Factors**] and option [**2:Zoom In**]. Option [4:Set Factors] is used to set the Zoom factors. These factors determine by how much the graph is magnified when you use option [2:Zoom In]. After you have pressed [ZOOM], set the factors to 10 by pressing

[4:Set Factors] **10** [ENTER] **10**.

Then return to the Home screen by pressing

[2nd] [QUIT].

Finding Solutions with Zoom In

Example 1: Find a solution to $2.7x + 3.28 = 0$ with an error of at most 0.01.

The solution can be found graphically by Tracing to the point where the graph of $y = 2.7x + 3.28$ crosses the x-axis. The difficulty with this approach is that if we graph in the Standard window, the Trace cursor won't be accurate enough. We will need to magnify the portion of the graph that crosses the x-axis. We can magnify with Zoom In.

Clear the Y= menu and enter

$Y_1 = 2.7X + 3.28$.

Enter the following values for the Range variables: Xmin = –10, Xmax = 10, Xscl = 0.01, Ymin = –10, Ymax = 10, and Yscl = 0.01.

We use a value of 0.01 for Xscl and Yscl because the problem required an error of at most 0.01. Press [GRAPH]. The axes look double thick because the tick marks are so close together. The tick marks

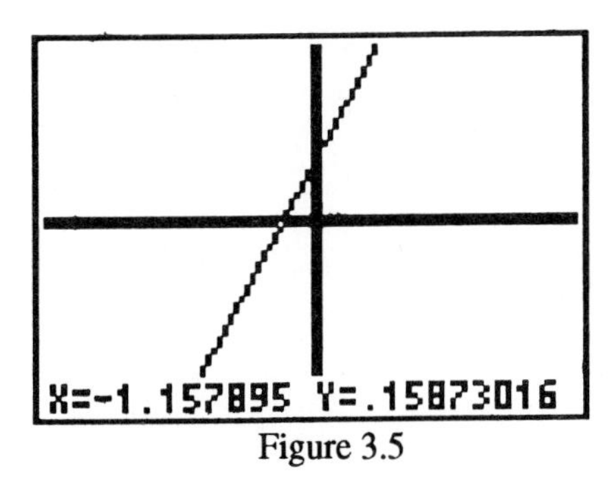

Figure 3.5

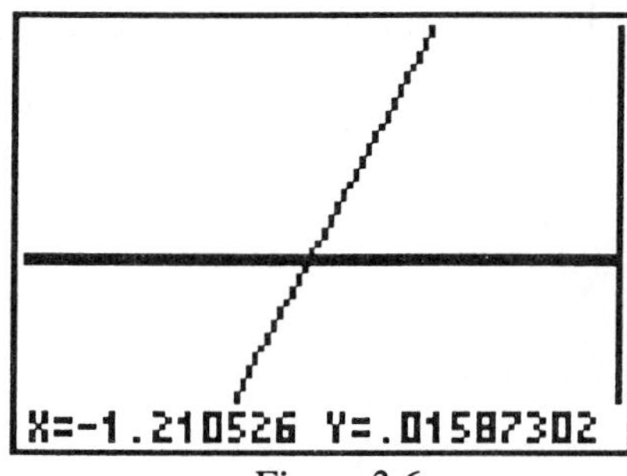

Figure 3.6

will spread out as we Zoom In. Now press

[ZOOM] [2:Zoom In].

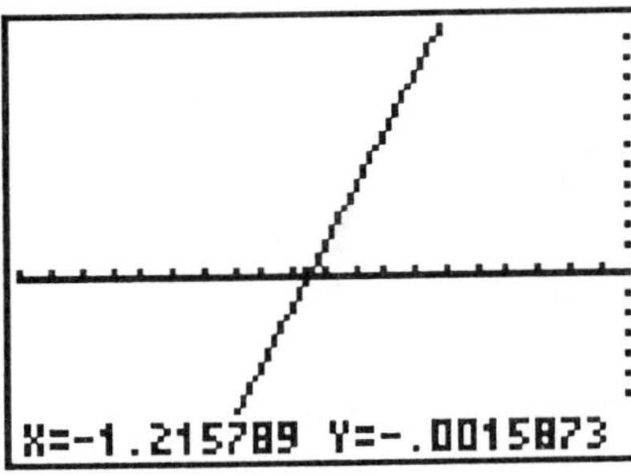

Figure 3.7

You should see the graph with the free-moving cursor at the origin. Use [◄] to move the cursor to the x-intercept of the graph. See Fig. 3.5. Press [ENTER]. The portion of the graph centered on the x-intercept is magnified by a factor of 10. Now the free-moving cursor appears slightly above the x-intercept. Use the cursor-movement keys to move the cursor back to the x-intercept. See Fig. 3.6. Press [ENTER]. The graph is magnified again. Move the cursor to the x-intercept one more time. See Fig. 3.7. Magnify the graph again by pressing [ENTER]. Now press [TRACE] and move the trace cursor to the x-intercept. See Fig. 3.8. The solution is approximately $X = -1.215$. We know the error is at most 0.01 because the graph crosses the x-axis between two tick marks that are 0.01 apart.

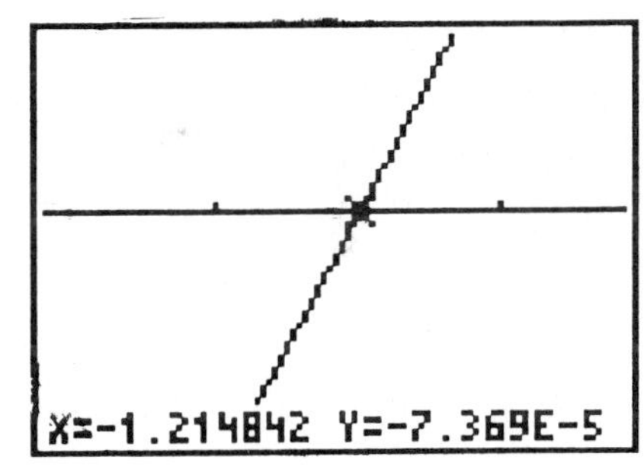

Figure 3.8

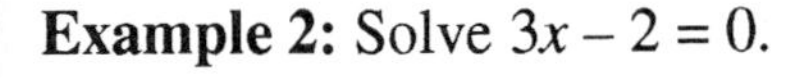

Example 2: Solve $3x - 2 = 0$.

Clear the Y= menu and enter $Y_1 = 3X - 2$. Enter the following values for the Range variables: Xmin = 0.6, Xmax = 0.7, Xscl = .01, Ymin= – 0.2, Ymax = 0.2, and Yscl = .1.

Graph and Trace to the intersection of the x-axis and the graph. See Fig. 3.9. (Note the tick marks for y appear on the edge of the screen since the y-axis is not included in the window.) The calculator solution is $X = .66631579$. We could Zoom In to get a better approximation, but no matter how many times we Zoom In, we will still only

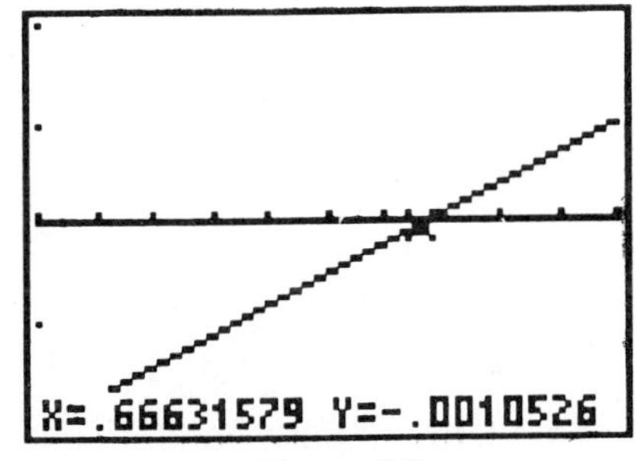

Figure 3.9

have an approximation. This is because the exact answer is 2/3, which has a repeating decimal representation. The grapher supports a solution of 2/3, but it does not confirm it.

How to Determine an Appropriate Window for a Graph

Example 3: Find a complete graph of $y = 32000 + 960x$.

Clear the Y= menu and enter $Y_1 = 32000 + 960X$. Graph in the [–5, 5] by [–5, 5] window. The Graph screen is blank because this window does not contain any portion of the graph. We can use [TRACE] to get a better idea of an appropriate window. Press [TRACE]. Even though you can't see the Trace cursor, you can see its coordinates. See Fig. 3.10. These coordinates tell you that the point (0, 32000) is either on or close to the graph. Based on this information, try the following values for the Range variables: Xmin = –5, Xmax = 5, Xscl = 1, Ymin = –1000, Ymax = 40000, and Yscl = 10000.

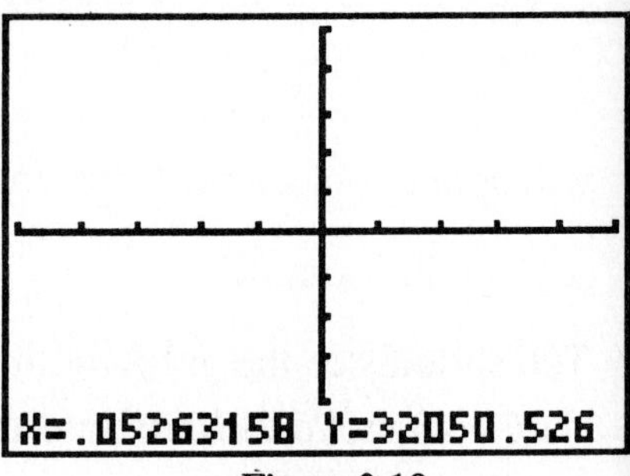

Figure 3.10

Press [GRAPH]. This window indicates the complete graph is a line with y-intercept of approximately 32,000. See Fig. 3.11.

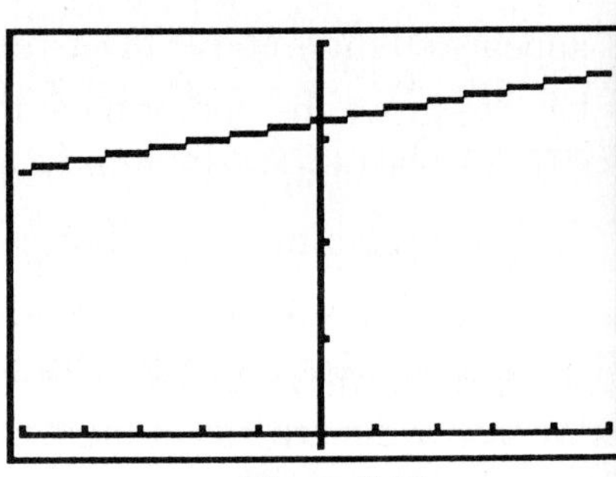
Figure 3.11

Section 3.3 Some Nonlinear Equations

Using Zoom with the Multigraph Method

Example 1: Solve $0.7\sqrt{(x-3)} = 0.3$ with the multigraph method.

In the multigraph method, we graph each side of the equation as a separate function. Then we Trace to the point where the two functions intersect. This point is a solution to the original equation. In our first attempt, we graph $Y_1 = 0.7\sqrt{(X-3)}$ and $Y_2 = 0.3$ in the Standard window, but only one graph appears. This is because the

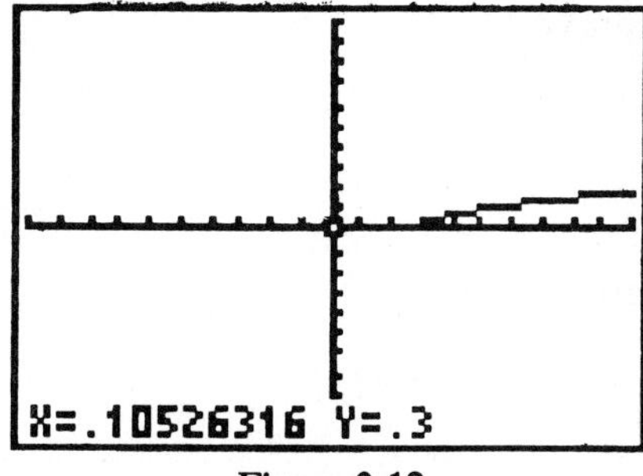

Figure 3.12

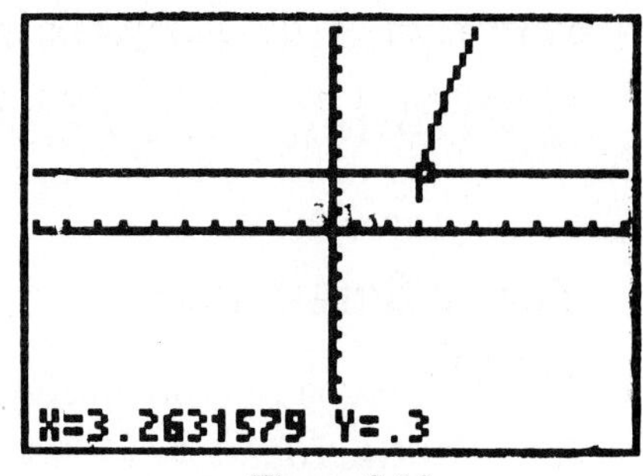

Figure 3.13

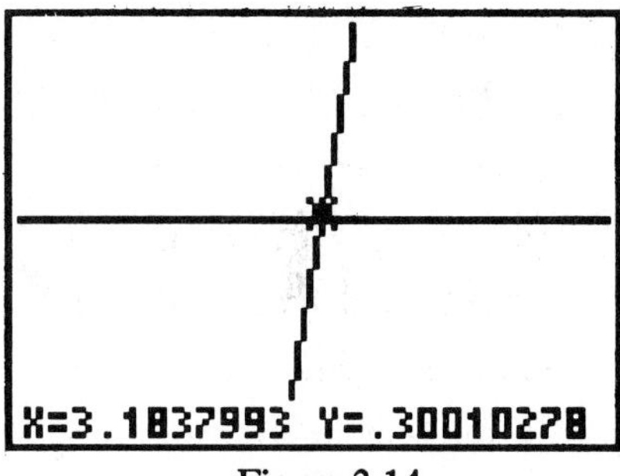

Figure 3.14

graph of Y_2 is too close to the x-axis to be seen in this window. Even though you can't see the graph of Y_2, you can still Trace it. Press [TRACE] and move the Trace cursor to the graph of Y_2 by pressing [▼]. See Fig. 3.12. A better viewing window would be [–10, 10] by [–1, 1]. You can see both graphs as well as their point of intersection in this window. See Fig. 3.13. Now you could Zoom In several times on the intersection to obtain a solution of $x = 3.18$. See Fig. 3.14.

Example 2: Solve $x^3 = x - 0.5$ with the multigraph method.

Graph $Y_1 = X \wedge 3$ and $Y_2 = X - 0.5$ in the Standard window. There appear to be two points of intersection. First we Zoom In on the point in the first quadrant. Press

[ZOOM] [2:Zoom In],

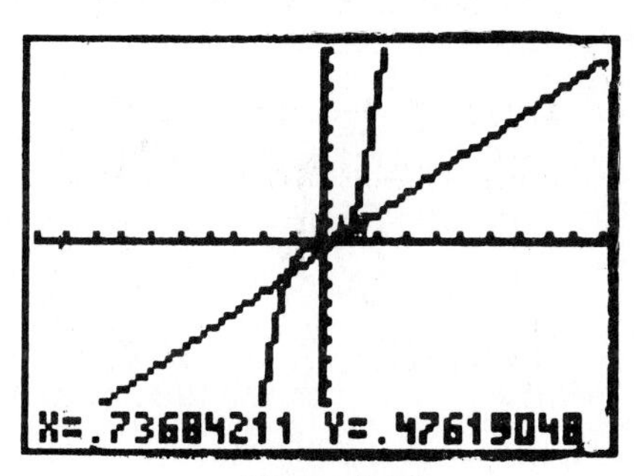

Figure 3.15

and move the free-moving cursor to this point. See Fig. 3.15. When you press [ENTER] to complete the Zoom In, you will see that the two graphs do not intersect after all, so there is no solution in the first quadrant. See Fig. 3.16. You could redraw the graphs in the Standard window and then Zoom In on the point of intersection in the third quadrant to obtain a solution of $x = -1.19$. Initially, it looked like there were two solutions to the example, but Zoom In helped us determine there was only one.

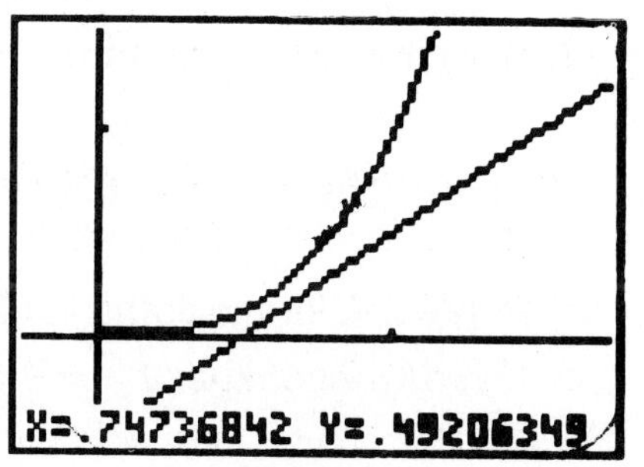

Figure 3.16

Section 3.4 Graphical Representations of Problem Situations

Motion Simulation

Parametric equations are useful in simulating moving objects. A simulation of two runners was described in Section 1.6. In Parametric mode, both the x- and y-coordinates of points on a graph are determined by a third variable, T. In motion simulations, T represents time.

Example 1: Simulate a ball that is thrown straight up with an initial velocity of 88 ft/sec.

The height of the ball is given by the equation

$h = -16t^2 + 88t.$

Select Parametric (Param) and **Dot** mode on the Mode screen. Now press [Y=] and erase any functions that may be in the menu. Then enter

$X_{1T} = 3$ and $Y_{1T} = -16T^2 + 88T.$

Press [RANGE] and enter the following values for the Range variables: Tmin = 0, Tmax = 5.5, Tstep = 0.1, Xmin = 0, Xmax = 5, Xscl = 1, Ymin = 0, Ymax = 150, Yscl = 10.

Press [GRAPH] to see the simulation. You can review the motion of the ball by pressing [TRACE]. See Fig. 3.17. This screen tells you that when $T = 2.7$ seconds, the ball has risen to a height of 120.96 feet. Press [▶] repeatedly to see the ball as time moves forward. Press [◀] to move backward in time through the simulation.

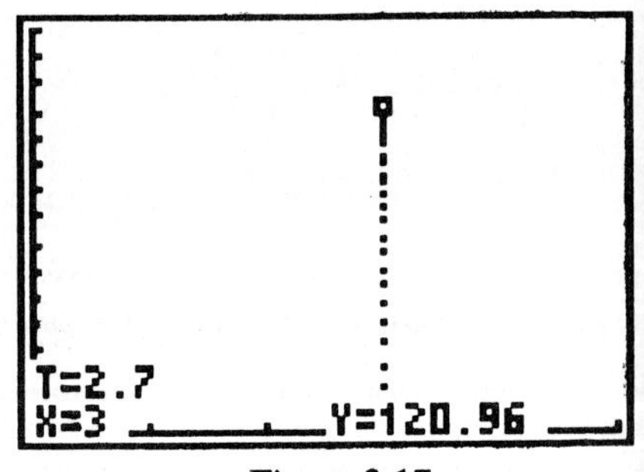

Figure 3.17

Plotting Points and Overlaying a Graph on the Points

Example 2: Plot the points (10, 50), (15, 33.33), (20, 25), (30, 16.67), and overlay the graph of $y = 500/x$ on the same screen.

Select the Function and **Connected** modes from the Mode screen. In Connected mode, the points that are calculated and plotted on the graph are connected with small line segments. This gives the appearance of a smooth graph. Now use [RANGE] to enter the [0, 60] by [0, 60] window. Clear the Y= menu and return to the Home screen. Plot the points given in the example statement with PT-On(or STAT. The PT-On(option of the DRAW menu and STAT options were described in Section 2.2. After you have plotted these points your screen should look like Fig. 3.18. Now we will use the DrawF option from the DRAW menu to overlay the graph of 500/*x*. Press

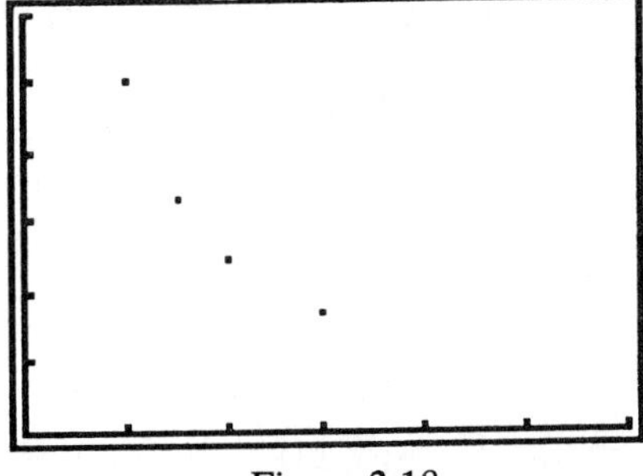

Figure 3.18

[2nd] [DRAW] [6:DrawF]

to copy the DrawF command on the Home screen, then press

500 [÷] [X|T] [ENTER].

See Fig. 3.19.

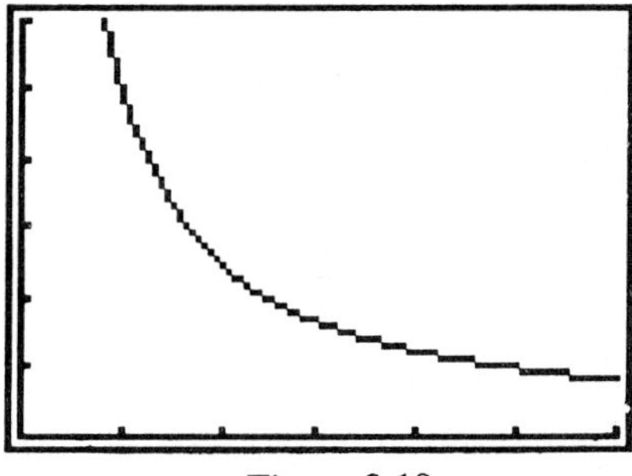

Figure 3.19

Finding an Integer Window that Satisfies [0, *X*max] by [0, *Y*max]

To find an Integer window that satisfies [0, *X*max] by [0, *Y*max], *X*max must be a multiple of 95, and *Y*max must be a multiple of 63. See Section 2.1.

Section 3.5 Solving Systems of Equations by Graphing

Using Trace on Two Functions

Example 2 in Section 3.1 describes how to Trace on two different graphs in the same window.

Solving Systems of Equations with both Zoom In and Zoom Out

We will be using Zoom Factors of *X*Fact = 5 and *Y*Fact = 5 in the next problem, so use the Set Factors option in the Zoom menu to set both factors to 5. This option is described in Section 3.2.

Why 5 and 10 Are Convenient *x* and *y* Zoom Factors

Five and ten are convenient Zoom factors because they cause large changes in the viewing window when used with Zoom In or Zoom Out.

Example 1: Solve the system $y = -2x + 175$ and $y = 3x - 320$.

Graph $Y_1 = -2X + 175$ and $Y_2 = 3X - 320$ in the Integer window, [–48, 47] by [–32, 31]. Since you don't see the graphs, we need a different window. We will change the window with **Zoom Out**. Zoom Out provides a more global view by enlarging the viewing window. Press

[ZOOM] [3:Zoom Out].

You should see the graph screen with the free-moving cursor blinking at the origin. When you press [ENTER], Zoom Out will enlarge the window centered on the free-moving cursor by a factor of 5. See Fig. 3.20. Now you can see the intersection of the two graphs. Next use the Zoom In option to find an accurate approximation of the point of intersection. Zoom In was also described in Section 3.2. The coordinates of the point of intersection are $x = 99$, $y = -23$. This is the solution to the system of equations.

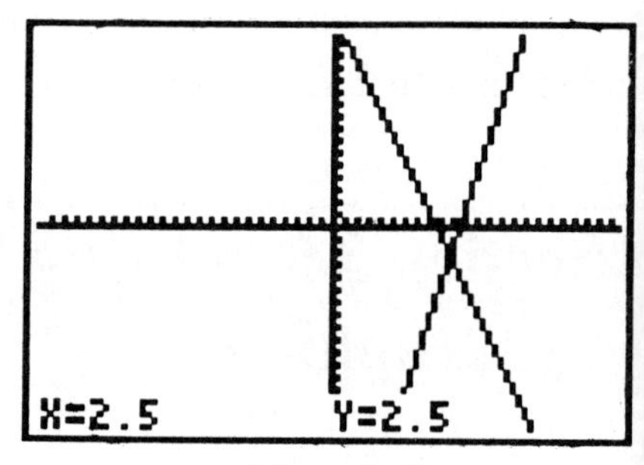

Figure 3.20

Chapter 4 Linear Inequalities and Systems of Linear Inequalities

Section 4.1 Solving Linear Inequalities Algebraically

See Section 1.3 to review Replay and Editing.

Section 4.4 Solving Absolute Value Equations and Inequalities

See Section 1.5 for an explanation of generating tables.

Section 4.5 Linear Inequalities in Two Variables

Shading Solutions to Inequalities in Two Variables

The **Shade** command is followed by three arguments that are separated by commas. The grapher shades all the points that are above the first argument and below the second. (The third argument refers to the x resolution.) If there is no third argument given, every pixel is turned on in the shaded portion. If the resolution is 2, every other pixel is turned on, 3 every third pixel and so on.

The following example employs two arguments.

Example 1: Graph the solution to $y > 4x - 5$.

Clear the Y= menu and select the Standard viewing window. From the Home screen enter the following keystroke sequence:

[2nd] [DRAW] [7:Shade(] **4** [X|T] [−] **5** [ALPHA] [,] **10** [)].

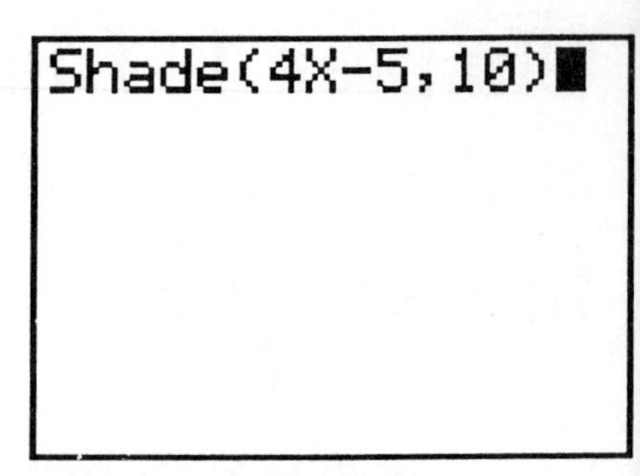

Figure 4.1

See Fig. 4.1. Press [ENTER] to see the solution. See Fig. 4.2. The solution is all points above the line $y = 4x - 5$. There is no upper bound for the solution, but the TI-81 requires us to provide an upper bound. We provided an upper bound of 10 so that the TI-81 would shade everything between the line and the top of the screen.

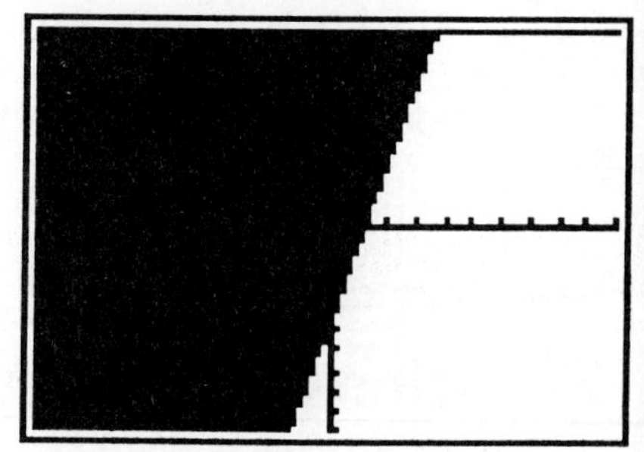
Figure 4.2

From the Home screen press

[2nd] [DRAW] [1:ClrDraw] [ENTER]

to clear the Graphics screen. Now use the DRAW menu to enter the command

Shade(–10, 4X – 5)

on the Home screen and press [ENTER] to see the solution to $y < 4x - 5$. See Fig. 4.3.

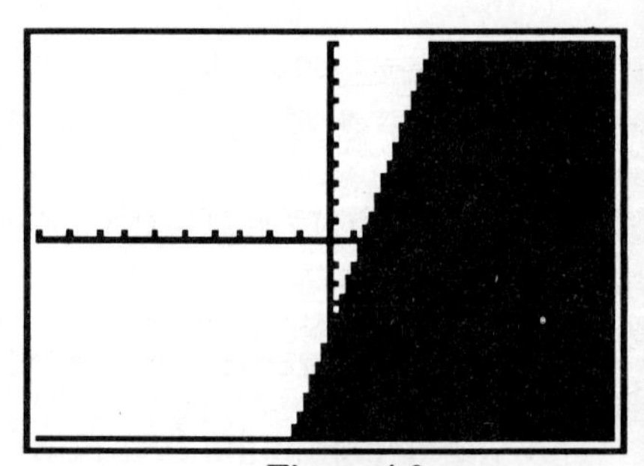
Figure 4.3

Section 4.6 Solving Systems of Inequalities

Graphing Systems of Inequalities Using Shading

Example 1: Graph the solution to the system $y < -x + 9$ and $y > 3x - 15$.

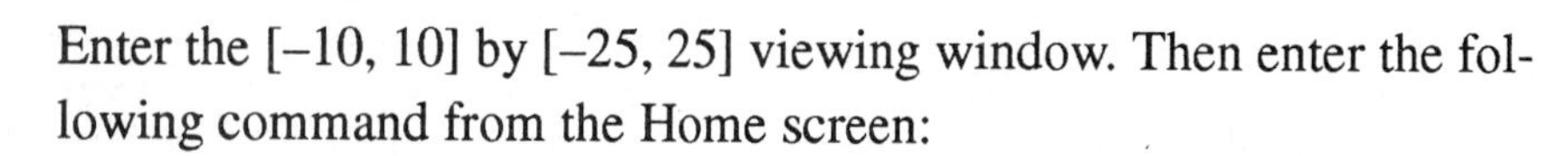

Enter the [–10, 10] by [–25, 25] viewing window. Then enter the following command from the Home screen:

Shade(3*X* – 15, –*X* + 9).

See Fig. 4.4. The lower boundary for the shaded area is the line $y = 3x - 15$, and the upper boundary is the line $y = -x + 9$.

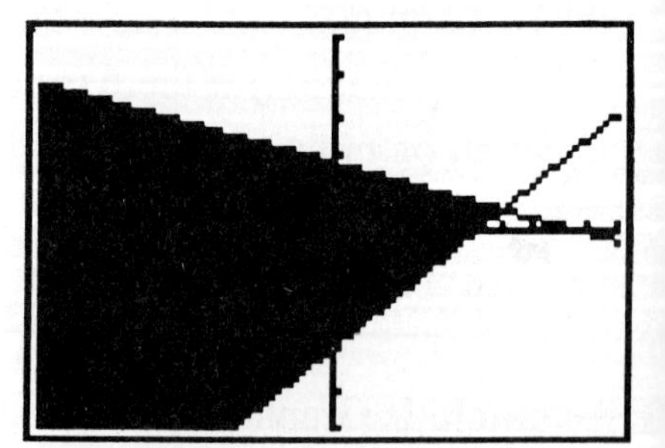
Figure 4.4

Chapter 5 Polynomials

Section 5.1 Exponents and Their Properties

Using Trace to Show Two Graphs that Appear to be Identical

Use [TRACE] to support rules of exponents by showing that two graphs appear to be identical.

Example 1: Determine graphically if $(2x^3)(3x^2) = 6x^5$ is correct.

Select Connected mode and enter

$Y_1 = (2X \wedge 3)(3X \wedge 2)$ and $Y_2 = 6X \wedge 5$.

Graph in the [–2, 2] by [–10, 10] window. Press [TRACE]. See Fig. 5.1. In Section 3.1, Example 2 we described how to move from one curve to the other by pressing [▼] and [▲]. When you press these cursor-movement keys, the cursor does not appear to move to another graph and the coordinates of the Trace cursor don't change. This is graphical evidence that both Y_1 and Y_2 are the same function.

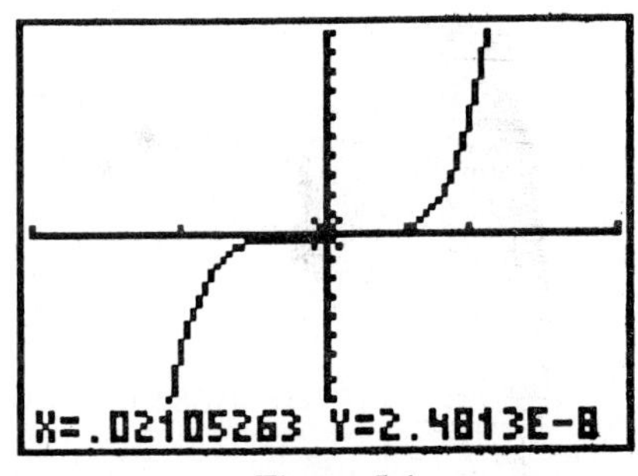

Figure 5.1

Scientific Notation and the [EE] Key

When you select **Sci** on the first line of the Mode screen, numbers are displayed in scientific notation. Press [MODE] and highlight 'Sci,' then return to the Home screen. Press

[CLEAR] **1230** [ENTER].

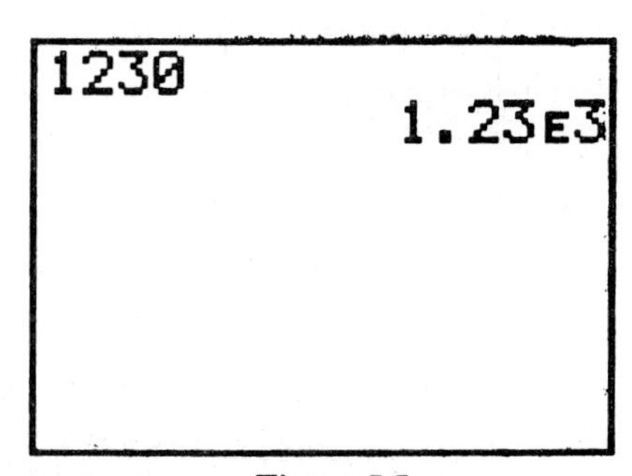

Figure 5.2

See Fig. 5.2. This notation means 1.23×10^3. You can enter numbers in scientific notation with [EE] whether the TI-81 is in Scientific mode or not. The [EE] key is on row six, column two. Press [MODE] and switch back to **Normal mode** by highlighting 'Norm.' Now return to the Home screen and press

[CLEAR] **1** [.] **23** [EE] **3** [ENTER].

The number was entered in Scientific mode but displayed in normal mode. See Fig. 5.3.

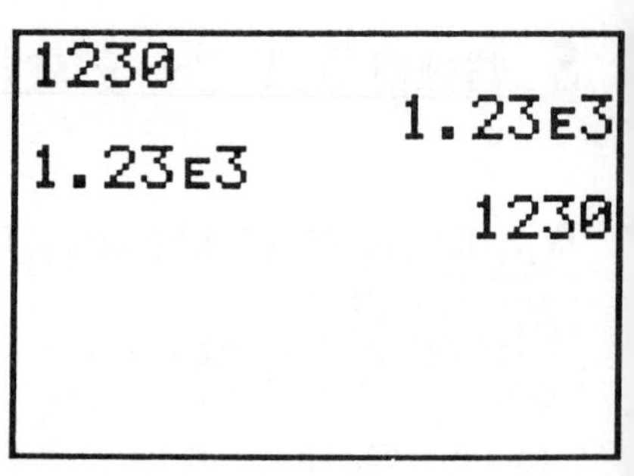

Figure 5.3

Large Numbers on the TI-81

The TI-81 displays numbers up to 10 digits long. If the number has more than 10 digits, the TI-81 converts the number to scientific notation. Enter

2 ^ 33

on the Home screen. The entire number is displayed since it is 10 digits long. Now enter

2 ^ 34

on the Home screen. See Fig. 5.4. This number has 11 digits so it is converted to scientific notation and only the first 10 digits are displayed.

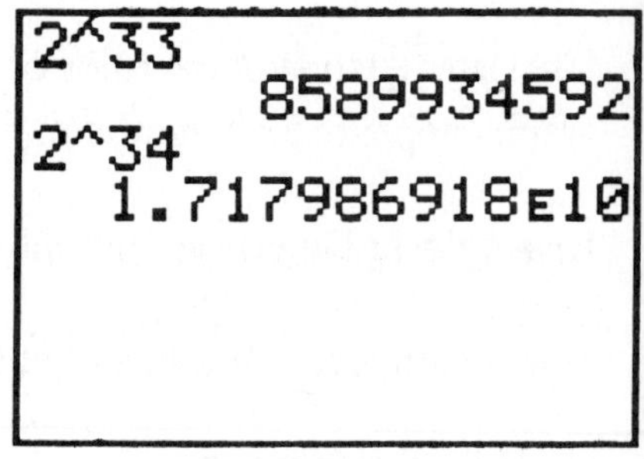

Figure 5.4

Small Numbers (Close to Zero) on the TI-81

Numbers that are close to zero are usually represented in scientific notation on the TI-81. Enter .000000000105 on the Home screen. See Fig. 5.5. The decimal is converted to scientific notation. This notation means 1.05×10^{-10}, which is the same as .000000000105. This number is very close to zero.

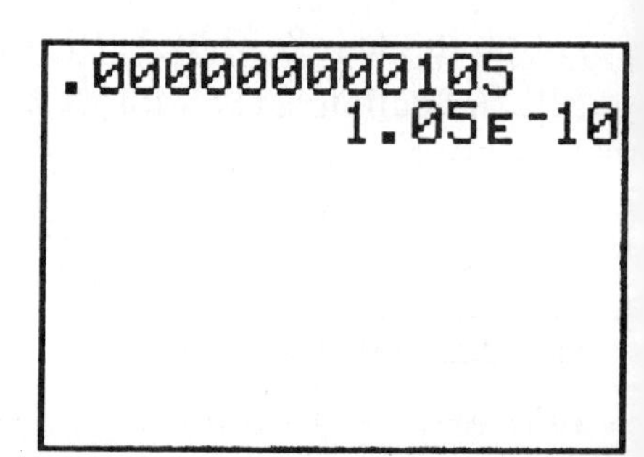

Figure 5.5

Graph $Y_1 = X$ in the Standard window. With the Zoom factors set at 5, Zoom In on the origin. Now Trace to the origin. See Fig. 5.6. The calculator has approximated zero for both coordinates.

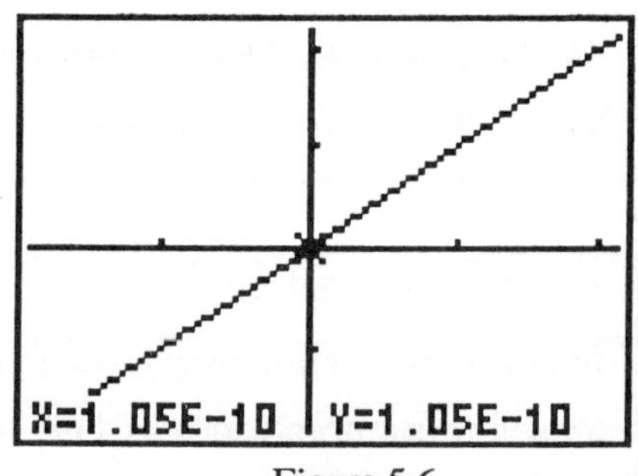

Figure 5.6

Section 5.2 Polynomial Expressions

A New Method of Evaluating a Polynomial Expression on a Grapher

Section 1.3 described one method of evaluating a polynomial expression at a particular value of x. Now we will describe another method to evaluate a polynomial that uses the Y= menu. We will enter the polynomial in Y_1, store the value of x in the TI-81 variable 'X,' and then print the value of Y_1 on the Home screen. We print the value of Y_1 by copying it from the **Y-vars menu** to the Home screen.

The Method of Entering the Expression in the Function-Defining Menu

Example 1: Evaluate $2x^3 - 4x^2 + 5x - 7$ at x = –1.

Enter $Y_1 = 2X \wedge 3 - 4X \wedge 2 + 5X - 7$ in the Y= menu, then exit to the Home screen and clear the Home screen. Store –1 in X by pressing

[(-)] **1** [STO▶] [X] [ENTER].

Now copy the variable Y_1 to the Home screen by pressing

[2nd] [Y-VARS][1:Y_1],

and then print the value of Y_1 by pressing [ENTER]. See Fig. 5.7. The TI-81 evaluates Y_1 at the current value of 'X.'

Table Building for Evaluating Polynomial Expressions

The TI-81 cannot make tables.

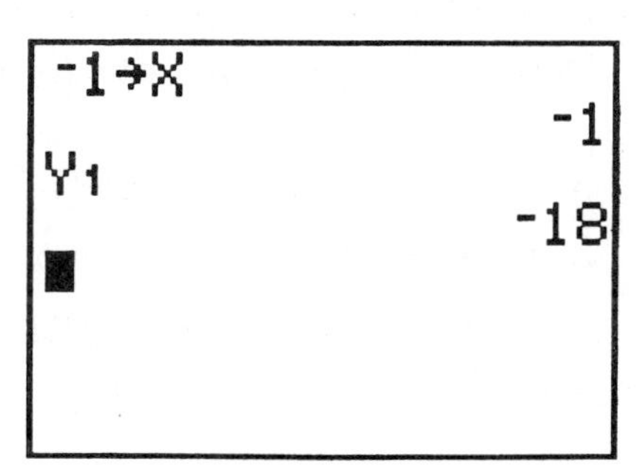

Figure 5.7

Graphical Support for the Addition of Polynomials

In Section 5.1 we used [GRAPH] and [TRACE] to provide visual support for rules of exponents. In this section we will use a similar method to provide support for addition of polynomials. We will enter the two polynomials in Y_1 and Y_2. Then we will enter $Y_3 = Y_1 + Y_2$ and enter the sum we derived algebraically in Y_4. Next we will deselect Y_1 and Y_2 and then graph Y_3 and Y_4 in the same window. If there appears to be only one graph in the window, we have support that Y_3 and Y_4 are the same. We can gather more support by Tracing both curves to see if they are the same.

Example 2: Find the sum of the polynomials

$6x^4 + 2x^3 + 2x^2 + 10x + 1$ and $-5x^4 - 7x^3 + 3x^2 - 3x - 8$,

and support your answer graphically.

The algebraic answer is $x^4 - 5x^3 + 5x^2 + 7x - 7$. Support this answer by clearing the Y= menu and entering

$Y_1 = 6X \wedge 4 + 2X \wedge 3 + 2X \wedge 2 + 10X + 1$

and

$Y_2 = -5X \wedge 4 - 7X \wedge 3 + 3X \wedge 2 - 3X - 8.$

Enter

$Y_3 = Y_1 + Y_2$

by placing the cursor next to 'Y_3 =' and pressing

[CLEAR] [2nd] [Y-VARS][1:Y_1] [+] [2nd] [Y-VARS][2:Y_2] [ENTER].

Then enter

$Y_4 = X \wedge 4 - 5X \wedge 3 + 5X \wedge 2 + 7X - 7.$

Deselect Y_1 by moving the cursor on to the '=' symbol beside Y_1 and pressing [ENTER]. This should unhighlight the '=' symbol. Deselect Y_2 the same way. See Fig. 5.8. Graph these polynomials in the [–5, 5] by [–10, 50] window. There appears to be just one graph. Reinforce this by pressing [TRACE] and moving the Trace cursor along the graph with the cursor-movement keys. As you move the Trace cursor along

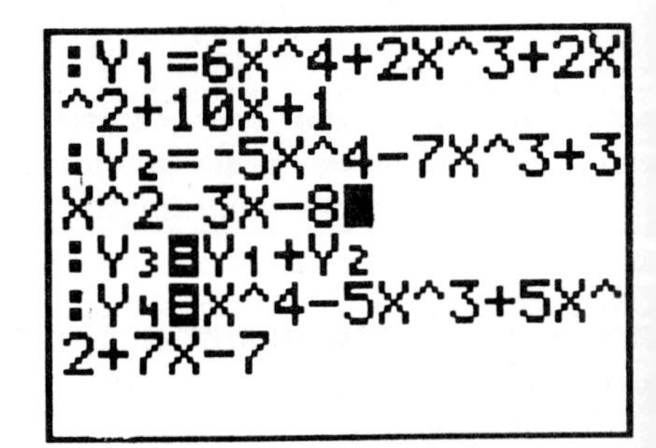

Figure 5.8

the graph, move from the graph of Y_3 to the graph of Y_4 by pressing [▼]. See Fig. 5.9. Since the coordinates don't change when you move from the graph of Y_3 to the graph of Y_4, we have support that the two polynomials are the same.

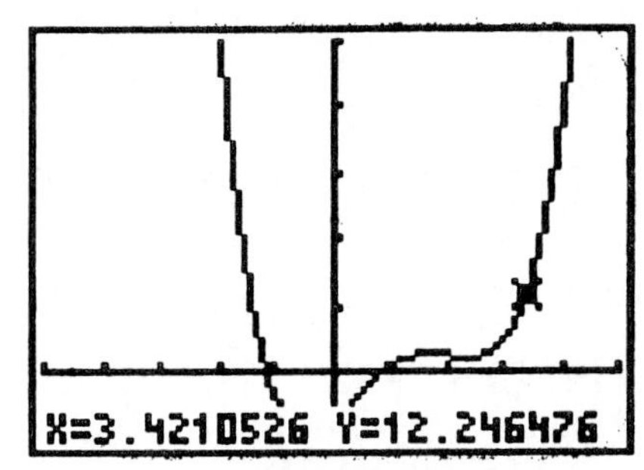

Figure 5.9

Evaluating Polynomials with Several Variables

We will use [STO▶] to assign values to the variables in the polynomial and then evaluate the polynomial. This method was introduced in Section 1.3.

Example 3: The volume of a tennis ball container is given as the polynomial

$\pi x^2 y + (2/3)\pi x^3$,

where x and y are dimensions of the container. Find the volume of the container when $x = 3.75$ and $y = 16.25$.

From the Home screen press

3 [.] **75** [STO▶] [X] [ENTER] **16** [.] **25** [STO▶] [Y] [ENTER].

Then press

[2nd] [π] [X|T] [x²] [ALPHA] [Y] [+] [(] **2** [÷] **3** [)] [2nd] [π] [X|T] [^] **3** [ENTER].

```
3.75→X
                3.75
16.25→Y
               16.25
πX²Y+(2/3)πX^3
           828.3496255
```

Figure 5.10

See Fig. 5.10. The volume is about 828.35.

Section 5.4 Polynomial Equations and Factoring

See Section 2.1 for a discussion of the 0.1 window.

Chapter 6 Rational Expressions and Equations

Section 6.1 Simplifying Rational Expressions

Using Parentheses in Rational Expressions

When evaluating and graphing rational expressions, you must use parentheses to enter the expression correctly.

Example 1: Graph $y = \frac{x-2}{x}$.

Enter $Y_1 = (X-2)/X$ and $Y_2 = X - 2/X$. Graph in the Standard window. See Fig. 6.1. Since there are two different graphs, we can see that Y_1 and Y_2 are not the same. The correct version is Y_1. This is because of the order of operations. Order of operations was discussed in Section 1.1. The parentheses in Y_1 force the subtraction to take place before the division. In Y_2 the division takes place before the subtraction since division has higher precedence than subtraction.

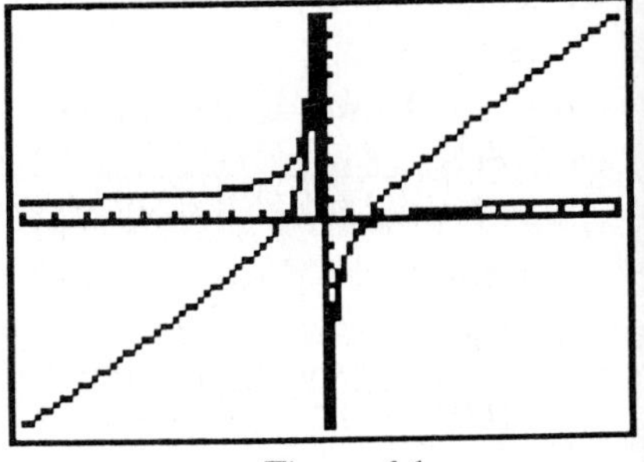

Figure 6.1

Math Errors with Rational Expressions

From the Home screen, press

[CLEAR] **0** [STO ▶] [X] [ENTER] **1** [÷] [X|T] [ENTER].

You should see the same MATH error that you saw in the Preliminary Chapter. This is because we have instructed the TI-81 to divide by zero. You may encounter this difficulty when evaluating other rational functions. For example, you cannot evaluate $1/(x-2)$ at $x = 2$ because division by zero is undefined.

Section 6.2 Multiplying and Dividing Rational Expressions

See Section 3.1 for a discussion of selecting and deselecting graphs.

Section 6.3 Adding and Subtracting Rational Expressions

Graphing Rational Expressions in Different Viewing Windows

Example 1: Graph $y = \dfrac{1}{x - 2}$.

Enter

$Y_1 = 1/(X - 2)$

and graph in the [–4.8, 4.7] by [–3.2, 3.1] window. See Fig. 6.2. Now regraph in the [–5, 5] by [–5, 5] window. See Fig. 6.3. The vertical line that you see in Fig. 6.3 is not really a part of the graph of $1/(x - 2)$. In some viewing windows, the bottom branch of the graph of a rational expression may be connected to the top branch by a vertical line. This vertical line is not a part of the correct graph. The occasional connecting of the branches has to do with the way the TI-81 graphs functions. It plots some of the points in the function and then connects those points to produce the final graph. In some windows, when the points are connected, a vertical line appears.

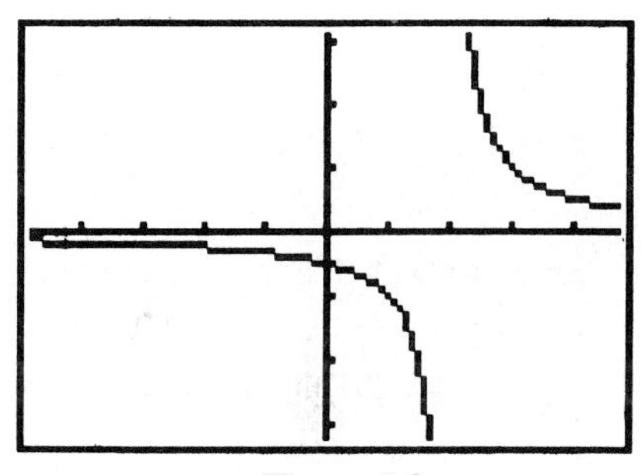

Figure 6.2

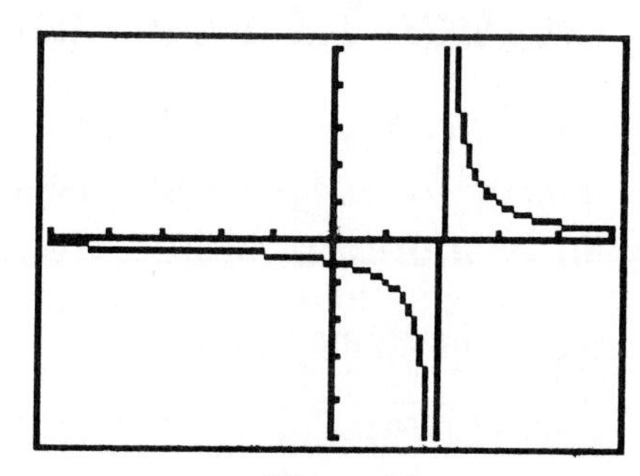

Figure 6.3

Section 6.4 Complex Fractions

Using Parentheses with Complex Fractions

Parentheses are required to enter expressions involving complex fractions. Make sure you have parentheses around the entire numerator and around the entire denominator.

Example 1: Evaluate $\dfrac{\frac{1}{2}+\frac{2}{3}}{\frac{5}{6}-\frac{3}{4}}$.

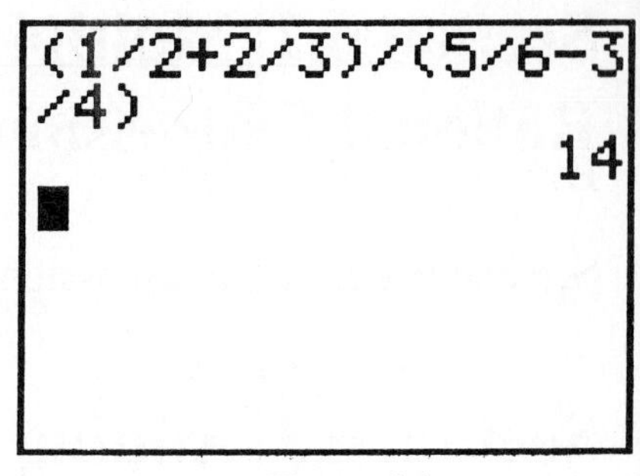

Figure 6.4

Clear the Home screen and enter (1/2 + 2/3)/(5/6 – 3/4). See Fig. 6.4. The answer is 14. The parentheses force the TI-81 to calculate the numerator and denominator first and then divide.

Section 6.6 Solving Equations Containing Rational Expressions

See Section 2.1 for a discussion of the 0.2 and Integer windows.

Chapter 7 Rational Exponents, Radicals, and Complex Numbers

Section 7.1 Roots and Radicals

Error Messages with Grapher Limitations of Evaluating Square Roots of Negative Numbers

From the Home screen press

[CLEAR] [2nd] [√] [(-)] **3** [ENTER].

An ERROR message appears indicating a MATH error. The MATH error occurs because the square root of negative three is not a real number. $\sqrt{-3}$ is not real because no real number squared is –3.

Evaluating Expressions with Fractional Exponents

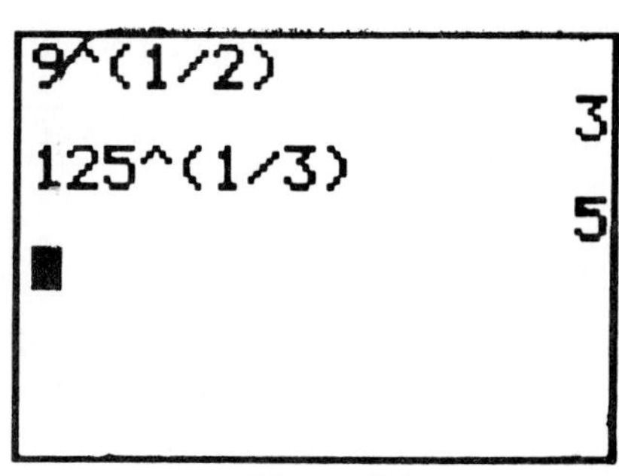

Figure 7.1

Example 1: Evaluate $9^{1/2}$ and $125^{1/3}$.

Go to the Home screen and press

[CLEAR] **9** [^] [(] **1** [÷] **2** [)] [ENTER] **125** [^] [(] **1** [÷] **3** [)] [ENTER].

See Fig. 7.1. You must place parentheses around the exponent.

Example 2: Compare the graphs of $y = x \wedge (1/4) \wedge 2$ and $y = x \wedge 2 \wedge (1/4)$.

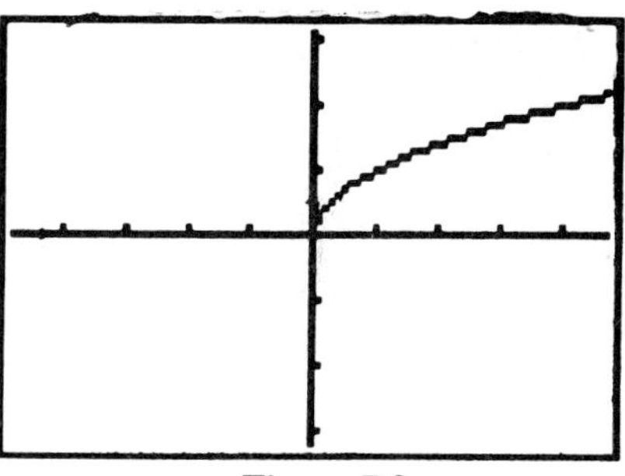
Figure 7.2

Enter $Y_1 = X \wedge (1/4) \wedge 2$ and graph in the Decimal window. See Fig. 7.2. Now erase Y_1 and graph $Y_1 = X\wedge 2\wedge (1/4)$ in the Decimal window. See Fig. 7.3. Both graphs appear in the first quadrant, but only the second graph also appears in the second quadrant where x is negative. This is because the TI-81 evaluates the expression from left to right. When it evaluates the first expression it takes the fourth root of x and then squares. The fourth root of a negative number is not real, so the first expression is not defined when x is negative. When the TI-81 evaluates the second expression it squares and then takes the fourth root. The square of a negative number is positive so the second expression is defined when x is negative.

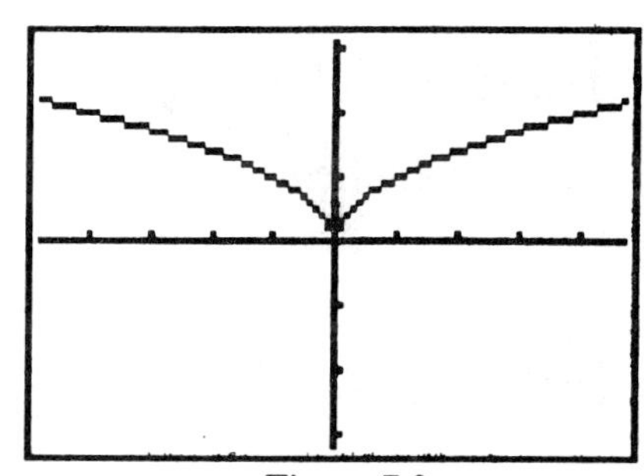
Figure 7.3

Finding Roots with Your Grapher

The square roots can be evaluated with [√], and cube roots can be evaluated with the fourth option in the MATH menu, but other roots must be evaluated by converting the root to a fractional exponent.

Example 3: Evaluate $\sqrt[4]{8}$.

The fourth root of eight is the same as eight raised to the one-fourth power.

Evaluate the fourth root of eight by pressing

[CLEAR] **8** [^] [(] **1** [÷] **4** [)] [ENTER].

Then evaluate the cube root of –8.75 by pressing

[MATH] [4: $\sqrt[3]{\ }$] [(-)] 8 [.] 75 [ENTER].

See Fig. 7.4.

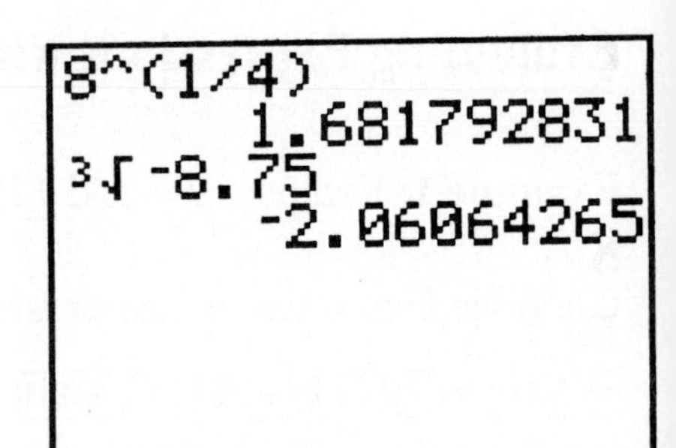

Figure 7.4

Adjusting the Number of Decimal Places Displayed by the Grapher

We will adjust the number of decimal places displayed with the Fixed mode option in the Mode screen. This was discussed in Section 1.1.

Example 4: Evaluate $\sqrt[4]{8}$ using a floating decimal display setting and then a fixed decimal display setting of 2 digits.

Select Floating Decimal mode and enter

8 ^ (1/4)

on the Home screen. Next, select Fixed Decimal mode by highlighting the '2' on the second line of the Mode screen, and enter

8 ^ (1/4)

on the Home screen. See Fig. 7.5. In Fixed Decimal (2) mode, only two digits to the right of the decimal are displayed.

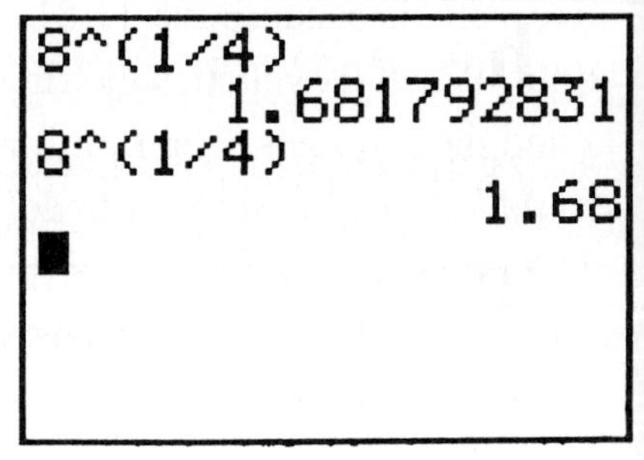

Figure 7.5

Section 7.2 Rational Exponents and Radicals

See Section 7.1 for a discussion of evaluating an expression with rational exponents.

Section 7.6 Complex Numbers

Using the Grapher to Work with Complex Numbers

The TI-81 cannot work with complex numbers.

Chapter 8 Quadratic Functions

8.4 Sketching Graphs of Quadratic Functions

See Section 2.1 to review the 0.2 window.

8.6 Vertex and Symmetry of a Parabola

See Section 1.1 to review the Frac command.

Chapter 9 Higher Order Systems of Equations and Matrices

Section 9.1 Solving Systems of Equations in Three Variables

Equations in three variables cannot be graphed on the TI-81. The only support the TI-81 can provide for a solution to a system of equations in three variables is numerical.

Section 9.2 Matrix Algebra

Entering a Matrix on the TI-81

The matrix key is on the fourth row, second column. Press [T][▶] to see the Matrix edit screen. See Fig. 9.1. You can use up to three matrices on the TI-81. The notation for these matrices is $[A]$, $[B]$, and $[C]$. The numbers to the right of each matrix name on the Matrix edit screen are the number of rows and columns in the matrix.

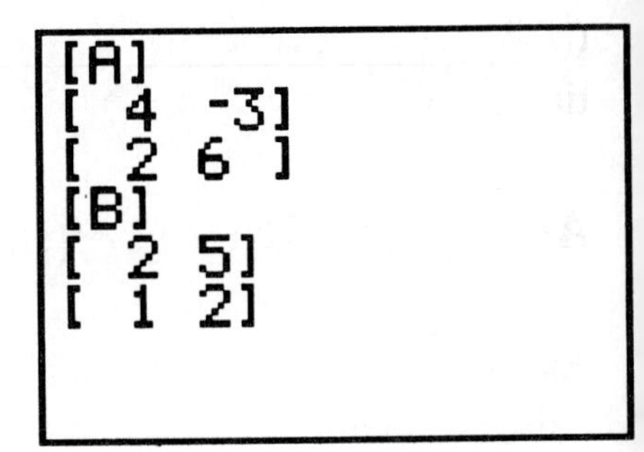

Figure 9.1

Example 1: Enter the 2×2 matrix

$$A = \begin{bmatrix} 4 & -3 \\ 2 & 6 \end{bmatrix}$$

and the 2×2 matrix

$$B = \begin{bmatrix} 2 & 5 \\ 1 & 2 \end{bmatrix}.$$

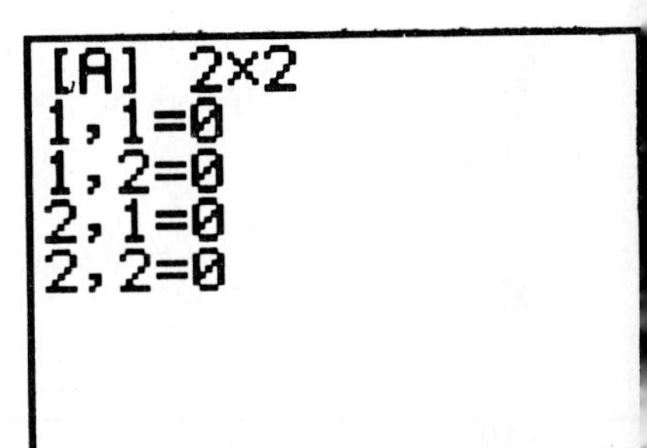

Figure 9.2

We must first change the dimension of matrix $[A]$ to 2 rows and 2 columns. From the Matrix edit screen press

1 2 [ENTER] **2** [ENTER].

See Fig. 9.2. Now enter the values in the matrix by pressing

4 [ENTER] [(-)] **3** [ENTER] **2** [ENTER] **6.**

[A] 2×2
1,1=4
1,2=-3
2,1=2
2,2=6

Figure 9.3

See Fig. 9.3. Now edit matrix $[B]$ by pressing

[T][▶] **2 2** [ENTER] **2** [ENTER] **2** [ENTER] **5** [ENTER] **1** [ENTER] **2.**

[B] 2×2
1,1=2
1,2=5
2,1=1
2,2=2

Figure 9.4

See Fig. 9.4. Press

[2nd] [QUIT]

to return to the Home screen. You can print the values in matrix $[A]$ and matrix $[B]$ by pressing

[2nd] [A] [ENTER] [2nd] [B] [ENTER].

([A] is the second function of the [1] key and [B] is the second function of the [2] key.) See Fig. 9.5.

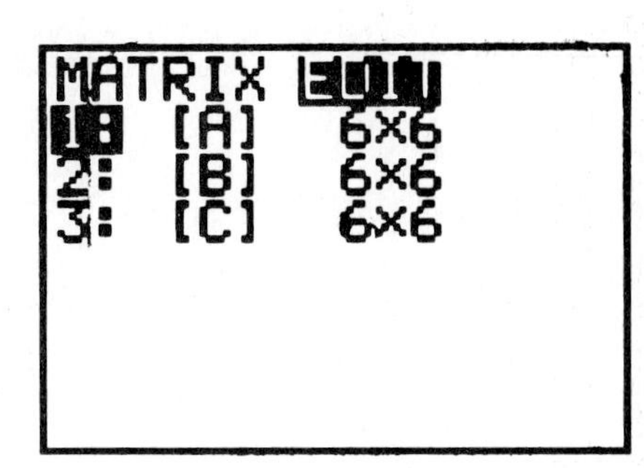

Figure 9.5

Adding and Subtracting Matrices

Example 2: Find $[A] + [B]$ and $[A] - [B]$, where $[A]$ and $[B]$ have the values assigned in Example 1.

Press

[CLEAR] [2nd] [A] [+] [2nd] [B] [ENTER] [2nd] [A] [−] [2nd] [B] [ENTER].

See Fig. 9.6.

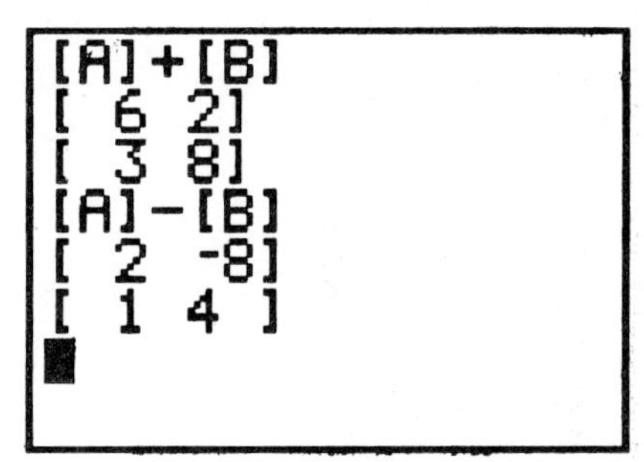

Figure 9.6

Scalar Multiplication of Matrices

Example 3: Find $3[A]$ where

$$[A] = \begin{bmatrix} 2 & 1 & 5 \\ 6 & -1 & 3 \end{bmatrix}.$$

Enter matrix $[A]$ by pressing

[T] [▶] **12** [ENTER] **3** [ENTER] **2** [ENTER] **1** [ENTER] **5** [ENTER] **6** [ENTER] [(−)] **1** [ENTER] **3**.

Figure 9.7

See Fig. 9.7. Notice you enter matrices one row at a time on the TI-81. Press

[2nd] [QUIT]

to return to the Home screen. Now perform the scalar multiplication by pressing

[CLEAR] **3** [2nd] [A] [ENTER].

See Fig. 9.8.

Figure 9.8

Section 9.3 Multiplication of Matrices

Matrix Multiplication

Once matrices have been entered on the TI-81, you can multiply them by simply entering them side-by-side on the Home screen.

Example 1: If

$$A = \begin{bmatrix} 3 & -4 \\ 2 & -8 \end{bmatrix} \quad \text{and} \quad B = \begin{bmatrix} 4 & -2 \\ -1 & 9 \end{bmatrix},$$

find $[A] \times [B]$.

Enter the matrices A and B with the Matrix edit screen. (Entering matrices was explained in Section 9.2.) Then from the Home screen press

[CLEAR] [2nd] [A] [2nd] [B] [ENTER].

See Fig. 9.9.

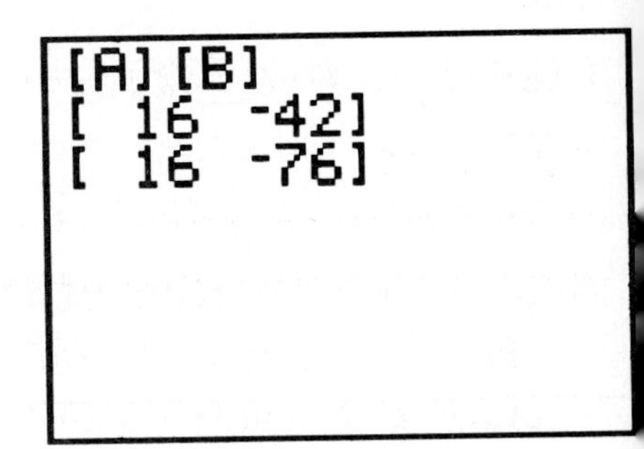

Figure 9.9

The Identity Matrix on the Grapher

The TI-81 does not have a built-in identity matrix. You must enter it using the Matrix edit screen. The Matrix edit screen was discussed in Section 9.2.

Inverse Matrices Using the [x^{-1}] Key

If a matrix can be inverted, you can usually find its inverse with [x^{-1}]. If the inverse doesn't exist, you will see an error message or an odd-looking matrix with entries that are very near zero.

Example 2: Find the inverse of

$$A = \begin{bmatrix} 5 & 1 \\ 4 & 1 \end{bmatrix}.$$

Enter matrix A with the Matrix edit screen, and then from the Home screen press

[CLEAR] [2nd] [A] [x^{-1}] [ENTER].

See Fig. 9.10.

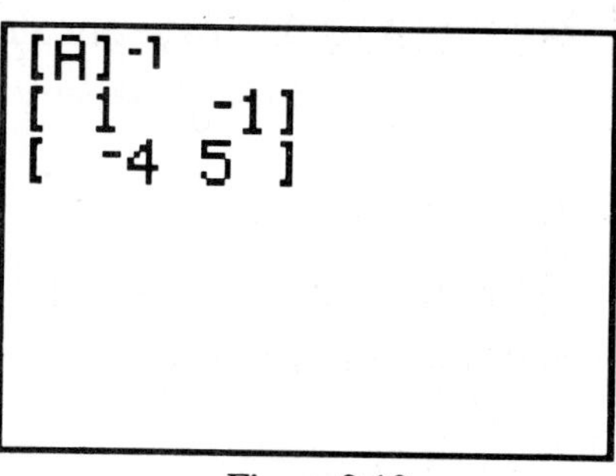

Figure 9.10

Example 3: Find the inverse of

$$A = \begin{bmatrix} 3 & 1 \\ 3 & 1 \end{bmatrix}.$$

Enter matrix [A] with the Matrix edit screen, and from the Home screen press

[CLEAR] [2nd] [A] [x^{-1}] [ENTER].

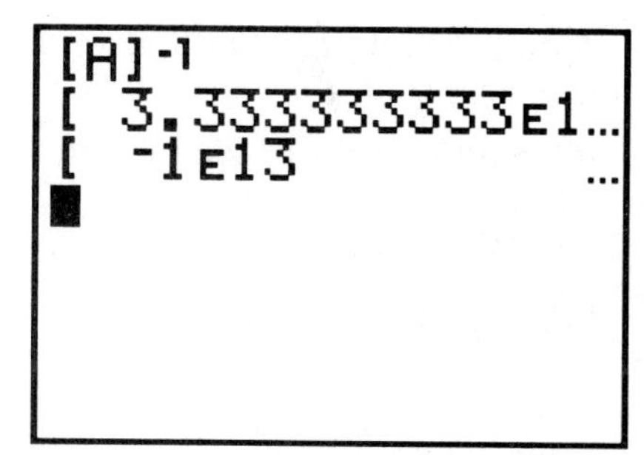

Figure 9.11

See Fig. 9.11. You can scroll to the right to see the rest of the inverse by pressing and holding [▶]. The entries in this calculator that produced the inverse are very near zero. In fact, the inverse of [A] does not exist. The TI-81 does not recognize this because it has approximated the inverse. If you see an error message or if all the entries are very near zero when you attempt to find the inverse of a matrix, the inverse may not exist.

Determinants of a Square Matrix

The option [5:det] (determinant) is the fifth option in the MATRIX menu. This feature can be used to find determinants of square matrices.

Example 4: Find the determinant of

$$A = \begin{bmatrix} 3 & 1 \\ 3 & 1 \end{bmatrix}.$$

Enter matrix [A] with the Matrix edit screen. Then from the Home screen press

[CLEAR] [T] [5:det] [2nd] [A] [ENTER].

See Fig. 9.12. The determinant is zero. The TI-81 approximates this determinant with a value very close to zero.

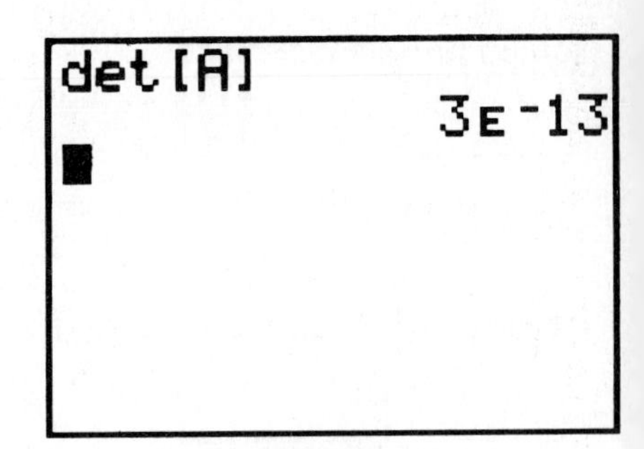

Figure 9.12

Section 9.4 Solving a System of Equations Using Matrices

How to Solve a System of Equations with Matrices on the Grapher

Systems of equations can be expressed with matrices. When this is done, the solution of the system is a matrix. We can solve for this solution matrix with the inverse matrix feature of the TI-81.

Example 1: Solve the system

$$\begin{cases} 4x - 3y = -26 \\ 15x + 11y = 36 \end{cases}$$

This system can be expressed in matrix form as $AX = B$, where

$$A = \begin{bmatrix} 4 & -3 \\ 15 & 11 \end{bmatrix}, \quad X = \begin{bmatrix} x \\ y \end{bmatrix}, \quad \text{and} \quad B = \begin{bmatrix} -26 \\ 36 \end{bmatrix}.$$

We then solve the matrix equation $AX = B$ for X, obtaining $X = A^{-1}B$. Matrix X contains the solution to the original system of equations. Enter the values just described for the 2×2 matrix A and the 2×1 matrix B with the Matrix edit screen. Don't forget to change the dimension of $[B]$ to two rows, one column. After you have entered A and B, return to the Home screen and find the product $A^{-1}B$ by pressing

[CLEAR] [2nd] [A] [x^{-1}] [2nd] [B] [ENTER].

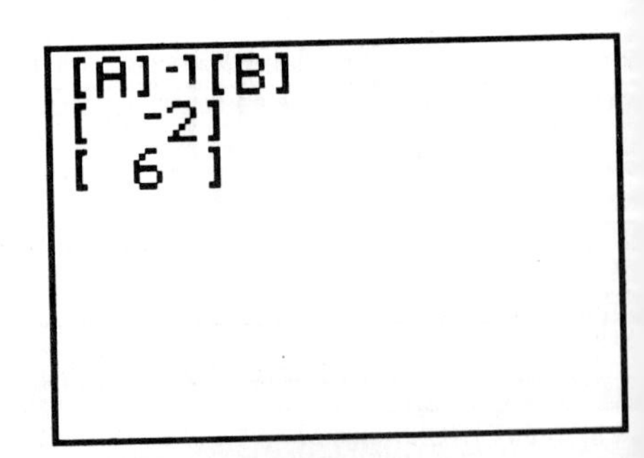

Figure 9.13

See Fig. 9.13. The solution is $x = -2$, $y = 6$. Matrix multiplication is not commutative, so you must enter $A^{-1}B$, not BA^{-1}.

Chapter 10 Conic Sections

Conic Sections

In general, when we graph conic sections on the TI-81, we must rewrite the conic equation as $y = \ldots$ and then graph. While solving for a conic equation for y, we usually obtain two new equations. A lengthy radical will probably be created in both equations. We will enter this radical into the Y_1 slot and enter the two equations in Y_2 and Y_3 with Y_1 used to represent the radical. Then we will deselect Y_1 and graph Y_2 and Y_3. There may be gaps between the graphs of Y_2 and Y_3, due to screen resolution. See Section 3.1 for a discussion of selecting and deselecting graphs since this skill will be used in this chapter.

Example 1: Graph the ellipse $(x - 1)^2/4 + (y + 2)^2/9 = 1$.

We solve for y and obtain $y = -2 \pm 3\sqrt{(4-(x-1)^2)}/2$. Enter

$Y_1 = 3\sqrt{(4-(X-1)^2)}/2$, $Y_2 = -2 + Y_1$, and $Y_3 = -2 - Y_1$. Deselect Y_1. See Fig. 10.1. Now graph in the Standard viewing window. See Fig. 10.2.

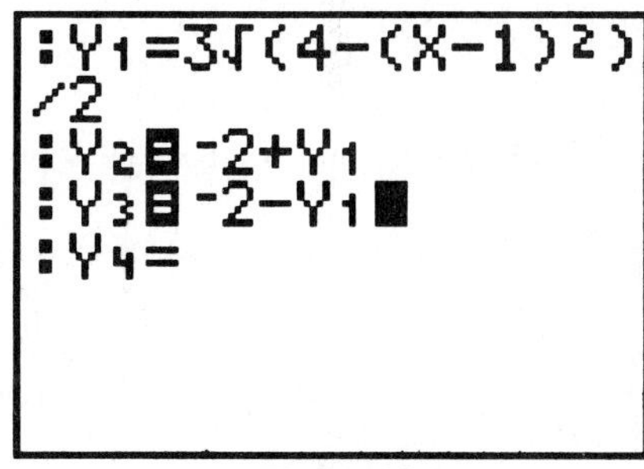

Figure 10.1

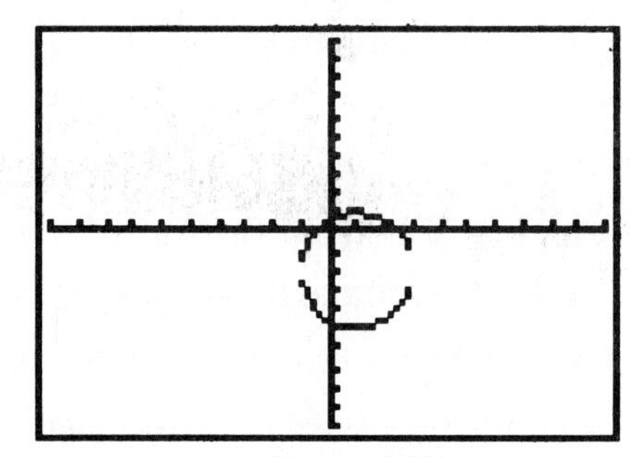

Figure 10.2

Section 10.1 Circles

Graph in the Square Window for a Circle

A circle must be graphed in a square window for the true shape of the circle to appear. In the Standard window, the circle will look distorted.

Example 1: Graph the circle $x^2 + y^2 = 25$.

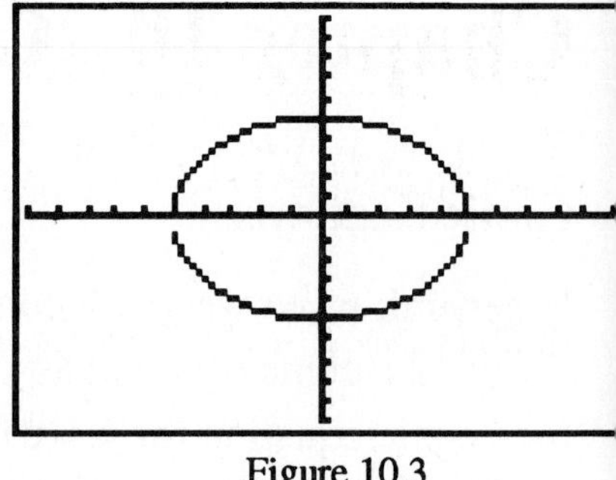
Figure 10.3

Enter $Y_1 = \sqrt{(25 - X \wedge 2)}$ and $Y_2 = -Y_1$. Graph in the Standard window. See Fig. 10.3. The circle looks like an ellipse. Now press ZOOM [5:Square].

See Fig. 10.4. The proportions of the circle are correct in a square window.

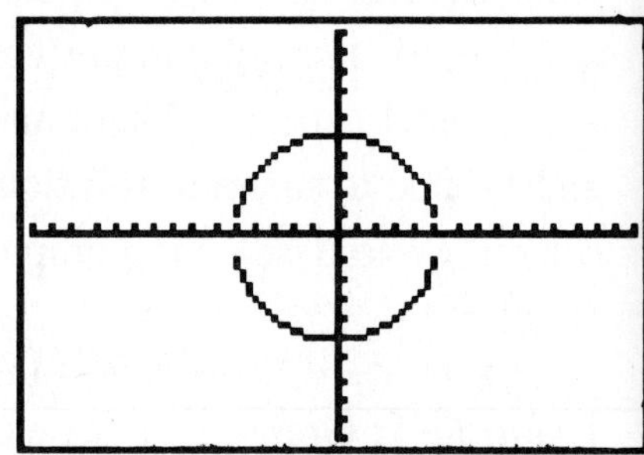
Figure 10.4

Section 10.5 Nonlinear Systems of Equations

A Review on Solving Systems of Equations Graphically

See Chapter 3 for a review of how to solve systems of equations graphically. Recall that we set the scale marks to 0.01 to assist with a solution with at most 0.01 error.

Example 1: Solve the system $x^2 + y^2 = 16$ and $y = 2x^2 - 2$.

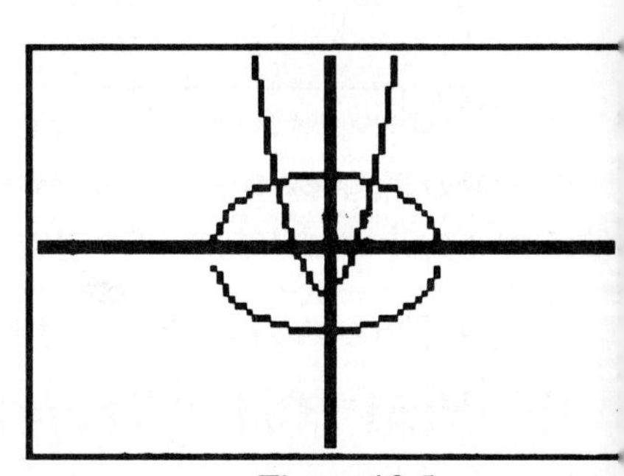
Figure 10.5

We must solve the first equation for y before we can solve graphically. This gives $y = \pm\sqrt{(16 - x^2)}$. Enter $Y_1 = \sqrt{(16 - X\wedge 2)}$ and $Y_2 = -Y_1$. Enter $Y_3 = 2X \wedge 2 - 2$ and graph all three in the [–10, 10] by [–10, 10] window with scale values of 0.01. See Fig. 10.5. Before we Zoom In, notice that Y_2 has no intersection points with Y_3, so we will deselect Y_2 for faster graphing. Now we Zoom In several times on the intersection in the first quadrant. See Fig. 10.6. Since the intersection is contained between the scale marks, we know that the error is at most 0.01. The other solution can be found by regraphing in the original window and then Zooming In on the intersection in the second quadrant.

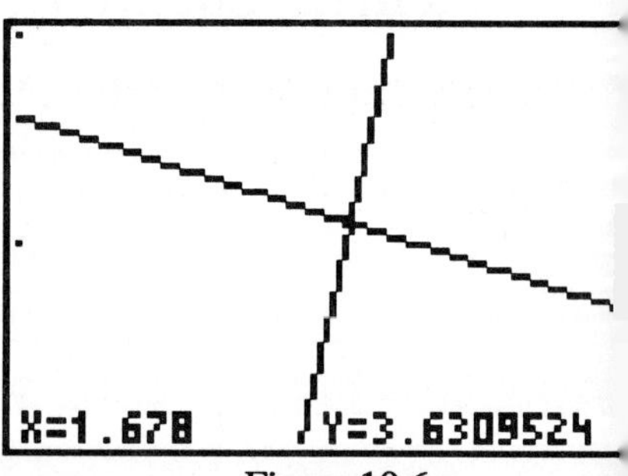

Figure 10.6

Chapter 11 Exponential and Logarithmic Functions

Section 11.3 Logarithmic Functions

The [LOG] and [LN] Keys

The base ten log key [LOG] is on the seventh column, first row. The natural log key [LN] is just below the base ten log key.

Example 1: Evaluate log 125 and ln 12.3.

From the Home screen press

[CLEAR] [LOG] **125** [ENTER] [LN] **12** [.] **3** [ENTER].

See Fig. 11.1.

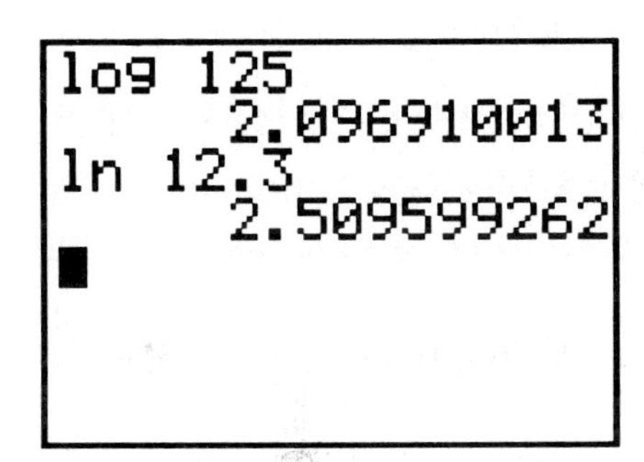

Figure 11.1

Section 11.5 Solving Logarithmic Equations

Some Graphers Cannot Evaluate Y_1 at 3 by Using the Notation $Y_1(3)$

The TI-81 cannot evaluate Y_1 at 3 by using the notation $Y_1(3)$. See Section 5.2 to review how to evaluate an expression on the TI-81.

Revisiting the Grapher-Generated Table

It is not possible to make tables on the TI-81.

Chapter 12 Sequences and Series and the Binomial Theorem

Section 12.1 Arithmetic Sequences and Series

Generating Sequences by Using a Sequence Command

It is not possible to generate sequences with a **seq(** command on the TI-81.

Displaying a Graph of a Sequence

Sequences can be graphed like functions if we use the Dot mode and graph in the Integer window [0, 95] by [0, 63].

Example 1: Display the graph of the sequence 23, 26, 29, . . . , and determine the value of the sixth term.

This sequence is given by the formula $23 + (X - 1)3$. Enter this formula in Y_1, change from Connected to Dot mode, and graph in the [0, 95] by [0, 63] window. Use x and y scales of 10. Now trace until $x = 6$. See Fig. 12.1. The sixth term is 38.

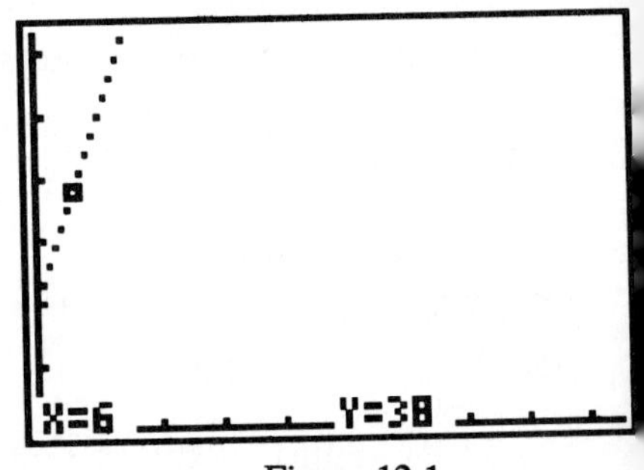

Figure 12.1

Section 12.2 Geometric Sequences and Series

See Section 12.1 for guidelines on graphing sequences.

Section 12.3 Binomial Expansion

Factorials

The Factorial function is the fifth option in the MATH menu.

Example 1: Evaluate 5! and 8!5!/(4!3!).

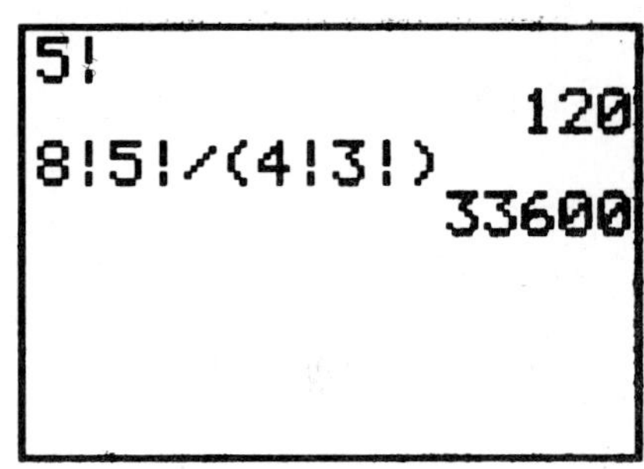

Figure 12.2

From the Home screen, press

CLEAR **5** MATH [5:!] ENTER

to evaluate the first expression and

8 MATH [5:!] **5** MATH [5:!] ÷ (**4** MATH [5:!] **3** MATH [5:!])
ENTER

to evaluate the second expression. See Fig. 12.2.

Evaluating a Binomial Coefficient $\binom{n}{r}$ on the Grapher

The notation $\binom{n}{r}$ means the same thing as ${}_nC_r$. The notation ${}_nC_r$ is used for a combination of n things taken r at a time. This option is found in the MATH Prob (probability) menu.

Example 2: Determine the value of the binomial coefficient $\binom{5}{3}$.
$\binom{5}{3} = {}_5C_3$.

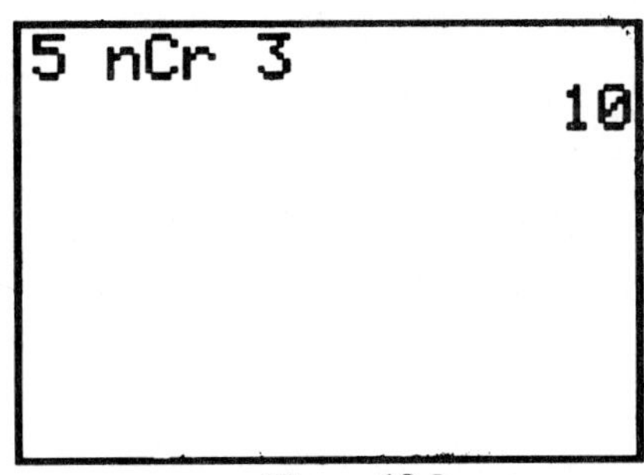

Figure 12.3

From the Home screen press

CLEAR **5** MATH ▶ ▶ ▶ [3:${}_nC_r$] **3** ENTER.

See Fig. 12.3. The result is 10.

Section 12.4 Counting Principle, Permutations, and Combinations

Evaluating a Permutation ${}_nP_r$

This option is also found in the MATH [Prob] menu. The notation '${}_nP_r$' means a permutation of *n* things taken *r* at a time.

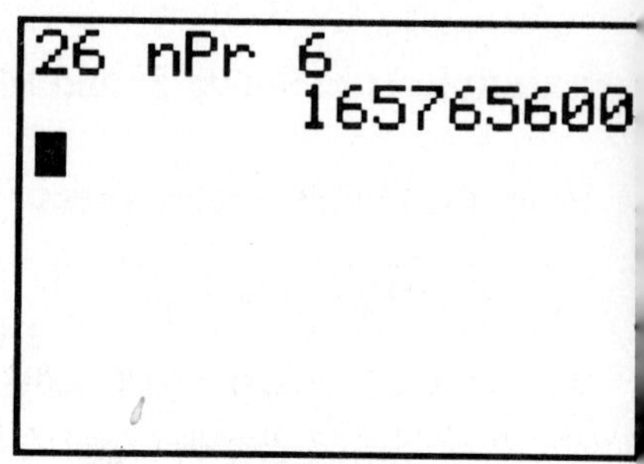

Figure 12.4

Example 1: Evaluate ${}_{26}P_6$.

From the Home screen, press

[CLEAR] **26** [MATH] [▶] [▶] [▶] [2:${}_nP_r$] **6** [ENTER].

See Fig. 12.4.

Evaluating a Combination ${}_nC_r$

This feature was described in Section 12.3.

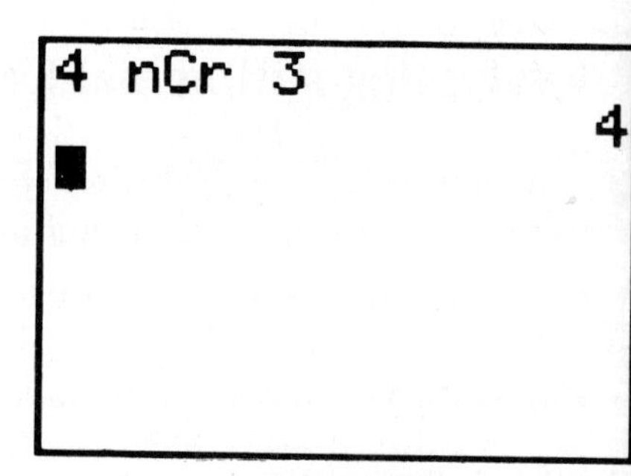

Figure 12.5

Example 2: Evaluate ${}_4C_3$.

Press

[CLEAR] **4** [MATH] [▶] [▶] [▶] [3:${}_nC_r$] **3** [ENTER].

See Fig. 12.5. The answer is 4.

Preliminary Chapter. Getting Started Using the TI-82

Section 1 The Keyboard

The keys on the TI-82 are organized by color and position. See Fig. 1. The grey keys on the lower part of the keyboard are used to enter numbers. The dark blue keys on the lower right-hand side of the keyboard are used to perform arithmetic. The black keys on the top row just under the screen are used to draw graphs of equations and make tables of numbers. The other black keys are used for scientific functions and menus. The dark blue keys with arrows in the upper right-hand corner of the keyboard are used to move the cursor on the screen. They are called the cursor-movement keys.

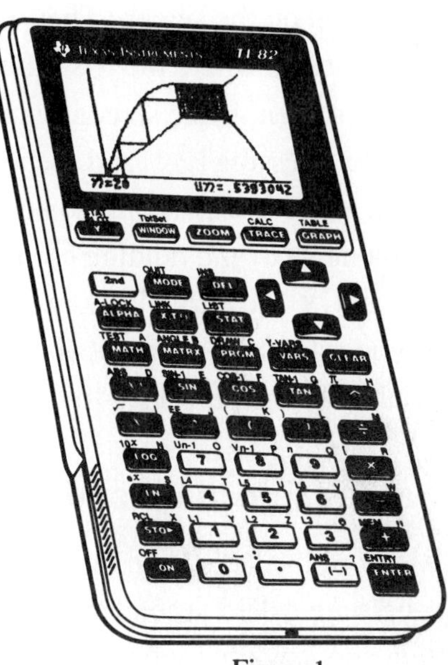

Figure 1

The [ON] and [OFF] Keys and the Automatic Off

The black key in the lower left-hand corner of the keyboard is used to turn the calculator on. Notice the key is labeled 'ON' and the word 'OFF' appears in light blue lettering above the key. Press [ON] to turn the calculator on. If you press [2nd] (the light blue key in the upper left-hand corner of the keyboard) and then press [ON], you will turn the calculator off. Pressing [2nd] lets you use the feature printed in light blue above the key. If you leave the TI-82 on for about five minutes without pressing any keys, it will turn itself off. This automatic off feature saves the batteries.

Section 2 The Cursor and Adjusting the Contrast

After you turn the calculator on, you should see a blinking rectangle displayed on the screen. This rectangle is called the cursor. The cursor indicates where the first number or letter you enter will be placed on the screen. If the screen seems too light, you can darken the display by pressing [2nd] and then holding [▲] until the contrast looks right. If the screen seems too dark, you can lighten the display by pressing [2nd] and holding [▼] until you have the contrast you want. ([▲] and [▼] are dark blue keys in the upper right-hand corner of the keyboard.) If you are not able to make the display dark enough, your batteries may need to be replaced.

Second Functions and the Alpha Key

Many of the keys have light blue labels above them. These blue labels are called second functions. You can access the second function above a key by first pressing [2nd] and then pressing the key. When you press [2nd] the cursor changes to '▮.' While the cursor looks like '▮,' you can access the second function above a key by pressing the key. Many of the keys also have a white letter or symbol above them. You can access the white letter above a key by first pressing [ALPHA] and then pressing the key. When you press [ALPHA] the cursor changes to '▮.' We say the cursor is in the Alpha mode. The role of [2nd] and [ALPHA] is similar to the shift key on a typewriter or computer keyboard.

Section 3 Keyboarding Instructions

When you see instructions to press a certain key, the instructions will include a box containing the label on the key. If the key is a number

key, however, the label will not be enclosed in a box. For example, if you follow the directions to turn on the calculator and press

[CLEAR] **2** [+] **4** [ENTER],

your screen should look like Fig. 2. When you are instructed to access a second function, the directions will include [2nd] followed by a box containing the second function label above the key rather than the label on the key. For example, notice the second function of [x^2] is [$\sqrt{\ }$]. ([x^2] is midway down the first column.) To calculate the square root of 2 you would press

[CLEAR] [2nd] [$\sqrt{\ }$] **2** [ENTER].

2+4
6

Figure 2

See Fig. 3. The reference to [$\sqrt{\ }$] in these directions means to press the key with x^2 on the key and $\sqrt{\ }$ above the key. The instructions to access a letter will include [ALPHA] followed by a box containing the letter above the key.

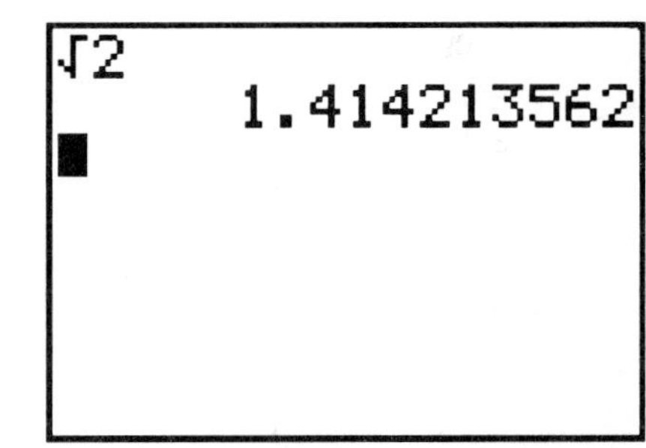

Figure 3

Section 4 Resetting

If you wish to reset the TI-82 to its factory settings and erase all previously entered data and programs, turn on the calculator, press [2nd], and then press [+]. ([+] is a dark blue key in the lower right-hand corner of the keyboard. Notice the word "MEM" in light blue lettering above [+].) After you press [2nd] and [+] you should see the **MEMORY** menu on the screen. The MEMORY menu should look like Fig. 4. Now press the [3] key to see the **MEMORY [2:Reset]** menu. This menu should look like Fig. 5. Press [2:Reset] to select **Reset**, the second item in the MEMORY [Reset] menu. The display should show the message 'Mem cleared.' Press

to clear the screen. ([CLEAR] is a black key on the fourth row, fifth column of the keyboard.) You may need to adjust the contrast after you reset your TI-82.

Figure 4

Section 5 The Home Screen, CLEAR, Menus, and QUIT

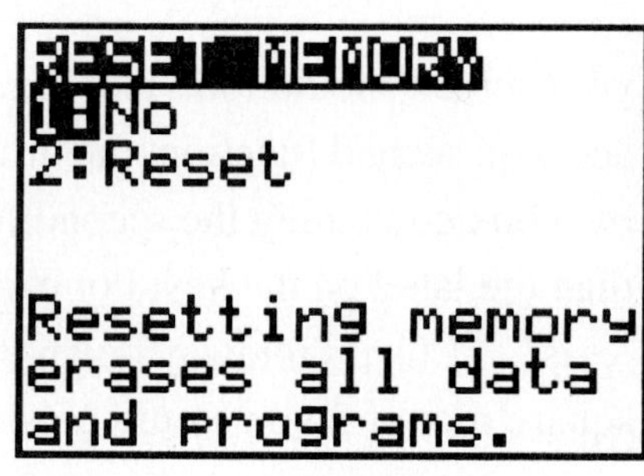

Figure 5

When you first turn on the TI-82, you see the **Home screen**. The calculations you performed in the Keyboarding Instructions Section were done on the Home screen. The CLEAR key you used before each calculation clears the Home screen of previous entries. If you make a mistake in a calculation you can start over by pressing [CLEAR]. You can access **Menu screens** from the Home screen. If you Reset your TI-82 in the Resetting Section, you used the MEMORY [Reset] menu. You can return from a Menu screen to the Home screen by pressing

[2nd] [QUIT].

([QUIT] is the second function of [MODE].) Sometimes you are automatically returned to the Home screen when you make a selection from the Menu. Press [MATH] to see the **MATH** menu. The [MATH] key is on the first column, fourth row of the keyboard. See Fig. 6. There are seven features visible on this menu, and the arrow by the seventh feature indicates there are more features available. To see the other features, press and hold [▼]. Press

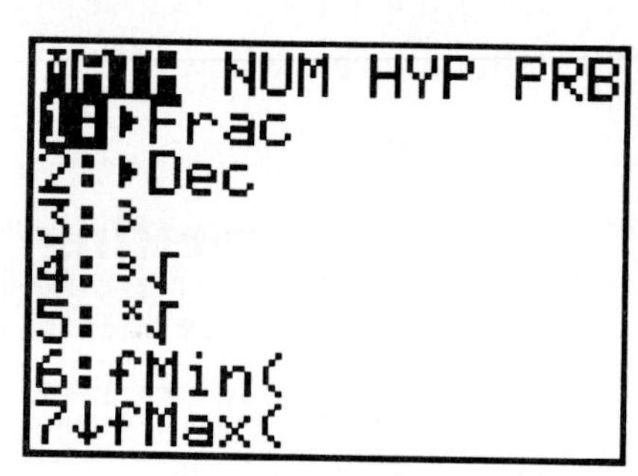

Figure 6

[2nd] [QUIT]

when you are ready to return to the Home screen.

You can calculate 7^3 by using a feature in the MATH menu. From the Home screen press

[CLEAR] **7** [MATH] [3:³] [ENTER].

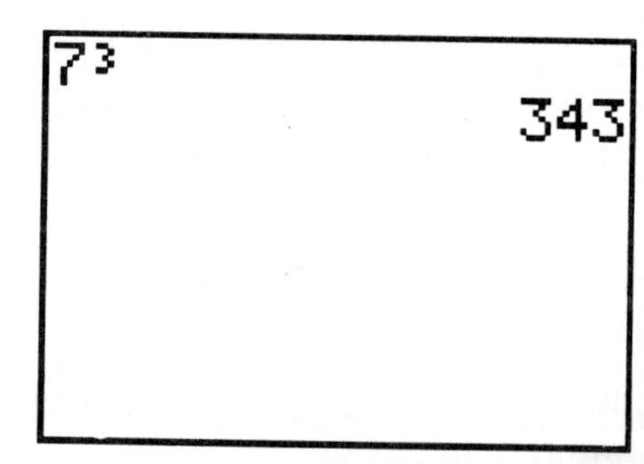

Figure 7

The result should be 343. See Fig. 7. The notation [3:³] in the keystroke instructions means to press 3 after you are in the MATH menu. Pressing the **3** key selects the third option in the menu. Did you notice that you were automatically returned to the Home screen after you selected the third option from the MATH menu? Selecting this option also caused the exponent of three to be printed on the Home screen.

The MODE Menu

From the Home screen press [MODE] (second row, second column). Your screen should look like Fig. 8. Each of the left-hand entries should be highlighted. We will discuss the meaning of these entries in later sections. If any of the left-hand entries are not highlighted on your calculator, use the cursor-movement keys to move the cursor to the unhighlighted item, and then press [ENTER] to highlight that item. Press [CLEAR] to return to the Home screen.

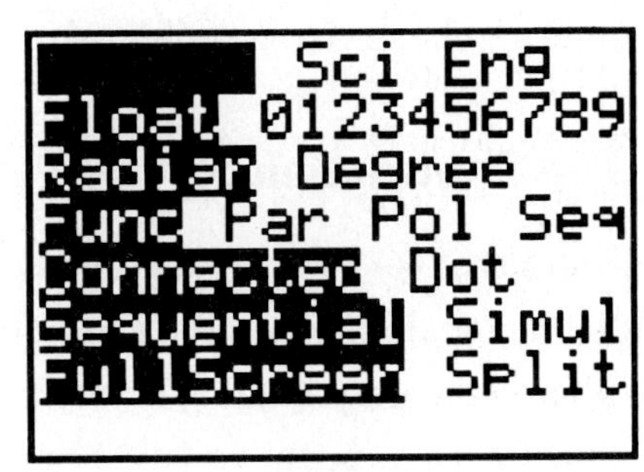

Figure 8

Section 6 Basic Operations Keys

The four dark blue keys [+] [−] [×] [÷] on the lower right-hand side of the keyboard are used for addition, subtraction, multiplication, and division. Press the following keys to perform the calculation 19 minus 27 plus 23:

[CLEAR] **19** [−] **27** [+] **23** [ENTER].

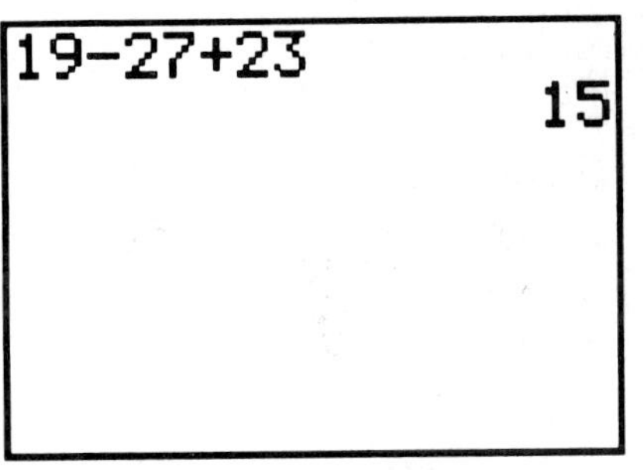

Figure 9

See Fig. 9. Notice you don't see the result of the calculation until you press [ENTER]. Then you see the expression on the left-hand side of the Home screen and the result of the calculation on the right-hand side of the next line. Press the following keys to multiply 18 by 145:

[CLEAR] **18** [×] **145** [ENTER].

The result is 2610. See Fig. 10. Notice pressing [×] produces the '*' symbol on the Home screen. Both '*' and '×' mean multiplication.

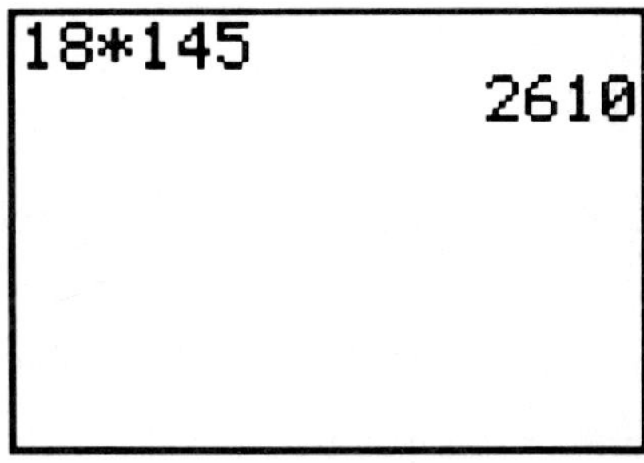

Figure 10

Now press

8 [÷] **156** [ENTER]

to divide 8 by 156. See Fig. 11. Notice pressing [÷] produces the '/' symbol on the Home screen. Both '÷' and '/' mean division. Also notice the previous calculation '18*145' remains on the screen. This is because you did not clear the screen by pressing [CLEAR] before entering the division problem.

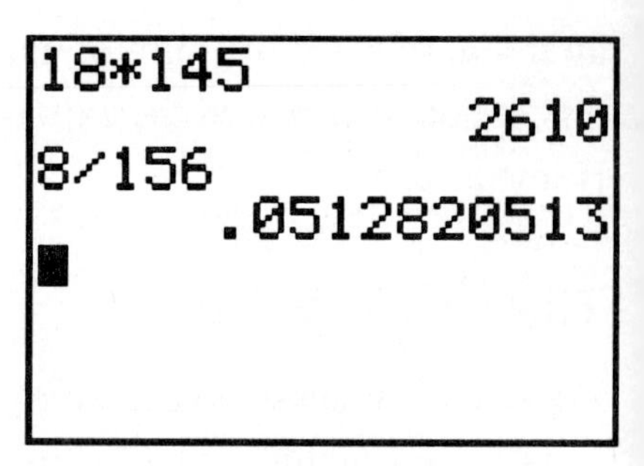

Figure 11

The Subtraction Key [−] versus the Opposite Key [(−)]

There are two keys which appear to be minus signs on the TI-82. One is the dark blue subtraction key [−] mentioned in Section 6. The other is the grey key [(−)] on the bottom row. This grey key is the *opposite of* or negative key. For example, –3 is denoted on the Home screen by pressing

[CLEAR] [(−)] **3** [ENTER].

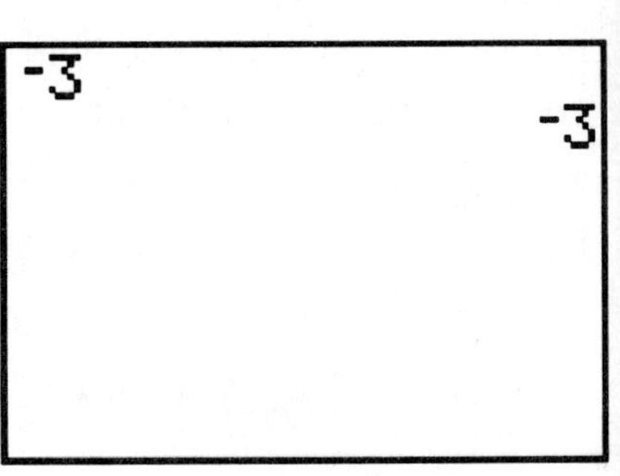

Figure 12

See Fig. 12. The number –3, when added to 3, is zero. To see this on your calculator, press

3 [+] [(−)] **3** [ENTER] or [(−)] **3** [+] **3** [ENTER].

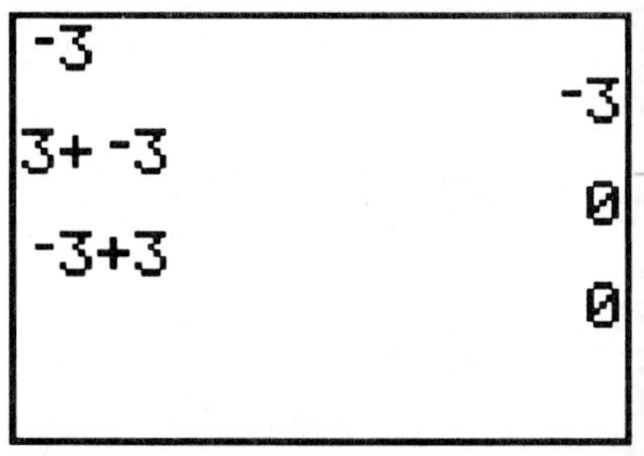

Figure 13

See Fig. 13. The negative key is different from the subtraction key. You can see the difference if you press

[CLEAR] [(−)] [−].

Observe the negative key screen symbol is smaller in length and higher than the subtraction symbol. See Fig. 14. Press [CLEAR] to clear the Home screen for the next example.

Figure 14

Section 7 Error Messages

If you want to see the result of 7 subtract 3 on the calculator, you must press

7 [−] **3** [ENTER].

If you press

7 [(–)] **3** [ENTER],

the TI-82 will give you a syntax error message. See Figs. 15 and 16. If you type **1** from the error message screen, the TI-82 will return you to the expression on the Home screen. The cursor blinking on the negative symbol tells you this is the mistake. You should have used the subtract key instead of the negative key. If you want to calculate 2(–3) then press

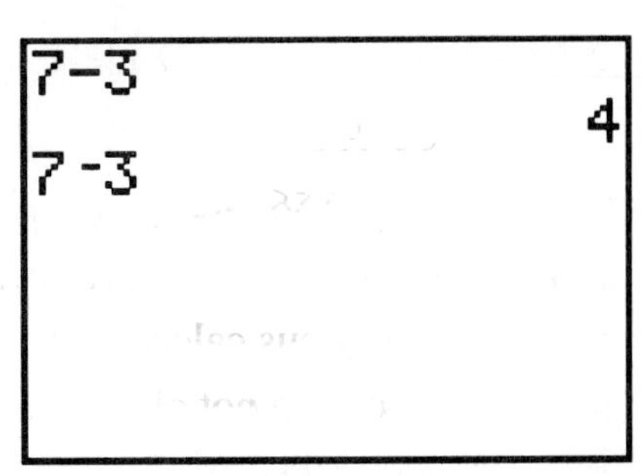

Figure 15

[CLEAR] **2** [×] [(–)] **3** [ENTER] or **2** [(] [(–)] **3** [)] [ENTER].

Figure 16

If you press

2 [(–)] **3** [ENTER]

the TI-82 will return a syntax error message. Press **2** from the error message screen to return to the Home screen. Why do you see an error when you press

2 [÷] **0**?

Section 8 Using the [x^2] and [x^{-1}] Keys

You can calculate squares with [x^2]. This key is in the first column. For example, the square of 5 can be calculated by pressing

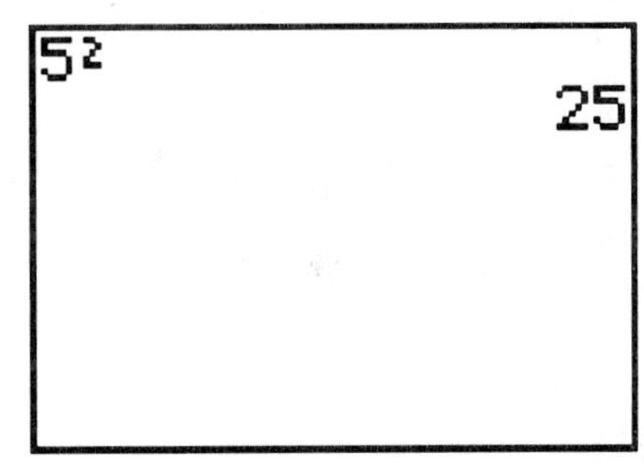

Figure 17

[CLEAR] **5** [x^2] [ENTER].

See Fig. 17. Notice when you pressed [x^2], the ^{2} symbol was printed on the Home screen next to the 5, but the answer did not appear until you pressed [ENTER]. You can calculate reciprocals with [x^{-1}]. This key is also in the first column. The reciprocal of 4 can be determined by pressing

4 [x^{-1}] [ENTER].

See Fig. 18.

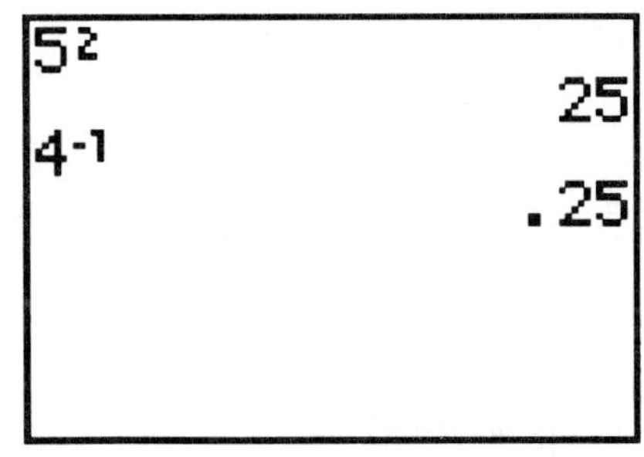

Figure 18

Section 9 Inserting and Deleting

You can edit expressions with the **Insert** key [INS] and the **Delete** key [DEL]. These keys are on the second row. Press

[CLEAR] **36** [+] **10**

but don't press [ENTER]. The cursor should be at the end of the expression. If you would like to change the expression to 316 + 10 by inserting a 1 between the 3 and the 6, press the left cursor-movement key until the cursor is blinking on the '6' in '36.' Now press

[2nd] [INS].

Notice the cursor changes from a blinking rectangle to a blinking underline. Press the **1** key. The expression changes to '316 + 10.' Press [ENTER] to evaluate this new expression. The result is 326. Now we will use the [DEL] key to delete a digit. Press

123 [−] **10**

but don't press [ENTER]. The cursor should be blinking at the end of this new expression. Press the left cursor-movement key until the cursor is blinking on the '2' in '123.' Now press [DEL]. The expression has been changed to '13 – 10.' Press [ENTER] to evaluate this new expression.

Chapter 1 Numerical Mathematics and the Graphing Calculator

Section 1.1 Real Numbers and a Graphing Calculator

Many of the computational skills you will need in this chapter are explained in the Preliminary Chapter, "Getting Started Using the TI-82." Be sure to read this chapter before beginning Chapter 1.

Changing Decimals to Fractions

Some decimals can be changed to equivalent fractions with the [1:▶Frac] option in the MATH menu.

Example 1: Convert 0.44 to an equivalent simplified fraction.

From the Home screen press

[CLEAR] **0** [.] **44** [MATH] [1:▶Frac] [ENTER].

See Fig. 1.1. The equivalent fraction is $^{11}/_{25}$.

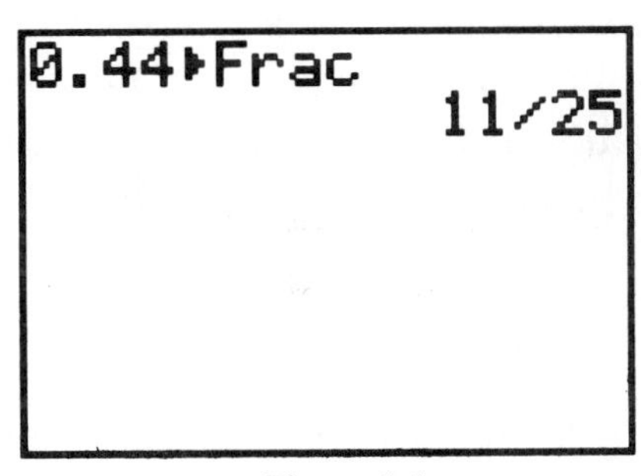

Figure 1.1

Decimal Representation of Fractions

Example 2: Convert the following fractions to decimals: $^{11}/_{25}$, $^{143}/_{999}$, $^{1}/_{17}$.

Press

[CLEAR] **11** [÷] **25** [ENTER] **143** [÷] **999** [ENTER] **1** [÷] **17** [ENTER].

See Fig. 1.2. The decimal representation of $^{11}/_{25}$ is exact because it is a terminating decimal. The TI-82 decimal representations of $^{143}/_{999}$ and $^{1}/_{17}$ are approximations because the correct representations of these fractions are repeating decimals. Although the screen shows an approximation to $^{143}/_{999}$, you can probably tell what the repeating pattern is from looking at your calculator result. $^{1}/_{17}$ is a repeating decimal but you cannot see the repeating block in the first 10 digits. The TI-82 can show only 10 decimals in a number. If the decimal repeats or terminates after the first 10 digits, you may not be able to determine the exact form from the TI-82 representation.

```
11/25
                .44
143/999
        .1431431431
1/17
        .0588235294
```

Figure 1.2

Floating Decimal Display versus Fixed-Point Display

When your calculator displayed the first 10 digits of the decimal representation of $^{143}/_{999}$ in the previous example, it was in Floating Decimal mode. If you only want to see the first two digits, you can switch to Fixed Decimal mode. Do this by pressing

.

See Fig. 1.3. Now press

CLEAR **143** ÷ **999** ENTER.

You should see the first two decimals in the answer, .14. Press MODE and change the number of decimals in Fixed Mode to four by using the cursor-movement keys to highlight the '4' on the second line of the Mode screen, then press

ENTER.

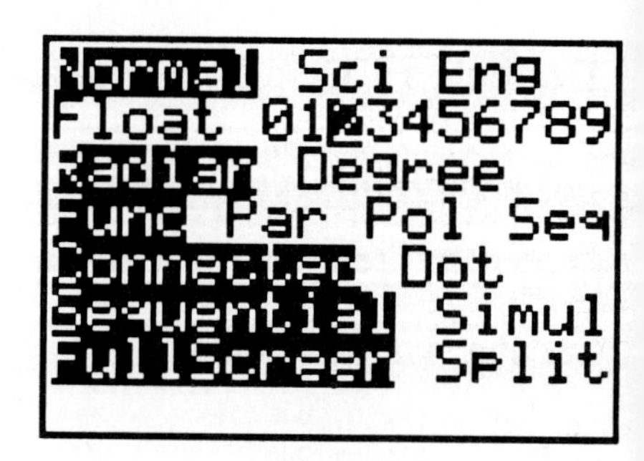

Figure 1.3

Press

CLEAR

to return to the Home screen and again divide 143 by 999. You should see the first four digits of the decimal representation, .1431. Now switch back to Floating Decimal mode by pressing MODE, highlighting 'Float,' and pressing

ENTER and CLEAR.

The Square Root Key and the Use of [2nd] Keys

The use of these keys was described in the Preliminary Chapter. The decimal representation of $\sqrt{2}$ shown in Fig. 4 is an approximation because $\sqrt{2}$ is a nonterminating, nonrepeating decimal; however, it is a very accurate approximation.

Powers

The Preliminary Chapter described how to square and cube numbers. Now we will see how to raise numbers to other powers. You can calculate powers with the power key on the fifth row, fifth column of the keyboard.

Example 3: Compute 3^5.

To see 3^5 press

[CLEAR] **3** [^] **5** [ENTER].

See Fig. 1.4.

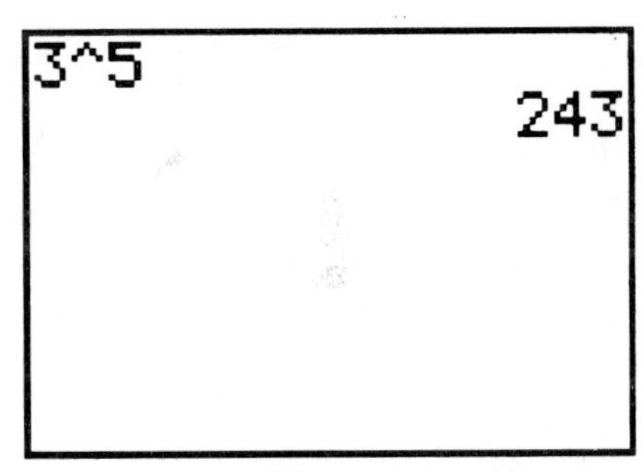

Figure 1.4

Example 4: Calculate -3^2 and $(-3)^2$.

Press

[CLEAR] [(-)] **3** [^] **2** [ENTER]

to see -3^2 and press

[(] [(-)] **3** [)] [^] **2** [ENTER]

to see $(-3)^2$. See Fig. 1.5. Notice the results of the two calculations are different. In the first calculation, the TI-82 squared three first and then negated. In the second calculation, the parentheses forced the TI-82 to negate three first and then square negative three. You can square numbers with [x^2] or with [^].

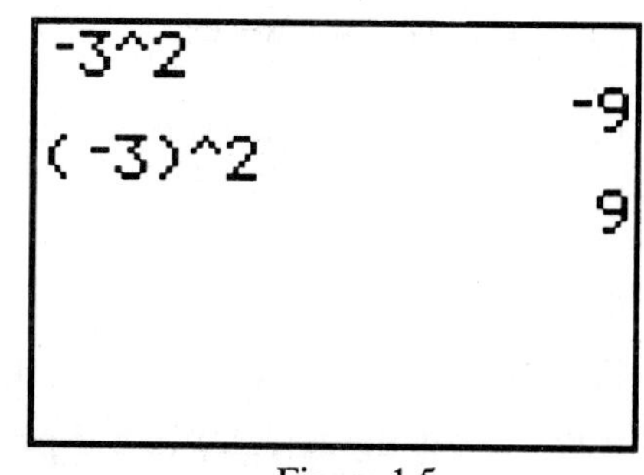

Figure 1.5

Example 5: $(2/5)^3$.

You must use parentheses to raise a fraction to a power. Press

[CLEAR] [(] **2** [÷] **5** [)] [^] **3** [ENTER].

The answer is .064.

Order of Operations

Example 6: Calculate $2 + 3 \times 4^2 - 6 \div 2$.

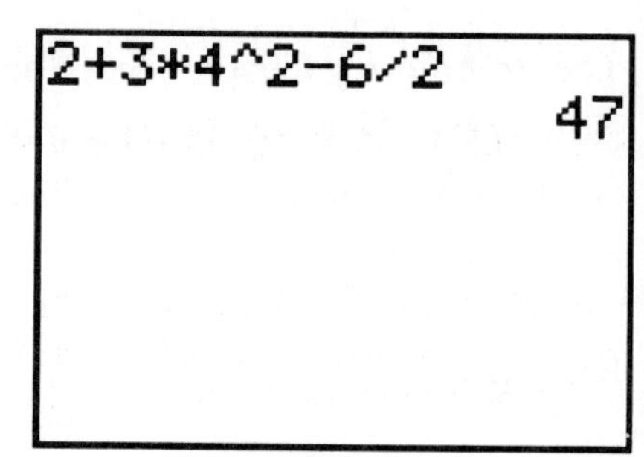

Figure 1.6

Press

[CLEAR] **2** [+] **3** [×] **4** [^] **2** [−] **6** [÷] **2** [ENTER].

See Fig. 1.6. The TI-82 does powers first, which reduces the expression to $2 + 3 \times 16 - 6 \div 2$. Next the calculator does multiplication and division, which simplify the expression further to $2 + 48 - 3$. Last it does addition and subtraction to yield a result of 47.

Comparing with the TEST Menu

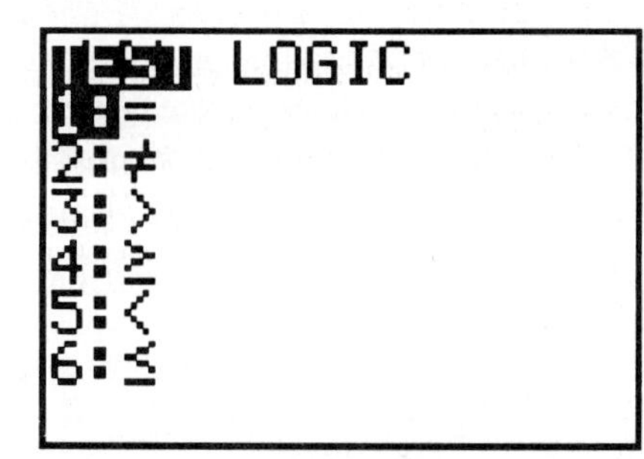

Figure 1.7

To see how the TI-82 interprets $3 < 5$, we will need to use the TEST menu. The [TEST] key is the second function of [MATH]. Before we evaluate the expression $3 < 5$, let's look at the TEST menu by pressing

[2nd] [TEST].

See Fig. 1.7. This menu contains equality and inequality symbols. We will indicate which symbol to select in our keystroke instructions by writing the menu number followed by the symbol. Press [CLEAR] to return to the Home screen and [CLEAR] again to clear the Home screen. Now evaluate the expression $3 < 5$ by pressing

3 [2nd] [TEST] [5:<] **5** [ENTER].

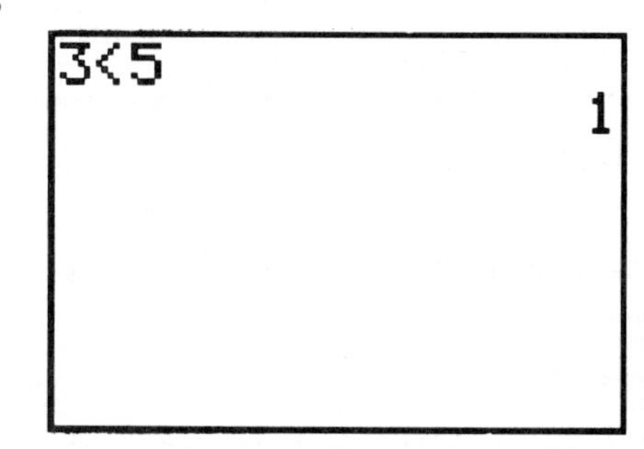

Figure 1.8

See Fig. 1.8. The TI-82 returns the value 1 because the expression is true. To see the result of $3 > 5$ press

3 [2nd] [TEST] [3:>] **5** [ENTER].

The calculator shows a result of 0 because this expression is false. You should also see a result of 0 when you press

[(-)] **1** [2nd] [TEST] [4:≥] [(-)] **0** [.] **5** [ENTER]

because –1 is not greater than or equal to –0.5.

Section 1.2 Properties of Real Numbers and the Basic Rules of Operations

Properties of Opposites and the Subtraction Key

These keys were discussed in the Preliminary Chapter.

The Absolute Value Key

Absolute value is the second function of [x^{-1}].

Example 1: Calculate | 7 |, | –7 |, and –| –7 |.

Press

[CLEAR] [2nd] [ABS] **7** [ENTER]

to see the absolute value of 7, and then press

[2nd] [ABS] [(-)] **7** [ENTER]

to see the absolute value of –7. Next press

[(-)] [2nd] [ABS] [(-)] **7** [ENTER]

to see –| –7 |. See Fig. 1.9.

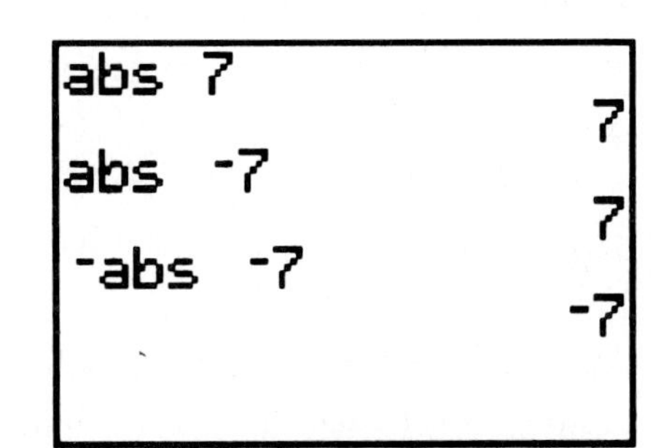

Figure 1.9

Example 2: Calculate | 5.2 – 17.8 |.

Press

[CLEAR] [2nd] [ABS] [(] **5** [.] **2** [–] **17** [.] **8** [)] [ENTER].

The result is 12.6. If you repeat the same keystroke sequence but leave out the parentheses, you will obtain an incorrect result. See Fig. 1.10. With the parentheses, the TI-82 subtracts 17.8 from 5.2 first and then finds the absolute value of the difference. Without the parentheses, the TI-82 takes the absolute value of 5.2 and then subtracts 17.8.

```
abs (5.2-17.8
                 12.6
abs 5.2-17.8
                -12.6
```

Figure 1.10

Section 1.3 Algebraic Expressions and Problem Situations

Using a Grapher to Evaluate an Expression by Using the [STO▶] Key and Replay

You can assign values to variables on the TI-82 and then use those variables in calculations. The TI-82 variables are the letters of the alphabet found above many of the keys. You assign values to variables with [STO▶]. This key is just above [ON]. After you press [STO▶] you must change the cursor to the Alpha mode so you can access the letters on the keyboard. When the TI-82 is in Alpha mode and you press a key, the letter above the key will be printed. You change the cursor to the Alpha mode by pressing [ALPHA].

Example 1: Evaluate x^3+17 at $x = 25$, 39, and 52.

Store the value of 25 in *X* by pressing

[CLEAR] **25** [STO▶] [ALPHA] [X] [ENTER].

The reference to [X] in these instructions means to press the key with 'X' above it. This key happens to be [STO▶]. Now press

[ALPHA] [X] [^] **3** [+] **17** [ENTER].

The screen should show you that $x^3 + 17 = 15{,}642$ when x is 25. See Fig. 1.11. We will evaluate $x^3 + 17$ at $x = 39$ and 52 by a different approach. Instead of storing 39 in *X* and then evaluating the expression, we will enter 39 directly into the expression. Press

[CLEAR] **39** [^] **3** [+] **17** [ENTER].

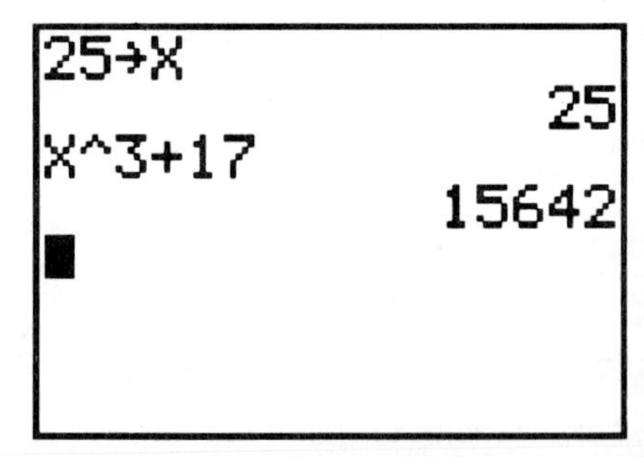

Figure 1.11

The answer is 59,336. Now we will use the **Replay** feature to edit the expression and replace 39 with 52. You can Replay an expression by using [ENTRY], which is the second function of [ENTER]. Recall the expression by pressing

[2nd] [ENTRY].

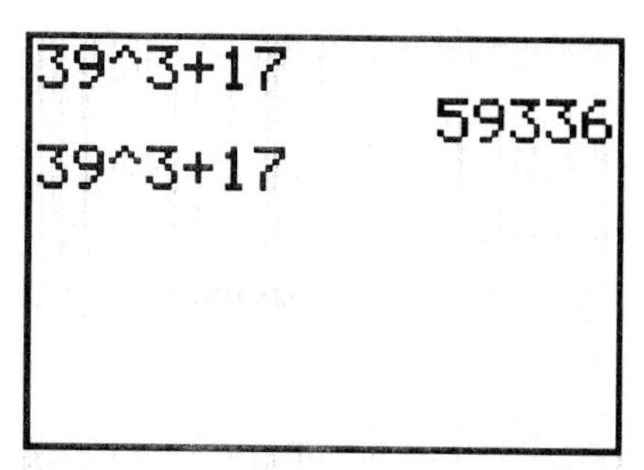

Figure 1.12

The previous expression is copied on the next line, and the cursor appears at the end of the line. See Fig. 1.12. Now press the left cursor-movement key [◀] until the cursor is on the '3' in '39.' Replace '39' with '52' by simply typing 52. See Fig. 1.13. Finish the calculation by pressing

[ENTER]

and the answer is $52^3 + 17 = 140{,}625$.

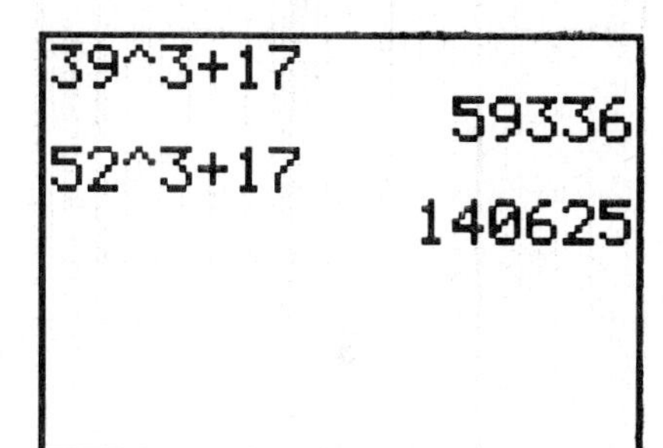

Figure 1.13

Example 2: Evaluate $4T - 3S^3/5R$ at $R = 13.5$, $S = -23.7$, and $T = 3.28$.

Assign values to the variables by pressing

[CLEAR] **13** [.] **5** [STO▶] [ALPHA] [R] [ENTER] [(-)] **23** [.] **7** [STO▶] [ALPHA] [S] [ENTER] **3** [.] **28** [STO▶] [ALPHA] [T] [ENTER].

Now evaluate the expression by pressing

4 [ALPHA] [T] [−] **3** [ALPHA] [S] [MATH] [3:³] [÷] [(] **5** [ALPHA] [R] [)] [ENTER].

See Fig. 1.14.

```
                13.5
-23.7→S
               -23.7
3.28→T
                3.28
4T-3S³/(5R)
            604.7668
```

Figure 1.14

Concatenation

Several calculator steps can be combined with the concatenation key [:]. [:] is the second function of [.]. Concatenate means to combine. The advantage of combining steps is that when you Replay the last expression it really recalls several expressions. This feature is illustrated in the next example.

Example 3: Evaluate $x^3 + 17$ at $x = 25$ and 39.

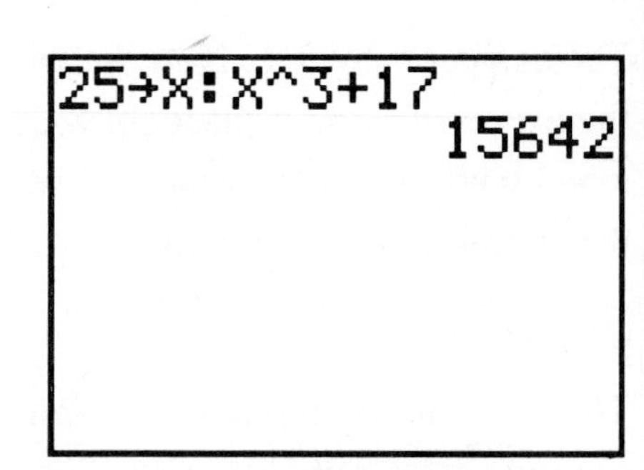

Figure 1.15

In Example 1 we evaluated the polynomial at $x = 25$ in two steps. The first step was to store the value of 25 in X, and the second step was to enter the polynomial. Now we concatenate these steps by pressing

[CLEAR] **25** [STO▶] [ALPHA] [X] [2nd] [.] [ALPHA] [X] [^] **3** [+] **17** [ENTER].

See Fig. 1.15. Now recall the expression by pressing

[2nd] [ENTRY].

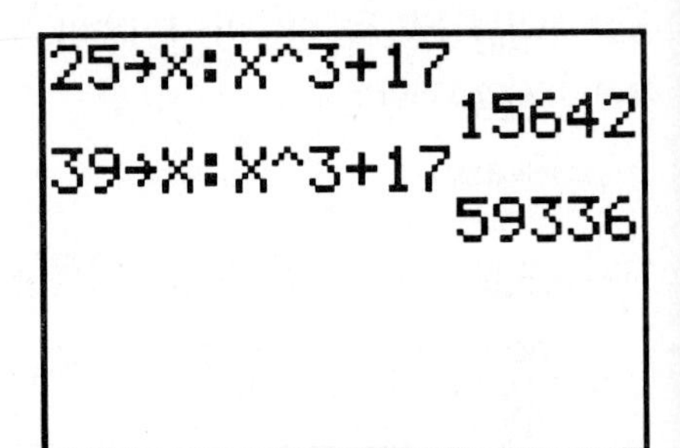

Figure 1.16

Move the cursor onto the '2' in '25,' replace the '25' with '39,' then press [ENTER]. See Fig. 1.16. The polynomial is reevaluated at $x = 39$. This is a quicker method to reevaluate the polynomial at various values.

Implied Multiplication

Press

[CLEAR] **3** [STO▶] [X/θ/T] [ENTER] **2** [X/θ/T] [ENTER].

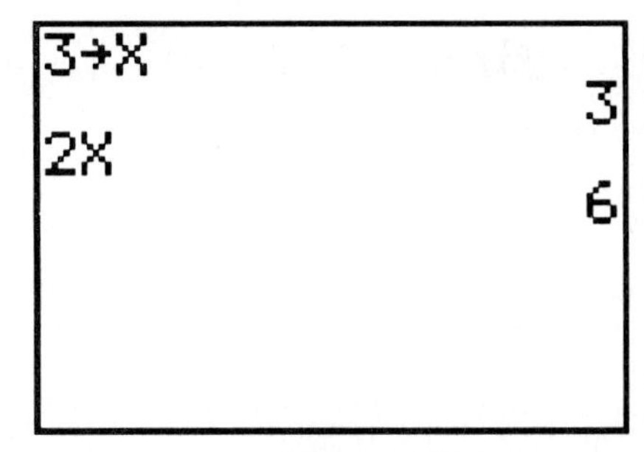

Figure 1.17

See Fig. 1.17. The [X/θ/T] key is a shortcut for pressing [ALPHA] [X]. The TI-82 multiplies the number 2 and the value of X even though there is no multiplication sign. This is called an implied multiplication because their position next to one another implies multiplication The expression 2*X is an explicit multiplication because there is a multiplication sign between the two factors. Now press

12 [÷] **2** [X/θ/T] [ENTER].

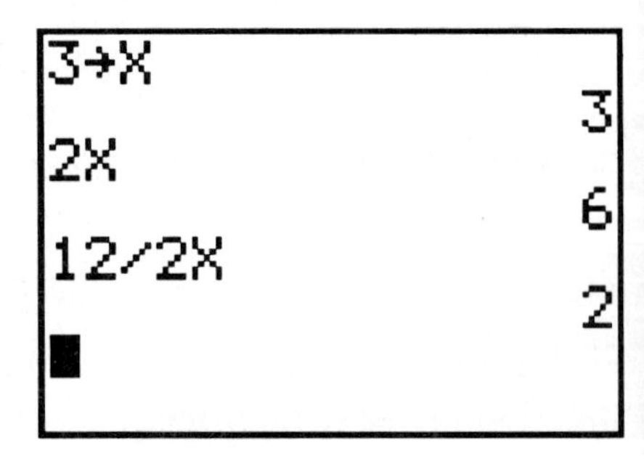

Figure 1.18

See Fig. 1.18. The result is 2 because the implied multiplication is done before the division. The TI-82 carries out an implied multiplication before an explicit multiplication or division.

Example 4: Evaluate $\left(\frac{5}{9}\right)(41 - 32)$.

You will need to use parentheses to enter this expression correctly in the TI-82. Press

[CLEAR] [(] **5** [÷] **9** [)] [(] **41** [−] **32** [)] [ENTER].

The answer is 5. See Fig. 1.19. The TI-82 performs the calculations inside both pairs of parentheses first and then multiplies the results. Now see what happens if you leave off the first set of parentheses. Press

5 [÷] **9** [(] **41** [−] **32** [)] [ENTER].

The result is different. See Fig. 1.19. This is because the TI-82 performs the implied multiplication between the 9 and the value in the parentheses before it does the division. The expression $^{5}/_{9}$ (41 – 32) is equivalent to 5/[9(41 – 32)].

```
(5/9)(41-32)
                5
5/9(41-32
      .0617283951
```

Figure 1.19

Now redo Example 2 and omit the pair of parentheses around the 5*R*. The result should be the same as with the parentheses because the implied multiplication between 5 and *R* is done before the division.

Section 1.5 Numerical Representations of Problem Situations

Tables

It is possible to make tables of values on the TI-82. This is helpful when investigating functions numerically. The function is stored in the **Y= menu**. Values for both the independent and dependent variables in the function can then be printed in a table.

Example 1: Make a table of values for the function $y = 2x$.

Enter the function in the Y= menu by pressing

[Y=] [CLEAR] **2** [X/θ/T].

See Fig. 1.20. Now set up the table by pressing

[2nd] [Tblset].

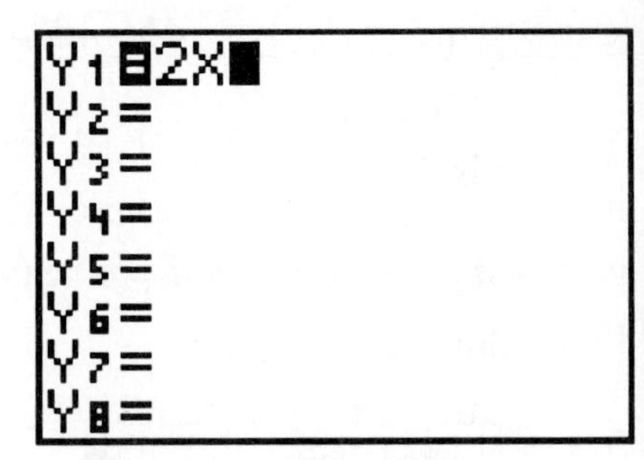

Figure 1.20

The [Tblset] key is on the top row. It is the second function of [WINDOW]. See Fig. 1.21. TblMin is the minimum value of the independent variable x. It should be zero. ΔTbl is the amount by which x will change from one row to the next in the table. It should be 1. If your Table Setup menu is different from Fig. 1.21, press

0 [ENTER] **1** [ENTER] [ENTER] [▼] [ENTER].

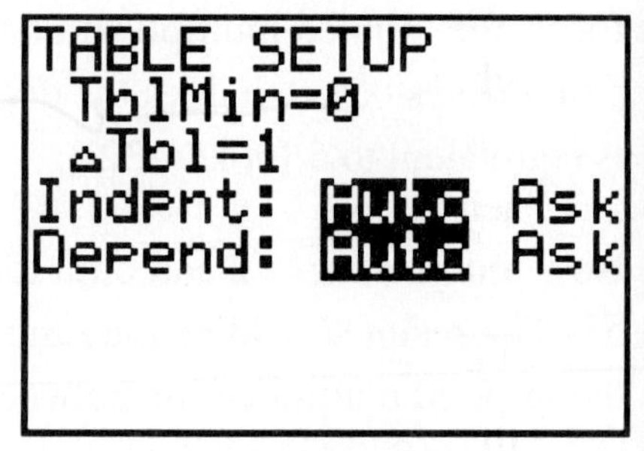

Figure 1.21

When your Table Setup menu matches Fig. 1.21, press

[2nd] [TABLE].

([TABLE] is the second function of [GRAPH] and [GRAPH] is on the top row.) See Fig. 1.22. Values of x are printed in the first column, and values of y are printed in the second column. You can use the cursor-movement keys to move to different cells in the table. You can even scroll down below the bottom of the current table if you continue to press [▼].

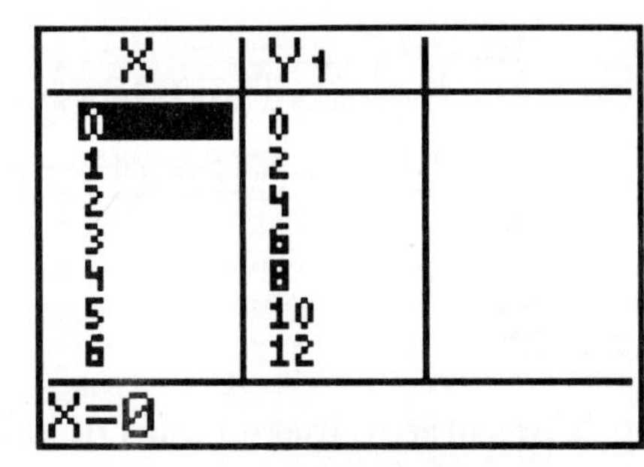

Figure 1.22

Example 2: A rectangle is to be constructed with width = x and length = $24 - 2x$. The area of the rectangle is length × width. Make a table with columns for width, length, and area to determine how the length and area are related to the width.

Erase the previous equation in Y_1 by pressing

[Y=] [CLEAR].

Enter the equations for length and area in the Y= menu by pressing

24 [−] **2** [X/θ/T] [ENTER] [CLEAR] [X/θ/T] [(] **24** [−] **2** [X/θ/T] [)].

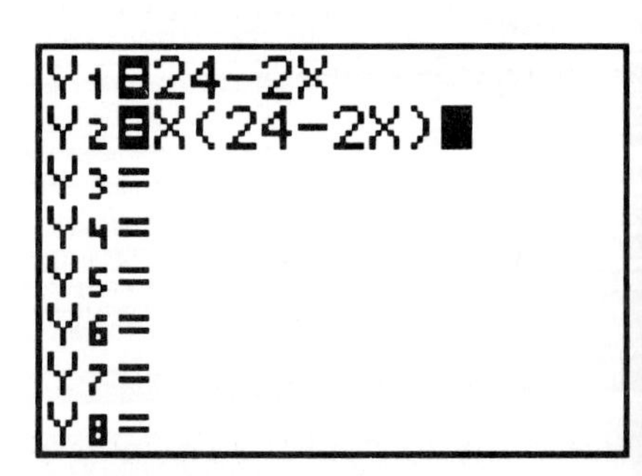

Figure 1.23

See Fig. 1.23. Set up the table by pressing

[2nd] [Tblset] **2** [ENTER].

View the table by pressing

[2nd] [TABLE].

See Fig. 1.24. The first column is width, the second is length, and the third is area. It appears from the table that the largest area occurs when x is 6.

X	Y_1	Y_2
2	20	40
3	18	54
4	16	64
5	14	70
6	12	72
7	10	70
8	8	64
X=2		

Figure 1.24

Example 3: The retail price of a book is equal to the wholesale price plus a 20% markup. Build a table that shows how the markup and retail price are related to the wholesale price.

Since we are dealing with money, we will only need to see two decimal places, so press [MODE] and change to Fixed Decimal mode by highlighting the '2' on the second line of the Mode screen. Don't forget to press [ENTER] after you highlight '2.' If the wholesale price is x, then the markup is $(0.20)x$ and the retail price is $x + (0.20)x$. Enter these equations in the Y= menu by pressing

[Y=] [CLEAR] [.] **2** [X/θ/T] [ENTER] [CLEAR] [.] **2** [X/θ/T] [+] [X/θ/T].

We will start the table with a wholesale price of $10, so set up the table by pressing

[2nd] [Tblset] **10** [ENTER] **1**.

Then view the table by pressing

[2nd] [TABLE].

See Fig. 1.25.

X	Y_1	Y_2
10.00	2.00	12.00
11.00	2.20	13.20
12.00	2.40	14.40
13.00	2.60	15.60
14.00	2.80	16.80
15.00	3.00	18.00
16.00	3.20	19.20
X=10		

Figure 1.25

Finding Percents

Example 4: Find the wholesale price for a book that sells retail for $17.10 if the markup is 20%.

We combine a formula with the method of *guess and correct*. The retail price is equal to the wholesale price plus 20% of the wholesale price. If we let x be the wholesale price, then the retail price is equal to $x + .20x$. We can make an initial guess for x and then check its correctness with the calculator. Since the numbers involved represent money, we are only interested in two decimal places. Use the

Mode screen to change from Floating Decimal mode to Fixed Decimal mode by highlighting the '2' on the second line of the Mode screen. Then press

[CLEAR]

to return to the Home screen. To check an initial guess of $15.00, press

[CLEAR] 15 [.] 00 [+] [.] 20 [x] 15 [.] 00 [ENTER].

Since the answer is $18.00, our initial guess is too high. To change the guess to $14.00, press

[2nd] [ENTRY]

to Replay the expression, and use the cursor-movement keys to replace each '5' with a '4.' This is done by placing the cursor on each '5' and then pressing 4. After you have changed the expression, press [ENTER]. This guess is too low. Use the Replay feature to change the guess to $14.50. Since the answer is $17.40, we are still high. Now Replay and change the expression to guess $14.25. This value works. See Fig. 1.26. Change to Floating Decimal mode when you are finished with this example.

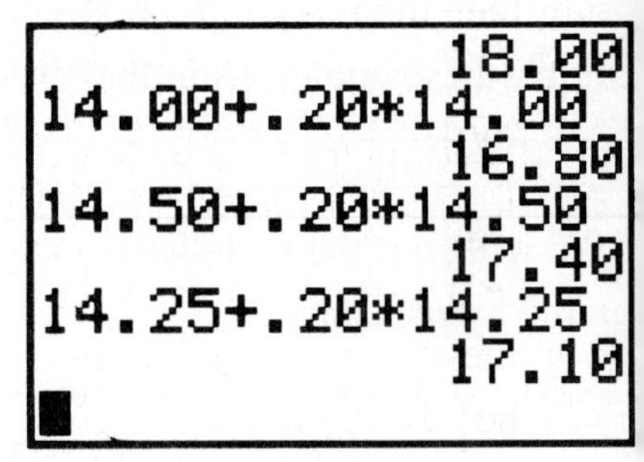

Figure 1.26

Section 1.6 Algebraic Representations of Problem Situations

Parametric Mode, Simultaneous Mode, and Parametric Simulations

Example 1: Suppose that at the instant Juan is running 30 feet per second, he is 10 feet behind Carlos, who is running at 28 feet per second. How long will it take Juan to catch Carlos? The runners in this example can be simulated by the TI-82 with the **Parametric Graphing mode** and **Simultaneous Graphing mode.** Press [MODE] and highlight 'Par' then 'Simul.' Don't forget to press ENTER after you move the cursor to each new mode selection. Next call up the **Window edit screen** by pressing WINDOW. The WINDOW key is on the top row. See Fig. 1.27. This screen is used to set up the viewing window for the simulation. The values on this screen can be changed by using the cursor-movement keys to move the cursor to the line you want to change and then typing in the new value. After you change a window value, you can move to the next line by pressing ▼ or ENTER. Change the values of the Window variables to the following: $T\text{min} = 0$, $T\text{max} = 5.5$, $T\text{step} = 0.01$, $X\text{min} = 0$, $X\text{max} = 150$, $X\text{scl} = 10$, $Y\text{min} = 0$, $Y\text{max} = 8$, and $Y\text{scl} = 1$. See Fig. 1.28. Now press Y= (also on the top row). We will enter the distance equations on this screen. Remember that Distance = Rate × Time. Press

30 X/θ/T ENTER **5** ENTER **28** X/θ/T + **10** ENTER **2** ENTER.

See Fig. 1.29. In Parametric mode, X/θ/T is a shortcut for ALPHA T. To see the simulation, press GRAPH (on the top row). The two moving lines represent the runners. You can trace their progress by pressing the TRACE key, which is next to the GRAPH key. The Trace cursor appears at the beginning of the top line. The value of T printed at the bottom of the screen is the elapsed time, and the value of X is the distance traveled by Juan. Press ▶ repeatedly to see Juan's progress each 0.01 second along the way. To see the distance traveled by Carlos at the same time, press ▼. Press ▲ to move the cursor back to the top graph. When $T = 5$, both runners have traveled a distance of 150 feet. See Fig. 1.30. Juan will catch Carlos after 5 seconds. Reselect Sequential mode for future examples.

```
WINDOW FORMAT
 Tmin=0
 Tmax=6.2831853…
 Tstep=.1308996…
 Xmin=-10
 Xmax=10
 Xscl=1
↓Ymin=-10
```

Figure 1.27

```
WINDOW FORMAT
↑Tstep=.01
 Xmin=0
 Xmax=150
 Xscl=10
 Ymin=0
 Ymax=8
 Yscl=1
```

Figure 1.28

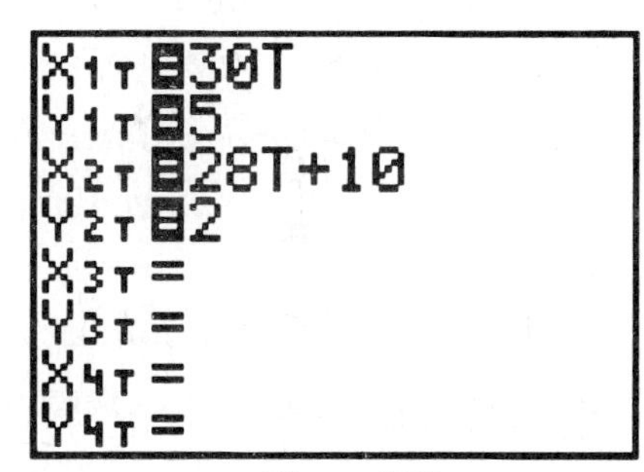

Figure 1.29

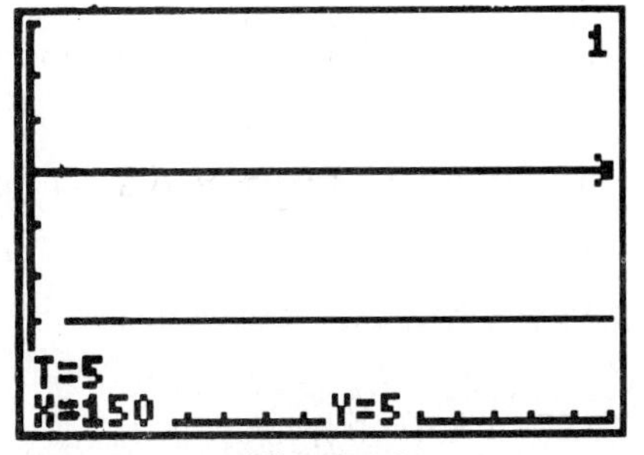

Figure 1.30

Chapter 2 Graphing Equations and Functions

Section 2.1 Rectangular Coordinate Plane

Displaying the Coordinate Plane on a Grapher

The TI-82 can use the screen to display a portion of the coordinate plane. This portion is called a **viewing window**. Graphs of functions can be displayed in the viewing window. The black keys on the top row are used to define the viewing window and create graphs. Press

[MODE]

and select the Function mode before continuing with the following examples.

The WINDOW and Range Adjustments

The [WINDOW] key is used to determine which portion of the coordinate plane will be displayed in the viewing window. Press [WINDOW] to see the Window edit screen. See Fig. 2.1. Your screen probably doesn't have the same numbers as those shown in Fig. 2.1. The variables on the Window edit screen are called Window variables. *X*min is the coordinate of the left side of the viewing window; *X*max is the coordinate of the right side of the window. *X*scl is the distance between the tick marks that are drawn on the *x*-axis. *Y*min is the coordinate of the bottom of the window. *Y*max is the coordinate of the top, and *Y*scl is the distance between the tick marks on the *y*-axis. You can change the values of your Window variables by moving the cursor to the value you would like to change and typing in the new value. Modify your Window edit screen until it looks like Fig. 2.1.

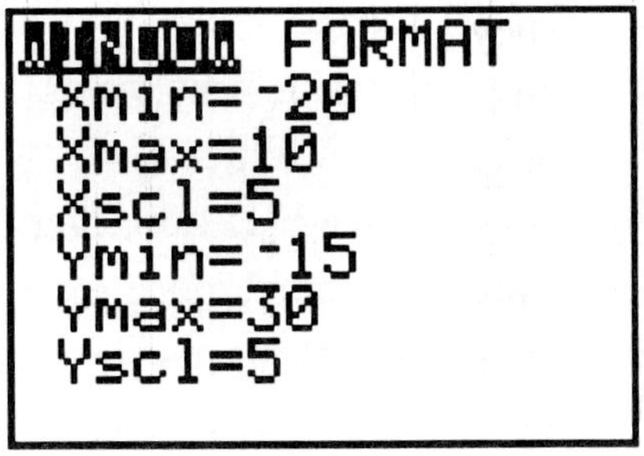

```
WINDOW FORMAT
Xmin=-20
Xmax=10
Xscl=5
Ymin=-15
Ymax=30
Yscl=5
```

Figure 2.1

Graphing and Setting Screen Coordinates (Range) and Defining Pixels

Clear the equations in the Y= menu by pressing [Y=] and then moving the cursor to each line with an equation and pressing [CLEAR]. Press

[2nd] [QUIT]

to return to the Home screen. Press [GRAPH] to see the viewing window. See Fig. 2.2. This viewing window represents points whose x-coordinates are between –20 and 10 and whose y-coordinates are between –15 and 30. The tick marks are each 5 units apart. We say that this is the window [–20, 10] by [–15, 30]. This notation means Xmin = –20, Xmax = 10, Ymin = –15, and Ymax = 30.

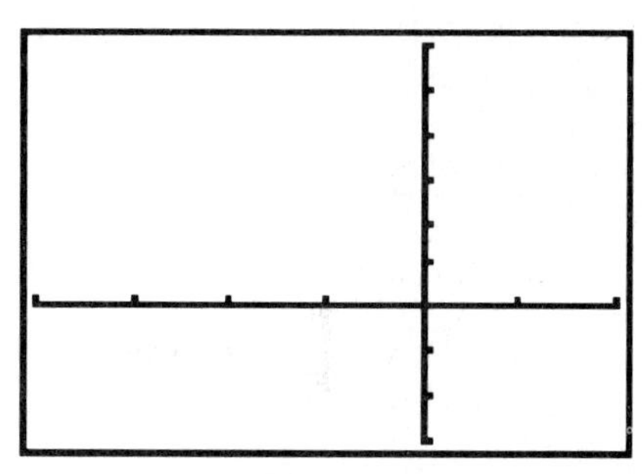

Figure 2.2

Now press [▶]. Pressing any cursor-movement key causes a cursor in the shape of a '+' with a blinking center to appear on the screen. This is called the **free-moving cursor**. The x- and y-coordinates of this cursor are printed at the bottom of the screen. See Fig. 2.3. You can move this cursor around the viewing window with the cursor-movement keys. The coordinates printed at the bottom of the screen are approximations to the actual coordinates of the free-moving cursor. Move the cursor to the origin. The coordinates should be (0,0) but they are slightly off. Press [WINDOW] and change the Window values to [–10, 10] by [–10, 10] with Xscl and Yscl values of 1, then press [GRAPH] and use the cursor-movement keys to move the free-moving cursor to the origin. Notice the cursor coordinates are correct this time. The coordinates of the cursor may be slightly different in different windows even though the cursor is on the same pixel in both windows.

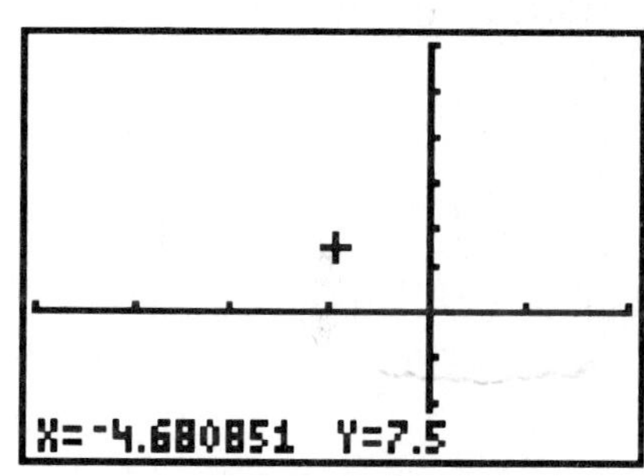

Figure 2.3

The Integer Windows

With the cursor at the origin of the [–10, 10] by [–10, 10] window, press

[ZOOM] [8:ZInteger] [ENTER].

Now use the cursor-movement keys to move the free-moving cursor. The *x*- and *y*-coordinates of the cursor are always integers. This is because the viewing window has been changed. Press [WINDOW] to see how the values of the Window variables have changed. They are [–47, 47] by [–31, 31]. This is called the **Integer window**. There are other windows where the cursor coordinates are integers, but since this window has the origin at the center, your textbook calls it the Integer window. You can find other Integer windows by moving the screen cursor away from the origin after you press

[ZOOM] [8:ZInteger]

but before you press [ENTER]. The Range [0, 94] by [0, 62] is also an Integer window, as is [0, 188] by [0, 124]. In an Integer window, *X*max – *X*min is a multiple of 94, and *Y*max – *Y*min is a multiple of 62.

The Decimal Window or the 0.1 Window

Press

[ZOOM] [4:ZDecimal]

and use the cursor-movement keys. The *x*- and *y*-coordinates are always tenths. This is called the **Decimal window** or **0.1 window**. Press [WINDOW] to see the values of the Window variables. They should be [–4.7, 4.7] by [–3.1, 3.1]. [–4.7, 4.7] by [–10, 10] is a window where the *x*-coordinates are always tenths, but not the *y*-coordinates. The **0.2 window** is [–9.4, 9.4] by [–6.2, 6.2]. Why is this called the 0.2 window?

Section 2.2 Equations in Two Variables and Their Graphs

Plotting Points with the Point-On (PT-On() and STAT Commands

Example 1: Plot the following solution pairs to $14x + 28y = 168$, using the points (–15, 13.5), (–10, 11), (0, 6),and (15, –1.5).

Use [WINDOW] to create a [–20, 20] by [–15, 15] viewing window. Press

[2nd] [QUIT]

to move from the Window edit screen to the Home screen. Now we will use the **point-on** option of the **DRAW [POINTS]** menu to plot these points. Press

[CLEAR] [2nd] [DRAW] [▶] [1:PT-On(].

The PT-On(command is copied to the Home screen. Now press

[(–)] **15** [,] **13** [.] **5** [)].

See Fig. 2.4. This command tells the TI-82 to plot the point (–15, 13.5). Press [ENTER] to see the point plotted on the viewing window. The point should be plotted in the upper left-hand corner of the viewing window. Now press [CLEAR] to return to the Home screen. Then press

[2nd] [DRAW] [▶] [1:PT-On(] [(–)] **10** [,] **11** [)] [ENTER]

to plot the second point. To plot the third point press

[CLEAR] [2nd] [DRAW] [▶] [1:PT-On(] **0** [,] **6** [)] [ENTER].

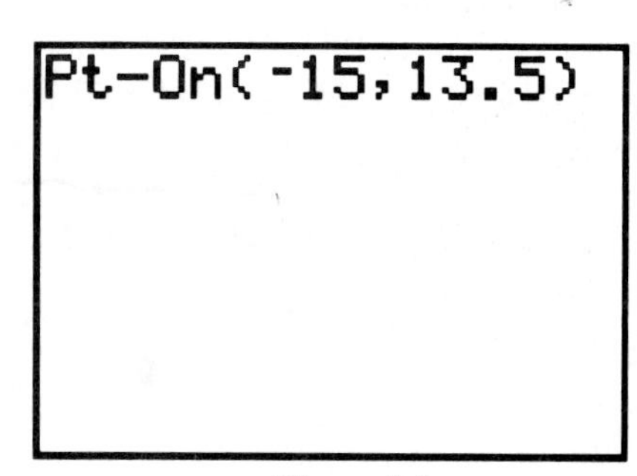

Figure 2.4

You don't see this point because it is on the *y*-axis. Plot the last point by pressing

[CLEAR] [2nd] [DRAW] [▶] [1:PT-On(] 15 [,] [(-)] 1 [.] 5 [)] [ENTER].

See Fig. 2.5. Press

[2nd] [DRAW] [1:ClrDraw]

to clear the viewing window.

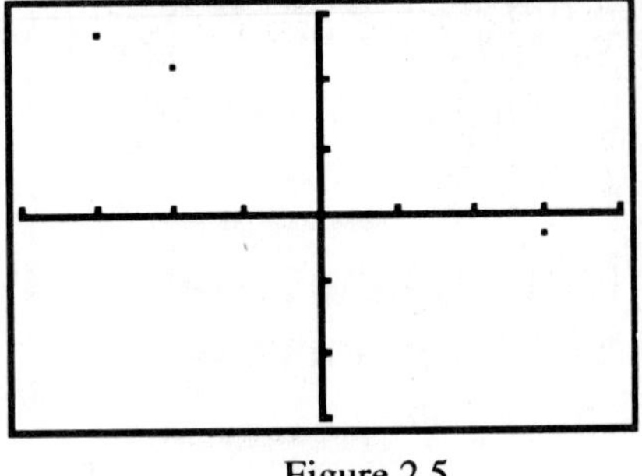

Figure 2.5

Another way to plot the points is with the Statistical features of the TI-82. You should still have the same values for the Window variables that were just used in the first solution to Example 1. The STATISTICAL menu is accessed with the key on the third row, third column. The first feature of the STATISTICAL menu we will access is the **Data Edit screen**. To do this press

[STAT] [1:Edit...].

See Fig. 2.6. We will enter the *x*-coordinates in the first list, L_1, and the *y*-coordinates in the second list, L_2. If your Data Edit screen does not look like this, you can erase the numbers in L_1 and L_2 by pressing

[STAT] [4:ClrList] [2nd] [L_1] [,] [2nd] [L_2] [ENTER].

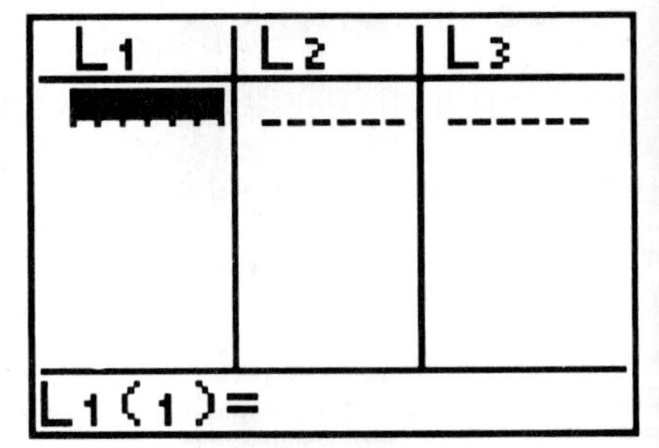

Figure 2.6

L_1 and L_2 are the second functions of the [1] and [2] keys. This will clear the lists and return you to the Home screen so you will need to return to the Data Edit screen by the keystroke sequence described above. Now enter the *x*-coordinates in the first list by pressing

[(-)] 15 [ENTER] [(-)] 10 [ENTER] 0 [ENTER] 15 [ENTER].

Next enter the *y*-coordinates in the second list by pressing

[▶] 13 [.] 5 [ENTER] 11 [ENTER] 6 [ENTER] [(-)] 1 [.] 5 [ENTER].

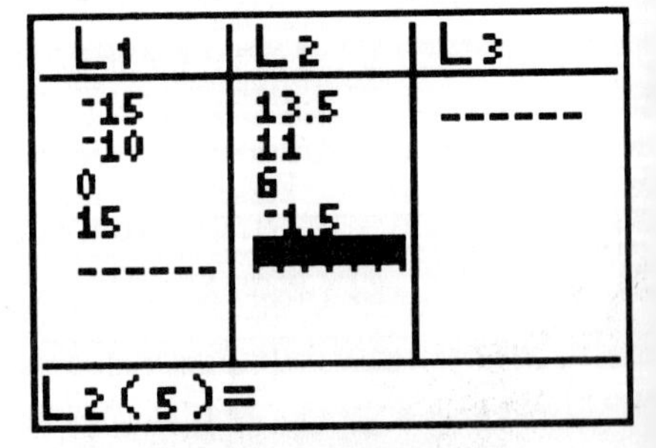

Figure 2.7

Your screen should look like Fig. 2.7. Next select the points for graphing by pressing

[2nd] [STATPLOT] [1:Plot1...].

STATPLOT is the second function of [Y=]. Press [ENTER] to select 'On.' This selects the points for graphing. Your screen should look like Fig. 2.8. This is the **Plot1 edit screen**. If the items highlighted on your screen don't match Fig. 2.9, select them the same way you would in the Mode screen. Use the cursor-movement keys to move the cursor to the item to be highlighted, and then press [ENTER]. 'Type' in the Plot1 edit screen refers to the type of graph. We have selected a **scatter plot graph**. '*X* list' and '*Y* list' designate the lists in the Data Edit screen that will be used for the *x*- and *y*-coordinates. The last row in the edit screen determines the symbol that will be used to plot each point. Now plot the points by pressing [GRAPH]. See Fig. 2.9. The points in this picture are the same ones you plotted previously with the PT On(option, but small boxes are used instead of dots. Press

Figure 2.8

[2nd] [STATPLOT] [1:Plot1...]

to return to the Plot1 edit screen, and then turn the plot off by pressing

[▶] [ENTER] [OFF].

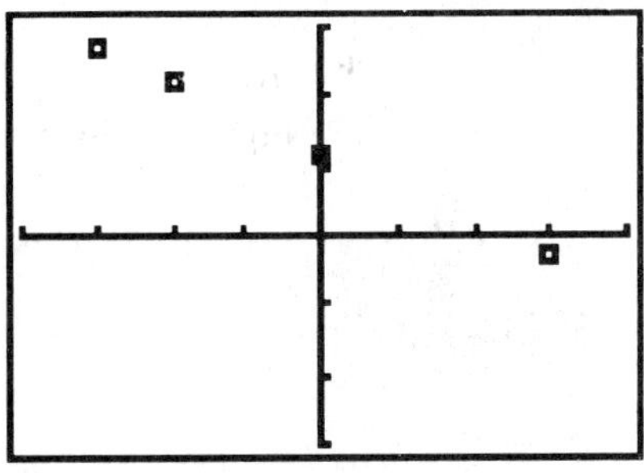
Figure 2.9

Press [CLEAR] to return to the Home screen. It is necessary to turn the Stat Plots off. All plots can be turned off at once by pressing

[2nd] Stat Plot [4:Plots Off].

Graphing Equations and Using TRACE and the Function-Defining Menu

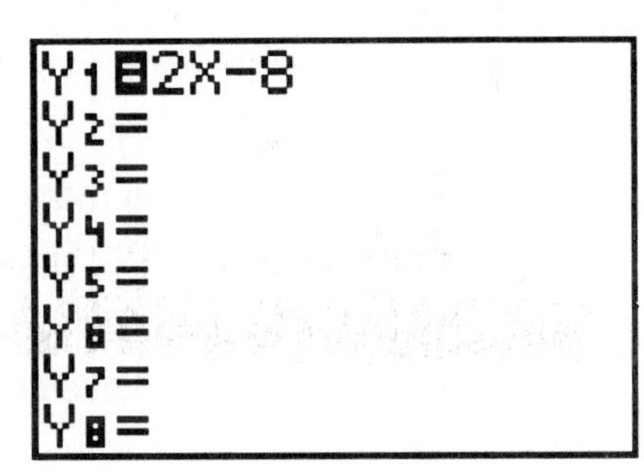

Figure 2.10

Example 2: Graph $y = 2x - 8$.

The [Y=] key on the top row is used to enter the equation to be graphed. Press [Y=]. You can enter 10 different equations. You can erase an old equation by pressing [CLEAR] when the cursor is on the equation. We will enter the equation in Y_1. Press

2 [X/θ/T] [–] **8**.

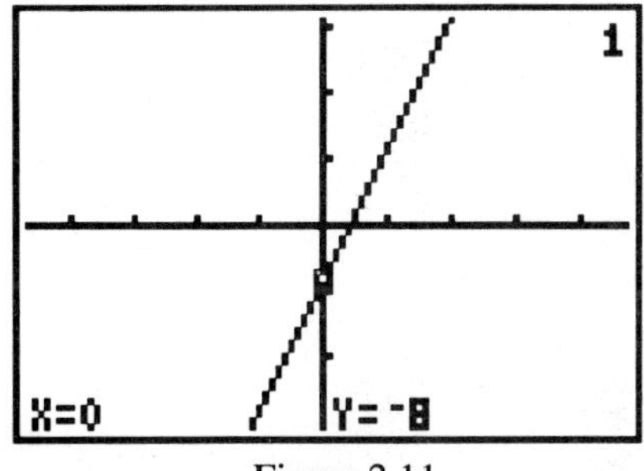

Figure 2.11

See Fig. 2.10. Now press [WINDOW] and enter a [–47, 47] by [–31, 31] viewing window. Press [GRAPH] to see the graph of $2x - 8$. Press [TRACE]. The Trace cursor appears on the graph. See Fig. 2.11. Notice the Trace cursor looks like a blinking 'X' with a box in the center. The coordinates at the bottom of the screen are the coordinates of the Trace cursor. Use the left and right cursor-movement keys to see the coordinates of points on the line. Each pair of coordinates you trace on the screen should be a solution to the equation $y = 2x - 8$.

Example 3: Graph $y = \frac{3}{4}x - 6$.

Press [Y=] and press [CLEAR] to erase the previous equation. Press [(] **3** [÷] **4** [)] [X/θ/T] [–] **6.**

The parentheses are needed here because of implied multiplication (Section 1.3). The viewing window from Example 2 will work here so there is no need to use the Window edit screen. Press [GRAPH] to see the graph. Trace to the point where $X = -9$ and $Y = -12.75$. See Fig. 2.12. This ordered pair is a solution to the equation $y = (^3/_4)x - 6$. The *y*-coordinate of the Trace cursor is calculated by replacing '*X*' from the equation in the *Y*= menu with the *x*-coordinate of the Trace cursor. Now we will move the free-moving cursor to the same point currently occupied by the Trace cursor. Press [GRAPH] to clear the Trace cursor from the viewing window and then use the cursor-movement keys to move the free-moving cursor until its *x*-coordinate is –9 and the cursor appears to be on the line. See Fig. 2.13. The *y*-coordinate of the free-moving cursor is not the same as the *y*-coordinate of the Trace cursor when it was on this point. This ordered pair (–9, –13) is not a solution to the equation. This is because the *y*-coordinate of the free-moving cursor is not calculated from the equation in the Y= menu.

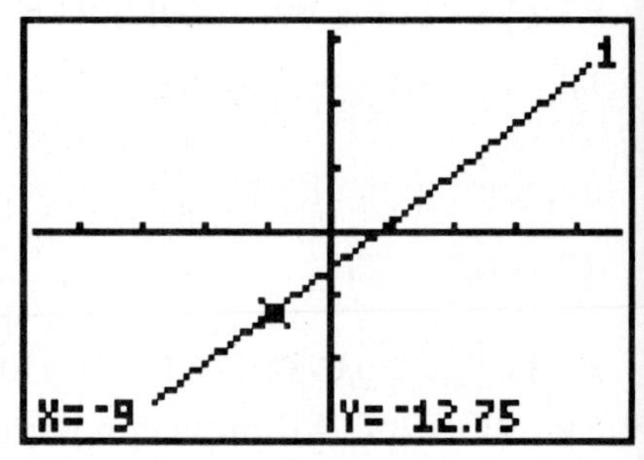

Figure 2.12

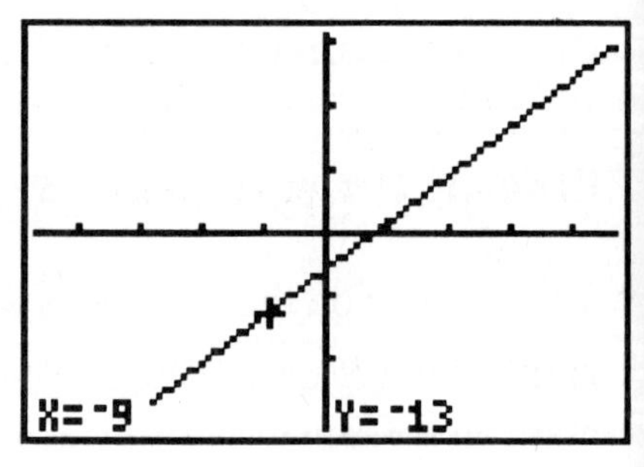

Figure 2.13

Section 2.3 Linear Equations and the Slope of a Line

Graphing Vertical Lines

Example 1: Graph $x = 3$.

Select the viewing window by pressing WINDOW and entering the following values for the Window variables: Xmin = –5, Xmax = 5, Xscl = 1, Ymin = –5, Ymax = 5, and Yscl = 1. Erase any functions in the Y= menu and return to the Home screen. The graph of $x = 3$ is a vertical line. This is not a function so we can't enter an equation in the Y= menu. Instead we will use an option in the DRAW menu. Press

2nd DRAW [4:Vertical]

to paste the Vertical command on the Home screen, then press

3 ENTER

to draw the vertical line $x = 3$. See Fig. 2.14.

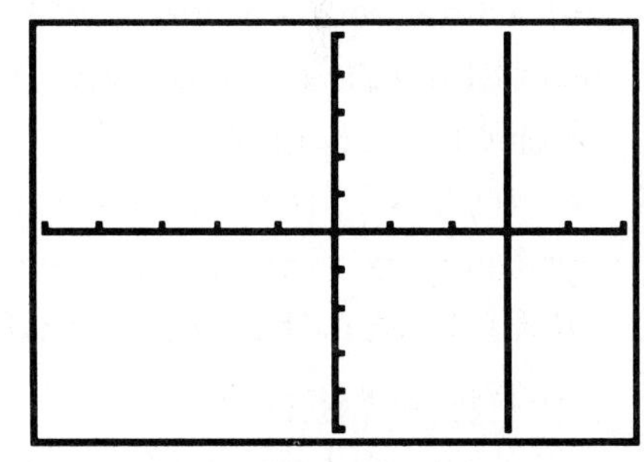

Figure 2.14

Section 2.4 The Slope-Intercept Form and Point-Slope Form of Linear Equations

Squaring Viewing Windows

Example 1: Graph $y = 0.5x + 2$ and $y = -2x - 3$ in a square viewing window.

Select Function mode and then press Y=. Erase any existing equations by pressing CLEAR on each line that has an equation. You can use the cursor-movement keys to move the cursor to each line. With

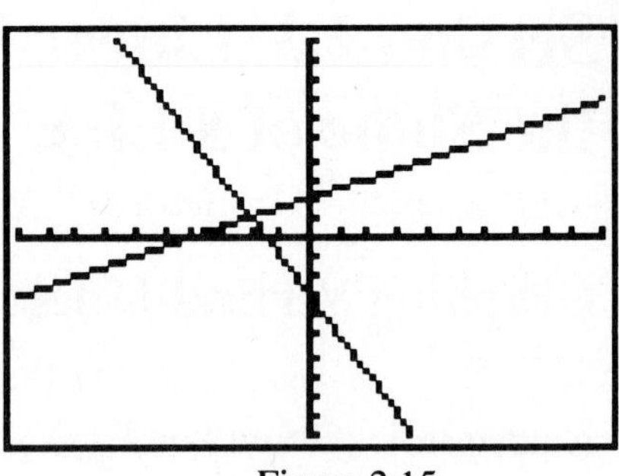
Figure 2.15

the cursor on the top line, enter the first equation in Y_1 by pressing

0 [.] **5** [X/θ/T] [+] **2** [ENTER],

and then enter the second equation in Y_2 by pressing

[(–)] **2** [X/θ/T] [–] **3**.

Press [WINDOW] and select a [–10, 10] by [–10, 10] window, then press [GRAPH]. The lines should be perpendicular but they don't look like they are. See Fig. 2.15. The window needs to be adjusted to make the lines look perpendicular. Press

[ZOOM] [5:ZSquare]

to make this adjustment. The ZSquare option in the ZOOM menu adjusts the values of the Window variables so the slopes and proportions of the picture are correct. Press [WINDOW] to see how these values have changed.

Section 2.5 Functions and Graphs of Functions

Functions on the TI-82 Grapher and the Standard Window

The method of graphing functions such as

$y = x^2$, $y = \sqrt{x}$, and $y = x^{-1}$

is identical to the one you used in Section 2.2 to find the graph of a line. You enter the equation in the Y= menu, select an appropriate window, and graph.

Example 1: Graph $y = -x^2 - 3x$.

Press [Y=] and use [CLEAR] and the cursor-movement keys to erase any equations that may be in the menu. Move the cursor to Y_1 and press

[(-)] [X/θ/T] [x²] [−] **3** [X/θ/T].

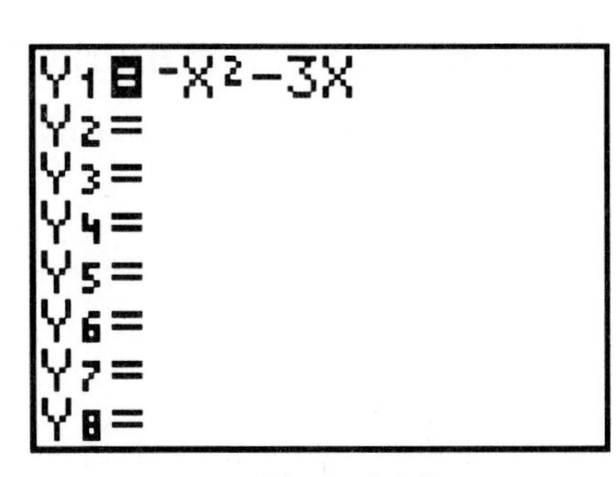

Figure 2.16

See Fig. 2.16. Most graphers use notation like $Y_1 = -X^2 - 3X$ to represent the function $f(x) = -x^2 - 3x$. Now press

[WINDOW]

and enter a [–5, 5] by [–10, 10] window, then press

[GRAPH].

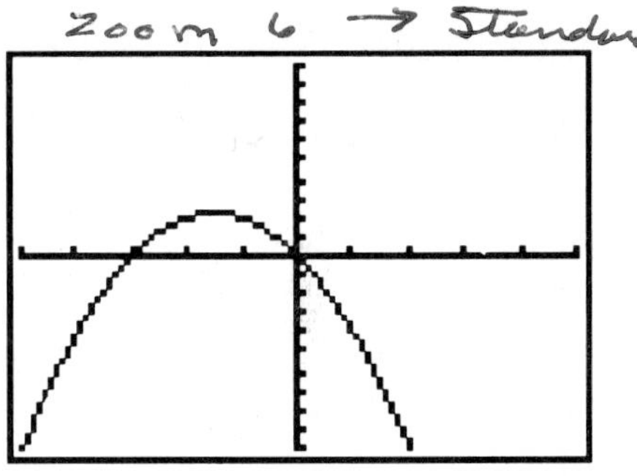
Figure 2.17

See Fig. 2.17. The graph is a parabola. Next we redraw the graph in the [–10, 10] by [–10 ,10] window. This window is called the **Standard window.** There is a shortcut to enter the Standard window. Instead of using [WINDOW], press [ZOOM] [6:ZStandard]. Notice the TI-82 redraws the graph in a new window as soon as you press '6.' See Fig. 2.18. Press [WINDOW] to verify this is the Standard window.

Example 2: Find the domain and range of $f(x) = \sqrt{(x-3)}$.

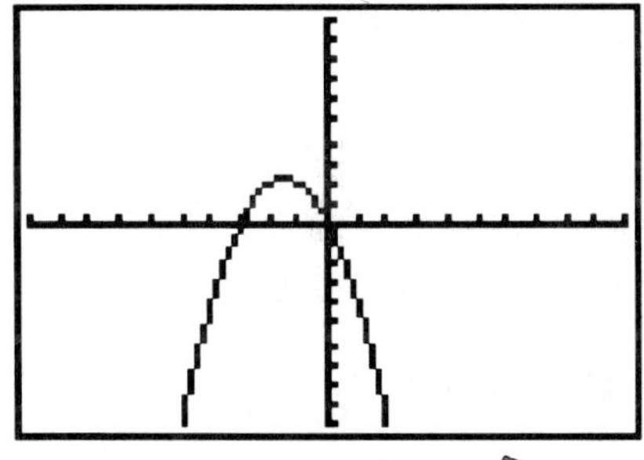
Figure 2.18

We will enter this equation in the Y= menu and graph in the Integer window. We are going to use a shortcut to enter the Integer window. Press [Y=] and erase existing functions from the menu. Move the cursor to Y_1 and press

[2nd] [√] [(] [X/θ/T] [−] **3** [)].

Now press [ZOOM] [8:ZInteger]. As soon as you press '8' you will see the graph drawn in the current window and the free-moving cursor blinking at the origin. Press [ENTER] to redraw the graph in the Integer window. This option in the ZOOM menu creates an Integer Window centered on the free-moving cursor. Now press [TRACE]. See Fig. 2.19. The X-coordinate is 0 and the Y-coordinate is blank. This is because the function is not defined for $X = 0$. Press [▶] three times. Now $X = 3$ and $Y = 0$. Press and hold [▶] to scroll to the right on the graph. The results of [TRACE] indicate the domain is $[3, \infty)$ and the range is $[0, \infty)$.

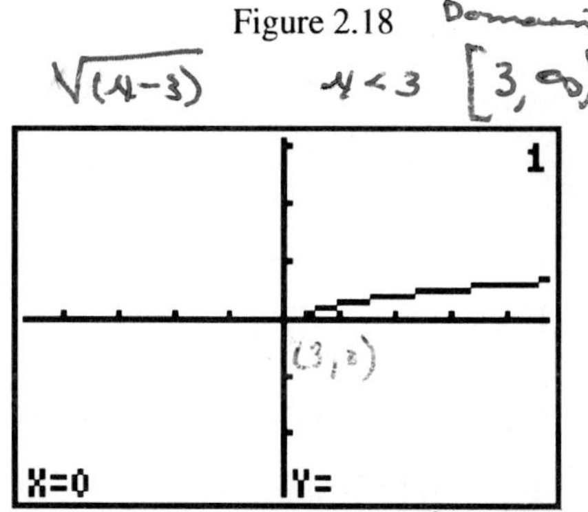

Figure 2.19

Using [TRACE] to Find the Domain and Range of a Function and Scrolling

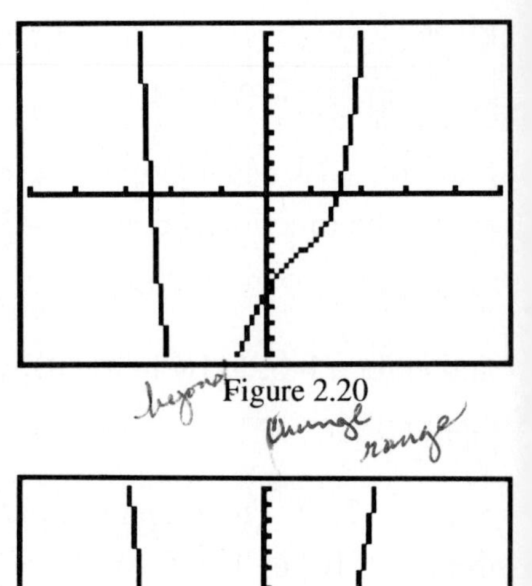
Figure 2.20

Figure 2.21

Example 3: Find a complete graph of $f(x) = x^4 - 3x^2 + 5x - 6$ and estimate its range.

Press [Y=] and erase existing functions. Move the cursor to Y_1 and press

[X/θ/T] [^] **4** [−] **3** [X/θ/T] [x²] [+] **5** [X/θ/T] [−] **6**.

Press [WINDOW] and enter a [–5, 5] by [–10, 10] window, then press [GRAPH]. See Fig. 2.20. This is not a complete graph because you can't see the bottom portion. Press [TRACE] and repeatedly press [▲]. Notice even when the Trace cursor moves off the screen, you can still see its coordinates. The smallest y-coordinate that is printed as the Trace cursor moves to the left is about –15.18. If you continue to press [▲], the Trace cursor will reappear on the graph. The results of [TRACE] give us an idea of a better viewing window. Press [WINDOW] and enter the [–5, 5] by [–20, 20] window with Xscl = 1 and Yscl = 2, then press [GRAPH]. See Fig. 2.21. This is a complete graph. The results of [TRACE] indicate the range is about [–15.2, ∞).

How to Graph a Relation that Is Not in Function Form

Example 4: Graph $y^2 = x$.

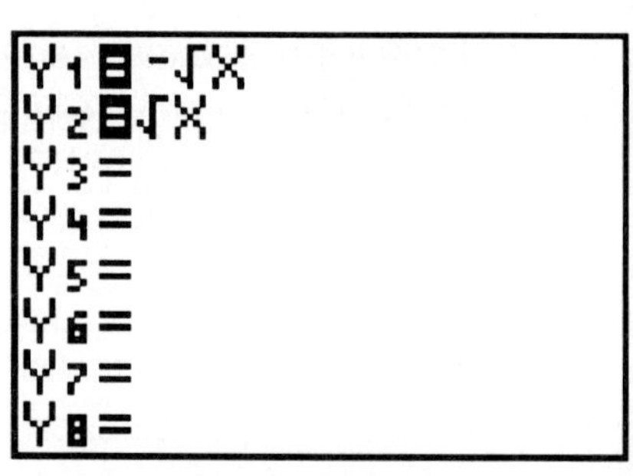

Figure 2.22

This relation is not a function and the TI-82 should be in Function mode. However, the relation can be split into two functions by solving for y. If you do this you get $y = -\sqrt{x}$ and $y = \sqrt{x}$. Enter these functions in Y_1 and Y_2 by pressing

[Y=] [CLEAR] [(−)] [2nd] [√] [X/θ/T] [ENTER] [CLEAR] [2nd] [√] [X/θ/T].

See Fig. 2.22. Graph in the Standard window by pressing

[ZOOM] [6:ZStandard].

The graph is a parabola. See Fig. 2.23.

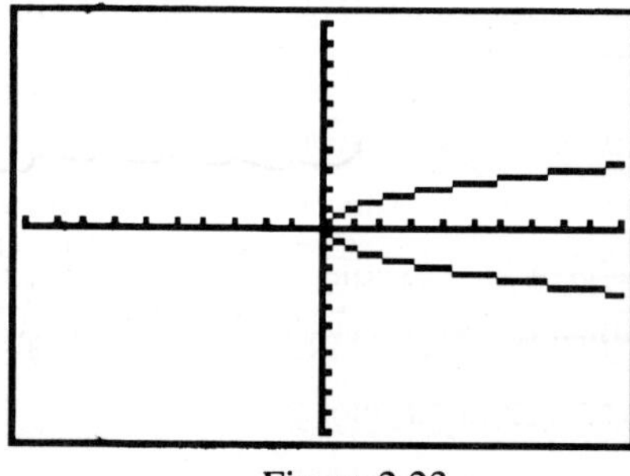
Figure 2.23

Chapter 3 Solving Equations and Systems of Equations

Section 3.1 Finding Graphical Solutions

Selecting and Deselecting Graphs and Sequential Versus Simultaneous Graphing

When several functions have been entered in the Y= menu, you can select some of them for graphing and deselect others so that they won't be graphed. If you select several functions to be graphed in the same window, you can graph them sequentially (one at a time) or simultaneously.

Example 1: Graph $y = 2x + 5$ and $y = 21$.

Clear the Y= menu and enter $Y_1 = 2X + 5$ and $Y_2 = 21$. Notice that the '=' signs next to both Y_1 and Y_2 are highlighted. This means both functions have been selected for graphing. Deselect Y_2 by using the cursor-movement keys to move the cursor onto the '=' next to Y_2 and then press

[ENTER].

Move the cursor off the '=' to see it is no longer highlighted. This means Y_2 is deselected. See Fig. 3.1. Now graph in the Integer window. You should see the graph of Y_1 but not Y_2. See Fig. 3.2. Now reselect Y_2 by pressing

[Y=] [▼] [◄].

This should place the cursor on the '=' beside Y_2. Highlight it by pressing [ENTER]. Notice that if the '=' is highlighted, pressing

[ENTER]

unhighlights it. If the '=' is not highlighted, pressing [ENTER] highlights it. Press [GRAPH]. Since both functions are highlighted, you see

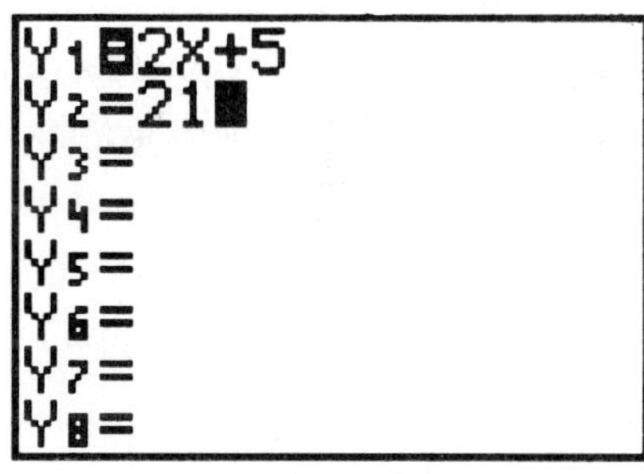

Figure 3.1

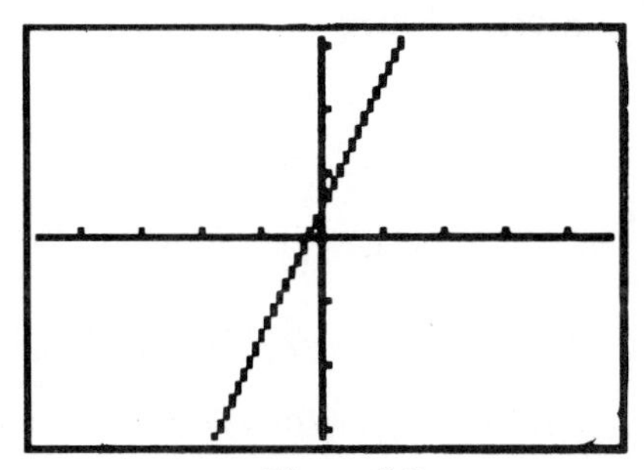
Figure 3.2

the graph of $y = 2x + 5$, followed by the graph of $y = 21$. See Fig. 3.3. When the TI-82 is in **Sequential mode**, the functions are graphed one at a time. Now press [MODE] and select **Simultaneous mode** by highlighting 'Simul' on the Mode screen. Don't forget to press [ENTER] after you move the cursor to 'Simul.' Press [GRAPH]. Both graphs are drawn at the same time.

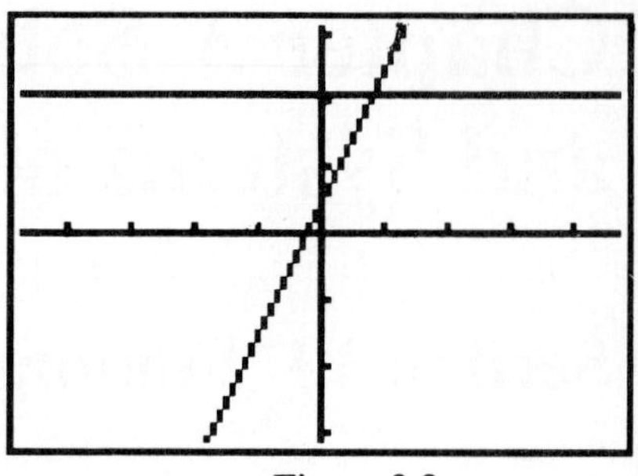
Figure 3.3

The Multigraph Methods of Solving Linear Equations Graphically

Example 2: Solve $17 - 1.5x = -10$.

Press [Y=] and use [CLEAR] and the cursor-movement keys to clear the Y= menu. Enter $Y_1 = 17 - 1.5X$ and $Y_2 = -10$. Graph in the Integer window. Now press [TRACE] and repeatedly press [▶] until the Trace cursor is on the intersection of the two lines. See Fig. 3.4. The '1' in the upper right-hand corner of the screen indicates that you are tracing on Y_1. If you press [▼] while tracing, the Trace cursor will move to the next graph on the screen. If you press [▼] when the Trace cursor is at the point of intersection, it won't appear to move because the Trace cursor is on both graphs, but the number in the upper right-hand corner changes. If you move the Trace cursor away from the point of intersection and then press [▼], you will be able to see the cursor move from one graph to the other.

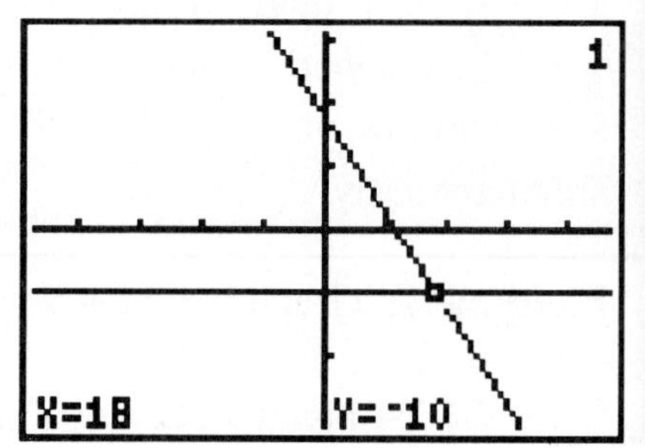

Figure 3.4

Section 3.2 Solving Equations Using Zoom In

Finding a Graphical Solution Using a 0.1 (Decimal) Window Setting

These window settings were described in Section 2.1. They are $X\text{min} = -4.7$, $X\text{max} = 4.7$, $X\text{scl} = 1$, $Y\text{min} = -3.1$, $Y\text{max} = 3.1$, and $Y\text{scl} = 1$.

Using the Grapher ZOOM Menu

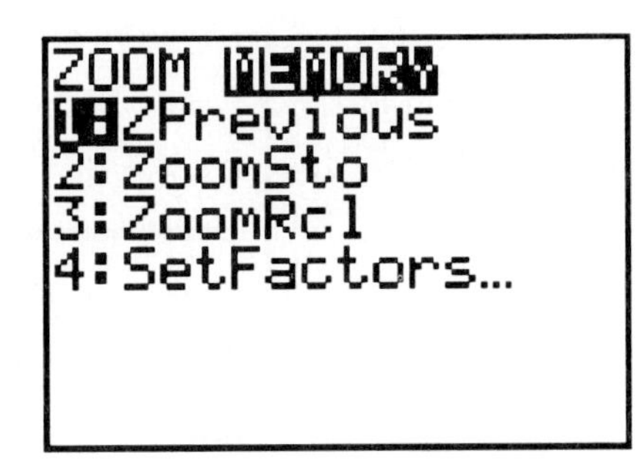

Figure 3.5

The **ZOOM** feature can be used to quickly magnify a portion of the screen. This is helpful when you need to find an accurate approximation to the solution of an equation. Press

to see the **ZOOM [MEMORY]** menu. See Fig. 3.5. In this section we will discuss option **[4:Set Factors]** in the ZOOM [MEMORY] menu and option **[2:Zoom In]** in the ZOOM menu.
[Option 4:Set Factors] is used to set the Zoom factors. These factors determine by how much the graph is magnified when you use option [2:Zoom In] from the ZOOM menu. After you have pressed ZOOM ▶, set the factors to 10 by pressing

[4:Set Factors] **10** ENTER **10.**

Then return to the Home screen by pressing

2nd QUIT.

Example 1: Find a solution to $2.7x + 3.28 = 0$ with an error of at most 0.01.

The solution can be found graphically by tracing to the point where the graph of $y = 2.7x + 3.28$ crosses the x-axis. The difficulty with this approach is that if we graph in the Standard window, the Trace cursor won't be accurate enough. We will need to magnify the portion of the graph that crosses the x-axis. We can magnify with Zoom In.

Clear the Y= menu and enter $Y_1 = 2.7X + 3.28$. Enter the following values for the Window variables: Xmin = –10, Xmax = 10, Xscl = 0.01, Ymin = –10, Ymax = 10, and Yscl = 0.01. We use a value of 0.01 for Xscl and Yscl because the problem required an error of at most 0.01. Press GRAPH. The axes look double thick because the tick marks are so close together. The tick marks will spread out as

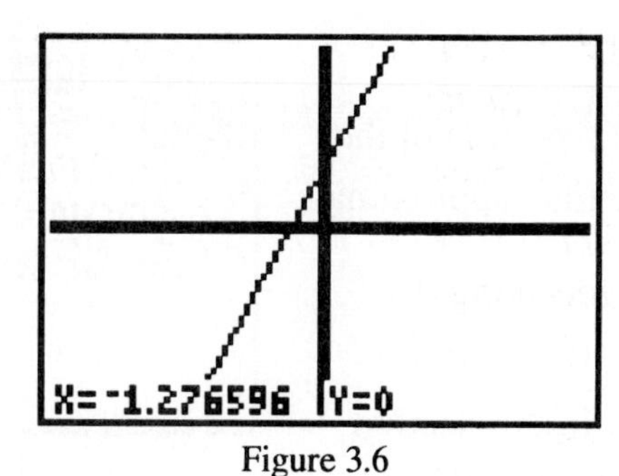

Figure 3.6

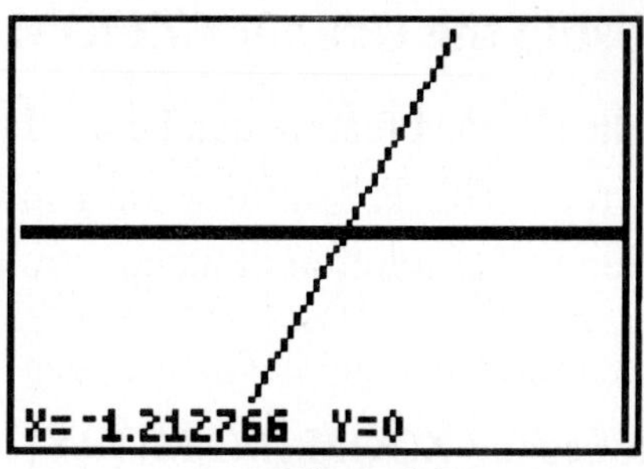

Figure 3.7

we Zoom In. Now press

[ZOOM] [2:Zoom In].

You should see the graph with the free-moving cursor at the origin. Use [▲] to move the cursor to the x-intercept of the graph. See Fig. 3.6. Press

[ENTER].

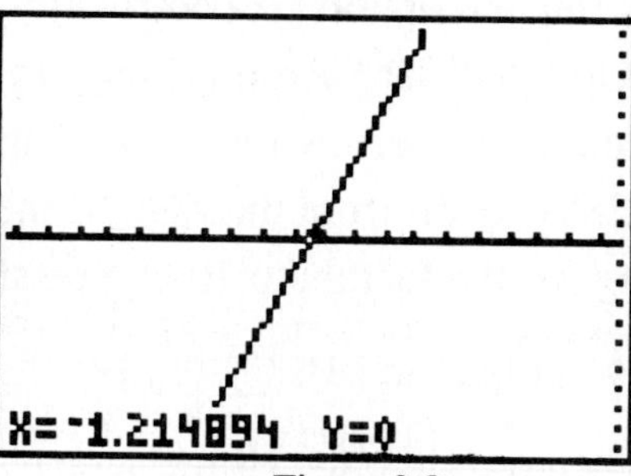

Figure 3.8

Use the cursor-movement keys to move the cursor back to the x-intercept. See Fig. 3.7. Press

[ENTER].

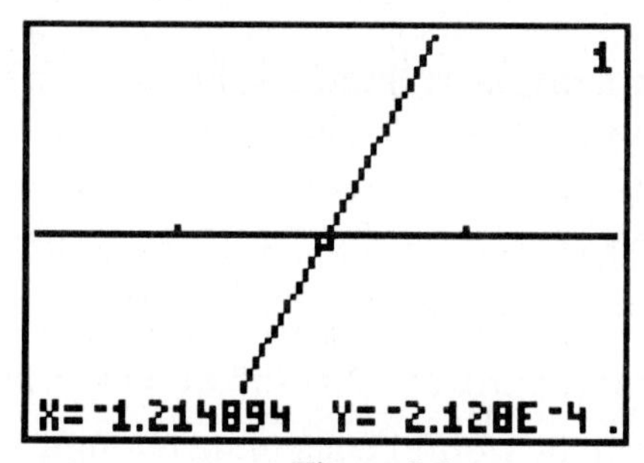

Figure 3.9

The graph is magnified again. Move the cursor to the x-intercept on more time. See Fig. 3.8. Magnify the graph again by pressing [ENTEF Now press [TRACE] and move the Trace cursor to the x-intercept. Se Fig. 3.9. The solution is approximately $X = -1.215$. We know the error is at most 0.01 because the graph crosses the x-axis between two tick marks that are 0.01 apart.

Problems Using the Zoom In Menu

Example 2: Solve $3x - 2 = 0$.

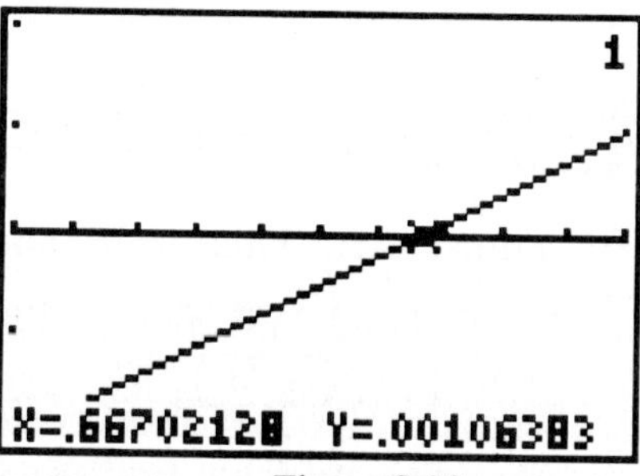

Figure 3.10

Clear the Y= menu and enter $Y_1 = 3X - 2$. Enter the following values for the Window variables: $X\text{min} = 0.6$, $X\text{max} = 0.7$, $X\text{scl} = .01$, $Y\text{min} = -0.2$, $Y\text{max} = 0.2$, and $Y\text{scl} = .1$. Graph and Trace to the inter section of the x-axis and the graph. See Fig. 3.10. (Note the tick marks for y appear on the edge of the screen since the y-axis is not included in the window.) The calculator solution is $X = .66702128$. We could Zoom In to get a better approximation, but no matter how many times we Zoom In, we will still only have an approximation. This is because the exact answer is $^2/_3$, which has a repeating decimal representation. The grapher supports a solution of $^2/_3$, but it does not confirm it.

How to Determine an Appropriate Window for a Graph with TRACE

Example 3: Find a complete graph of $y = 32000 + 960x$.

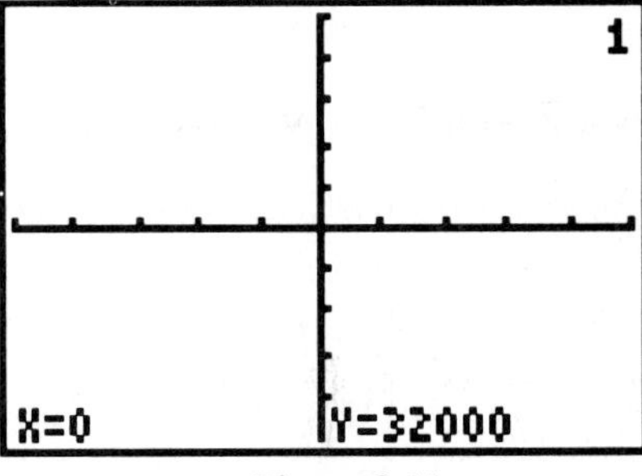

Figure 3.11

Clear the Y= menu and enter $Y_1 = 32000 + 960X$. Graph in the [–5, 5] by [–5, 5] window. The graph screen is blank because this window does not contain any portion of the graph. We can use TRACE to get a better idea of an appropriate window. Press TRACE. Even though you can't see the Trace cursor, you can see its coordinates. See Fig. 3.11. These coordinates tell you that the point (0, 32000) is either on or close to the graph. Based on this information, try the following values for the Window variables: $X\text{min} = -5$, $X\text{max} = 5$, $X\text{scl} = 1$, $Y\text{min} = -1000$, $Y\text{max} = 40000$, and $Y\text{scl} = 10000$. Press GRAPH. This window indicates the complete graph is a line with y-intercept = 32,000. See Fig. 3.12.

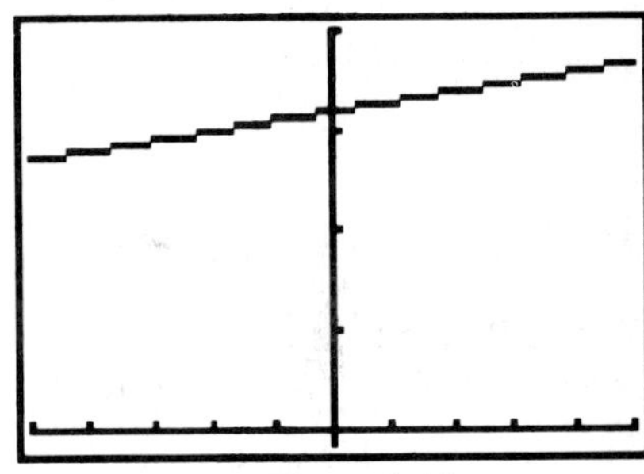
Figure 3.12

Section 3.3 Some Nonlinear Equations

Using Zoom with the Multigraph Method

Example 1: Solve $0.7\sqrt{(x-3)} = 0.3$ with the multigraph method.

In the multigraph method, we graph each side of the equation as a separate function. Then we trace to the point where the two func-

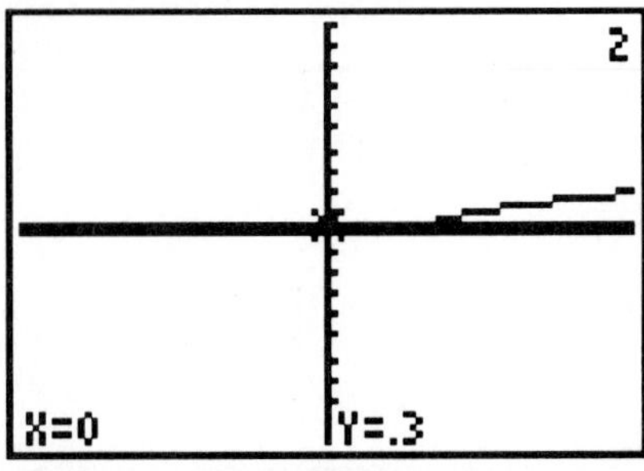

Figure 3.13

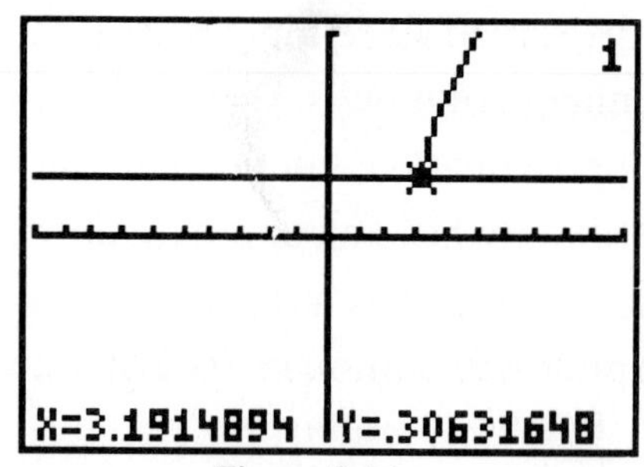

Figure 3.14

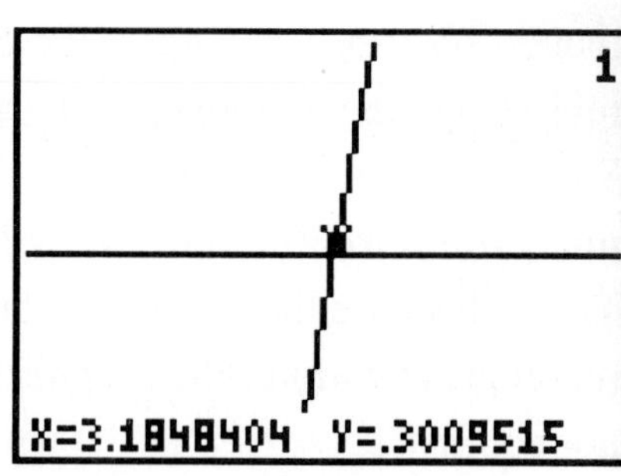

Figure 3.15

tions intersect. This point is a solution to the original equation. In our first attempt, we graph $Y_1 = 0.7\sqrt{(X-3)}$ and $Y_2 = 0.3$ in the Standard window. The graph of Y_2 is too close to the x-axis to be seen clearly in this window. Even though you can't see the graph of Y_2 very well, you can still trace it. Press [TRACE] and move the Trace cursor to the graph of Y_2 by pressing [▼]. See Fig. 3.13. A better viewing window would be [–10, 10] by [–1, 1]. You can see both graphs as well as their point of intersection in this window. See Fig. 3.14. Now you could Zoom In on the intersection three times using Zoom factors of 10 to obtain a solution of $x = 3.18$. See Fig. 3.15.

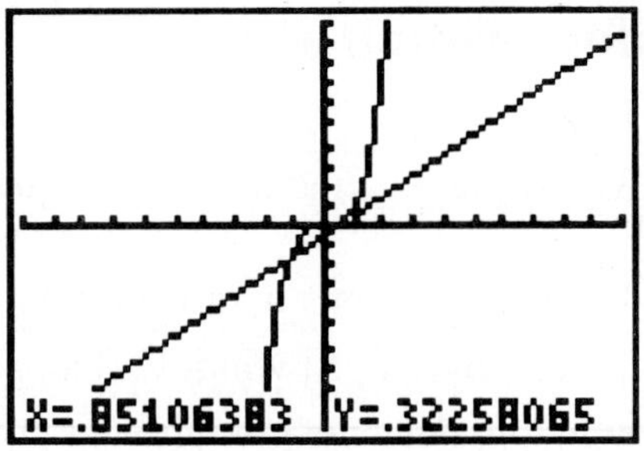

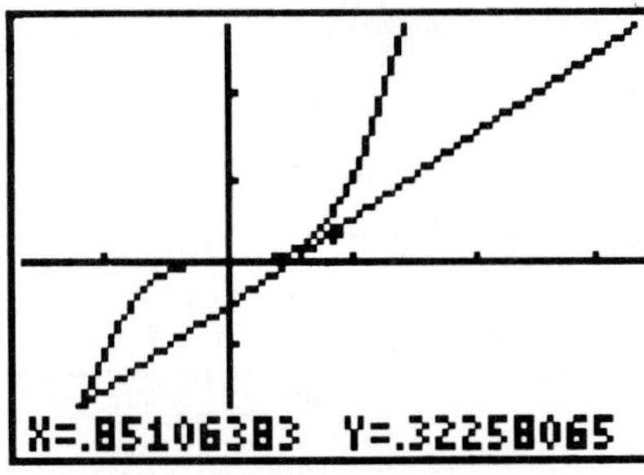

Figure 3.16

Example 2: Solve $x^3 = x - 0.5$ with the multigraph method.

Graph $Y_1 = X \wedge 3$ and $Y_2 = X - 0.5$ in the Standard window. There appear to be two points of intersection. First we Zoom In on the point in the first quadrant. Press

[ZOOM] [2:Zoom In]

and move the free-moving cursor to this point. See Fig. 3.16. When you press [ENTER] to complete the Zoom In, you will see that the two graphs do not intersect after all, so there is no solution in the first quadrant. See Fig. 3.17. You could redraw the graphs in the Standard window and then Zoom In several times on the point of intersection in the third quadrant to obtain a solution of $x = -1.19$. See Fig. 3.18. Initially, it looked like there were two solutions to the problem, but using Zoom In helped us determine that there was only one.

Figure 3.17

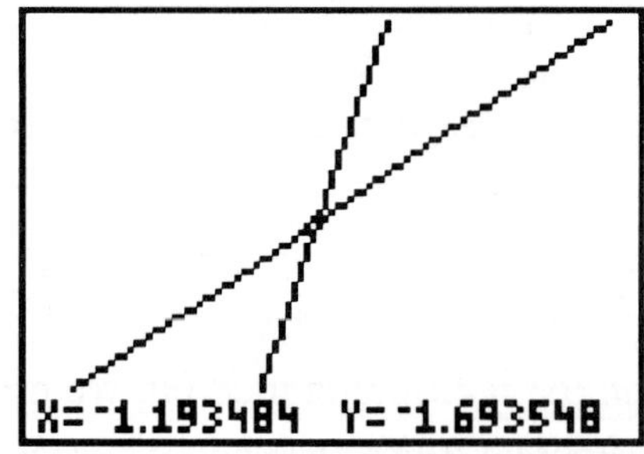

Figure 3.18

Section 3.4 Graphical Representations of Problem Situations

Motion Simulation

Parametric equations are useful in simulating moving objects. A simulation of two runners was described in Section 1.6. In Parametric mode, both the x- and y-coordinates of points on a graph are determined by a third variable, T. In motion simulations, T represents time.

Example 1: Simulate a ball that is thrown straight up with an initial velocity of 88 ft/sec.

The height of the ball is given by the equation $h = -16t^2 + 88t$. Select Parametric (Par) and **Dot** mode on the Mode screen. Now press [Y=] and erase any functions that may be in the menu. Then enter $X_{1T} = 3$ and $Y_{1T} = -16T^2 + 88T$. Press [WINDOW] and enter the following values for the WINDOW variables: Tmin = 0, Tmax = 5.5, Tstep = 0.1, Xmin = 0, Xmax = 5, Xscl = 1, Ymin = 0, Ymax = 150, and Yscl = 10. Press [GRAPH] to see the simulation. You can review the motion of the ball by pressing [TRACE]. The Trace cursor appears at the bottom of the screen. Press [▶] repeatedly to see the ball as time moves forward. See Fig. 3.19. This screen tells you that when T = 2.7 seconds, the ball has risen to a height of 120.96 feet. Press [◀] to move backwards in time through the simulation.

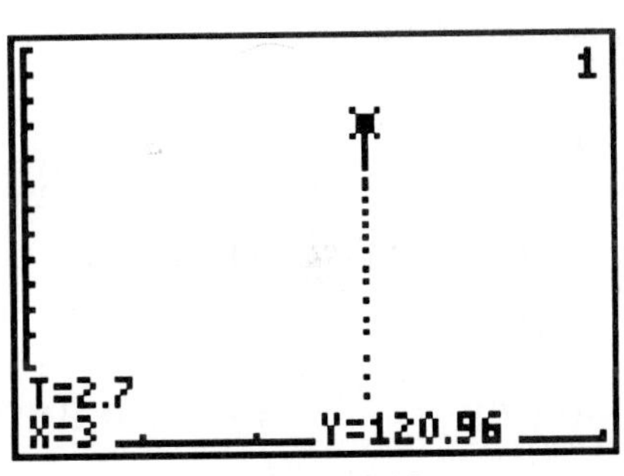

Figure 3.19

Plotting Points and Overlaying a Graph on the Points

Example 2: Plot the points (10, 50), (15, 33.33), (20, 25), and (30, 16.67), and overlay the graph of, $y = 500/x$ on the same screen.

Select Function and **Connected** modes from the Mode screen. In Connected mode, the points that are calculated and plotted on the graph are connected with small line segments. This gives the appearance of a smooth graph. Now use [WINDOW] to enter the [0, 60] by [0,

60] window. Clear the Y= menu and return to the Home screen. Plot the points given in the example statement with PT-On(or STAT. The Pt-On(option of the DRAW menu and STAT options were described in Section 2.2. After you have plotted these points your screen should look like Fig. 3.20. Now we will use the DrawF option from the DRAW menu to overlay the graph of 500/*x*. Press

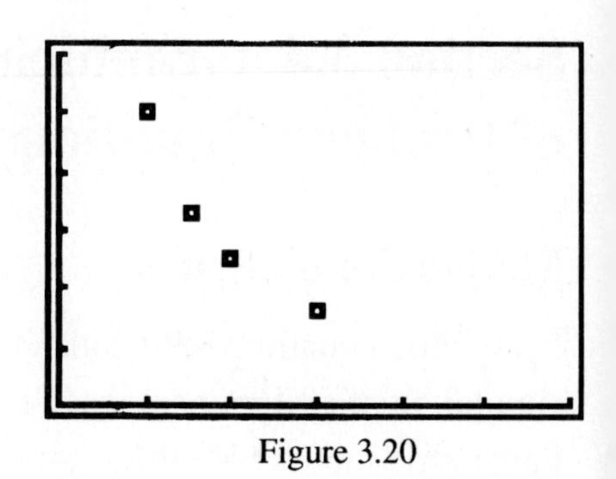

Figure 3.20

[2nd] [DRAW] [6:DrawF]

to copy the DrawF command on the Home screen, then press

500 [÷] [X/θ/T] [ENTER].

See Fig. 3.21.

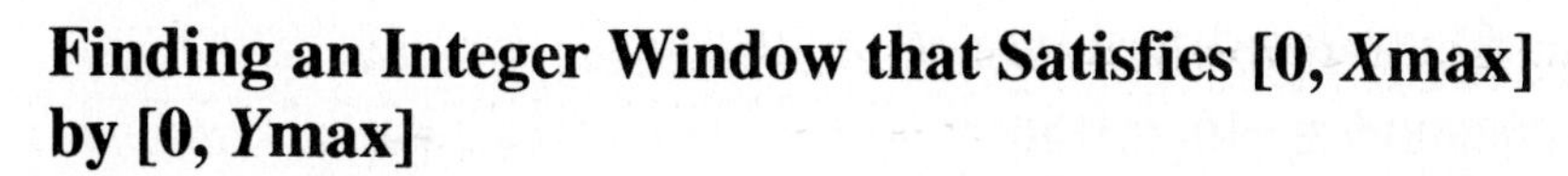

Finding an Integer Window that Satisfies [0, *X*max] by [0, *Y*max]

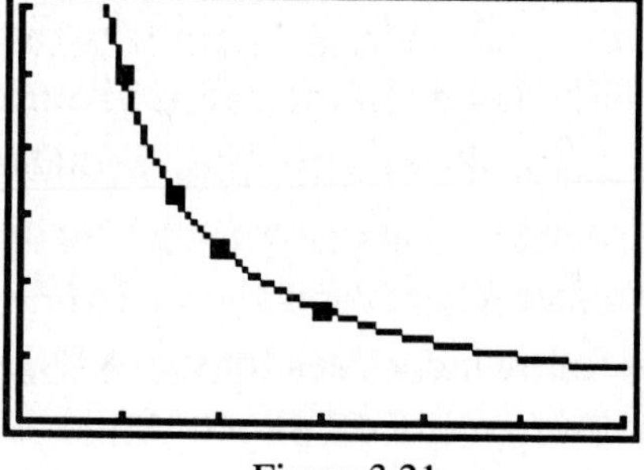

Figure 3.21

To find an Integer window that satisfies [0, *X*max] by [0, *Y*max], *X*max must be a multiple of 94, and *Y*max must be a multiple of 62. See Section 2.1.

Section 3.5 Solving Systems of Equations by Graphing

Using [TRACE] on Two Functions

The second example in Section 3.1 describes how to Trace on two different graphs in the same window.

Solving Systems of Equations with both Zoom In and Zoom Out

We will be using Zoom Factors of *X*Fact = 5 and *Y*Fact = 5 in the next problem, so use the Set Factors option in the ZOOM [MEMORY] menu to set both factors to 5. This option is described in Section 3.2.

Why 5 and 10 Are Convenient *X* and *Y* Zoom Factors

Example 1: Solve the system $y = -2x + 175$ and $y = 3x - 320$.

Graph $Y_1 = -2X + 175$ and $Y_2 = 3X - 320$ in the Integer window [−48, 47] by [−32, 31]. Since you don't see the graphs, we need a different window. We will change the window with **Zoom Out**. Zoom Out provides a more global view by enlarging the viewing window. Press

ZOOM [3:Zoom Out].

You should see the Graph screen with the free-moving cursor blinking at the origin. When you press ENTER, Zoom Out will enlarge the window centered on the free-moving cursor by a factor of 5. See Fig. 3.22. Now you can see the intersection of the two graphs. Next use the Zoom In option to find an accurate approximation of the point of intersection. Zoom In was also described in Section 3.2. The coordinates of the point of intersection are $x = 99$, $y = -23$. This is the solution to the system of equations.

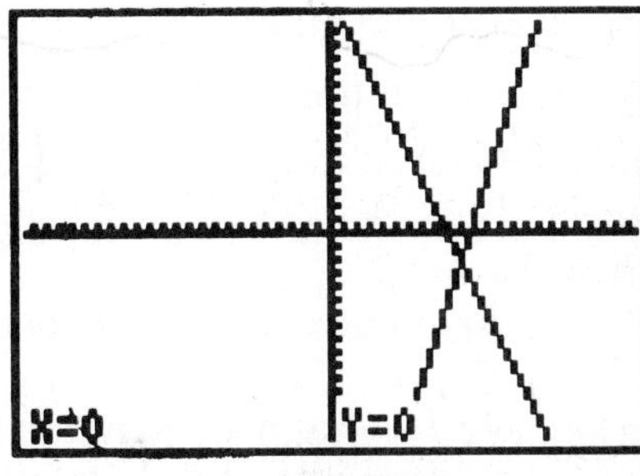

Figure 3.22

Chapter 4 Linear Inequalities and Systems of Linear Inequalities

Section 4.1 Solving Linear Inequalities Algebraically

See Section 1.3 to review Replay and Editing.

Section 4.4 Solving Absolute Value Equations and Inequalities

See Section 1.5 for an explanation of generating tables.

Section 4.5 Linear Inequalities in Two Variables

The Shade Command and Shading Solutions to Inequalities in Two Variables

The **Shade** command is followed by three arguments that are separated by commas. The grapher shades all the points that are above the first argument and below the second. (The third argument refers to the x resolution.) If there is no third argument given, every pixel is turned on in the shaded portion. If the resolution is 2, every other pixel is turned on, 3 every third pixel and so on.

The following example employs two arguments.

Example 1: Graph the solution to $y > 4x - 5$.

Clear the Y= menu and select the Standard viewing window. From the Home screen enter the following keystroke sequence:

[2nd] [DRAW] [7:Shade(] **4** [X/θ/T] [−] **5** [,] **10** [)].

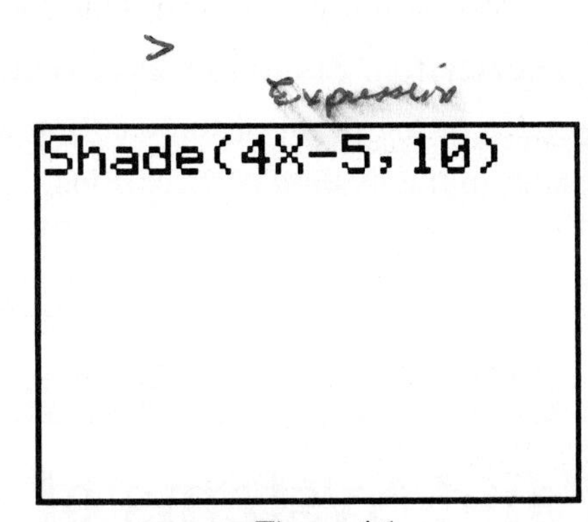

Figure 4.1

See Fig. 4.1. Press [ENTER] to see the solution. See Fig. 4.2. The solution is all points above the line $y = 4x - 5$. There is no upper bound for the solution, but the TI-82 requires us to provide an upper bound. We provided an upper bound of 10 so that the TI-82 would shade everything between the line and the top of the screen.

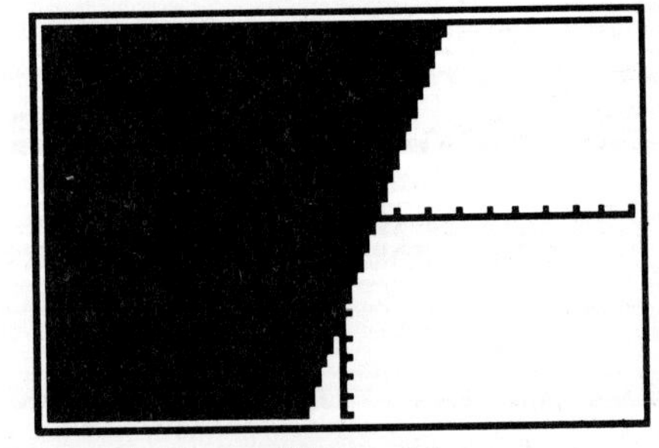

Figure 4.2

From the Home screen press

[2nd] [DRAW] [1:ClrDraw] [ENTER]

to clear the Graphics screen. Now use the DRAW menu to enter the command

Shade(–10, 4X–5)

on the Home screen and press [ENTER] to see the solution to $y < 4x - 5$. See Fig. 4.3.

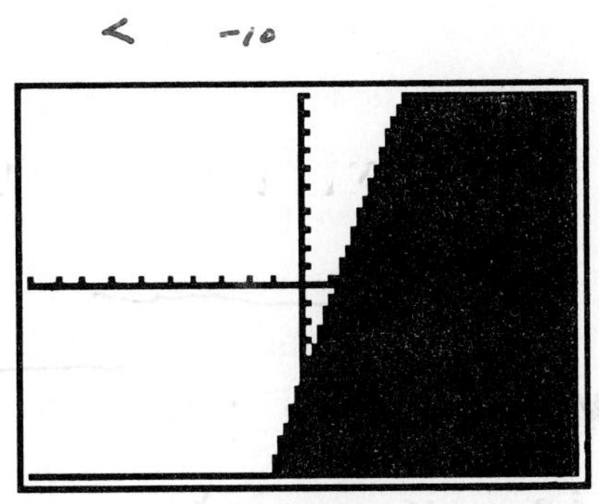

Figure 4.3

Section 4.6 Solving Systems of Inequalities

Graphing Systems of Inequalities Using Shading

Example 1: Graph the solution to the system $y < -x + 9$ and $y > 3x - 15$.

Enter the [–10, 10] by [–25, 25] viewing window. Then enter the following command from the Home screen: Shade(3X–15, –X+9). See Fig. 4.4. The lower boundary for the shaded area is the line $y = 3x - 15$ and the upper boundary is the line $y = -x + 9$.

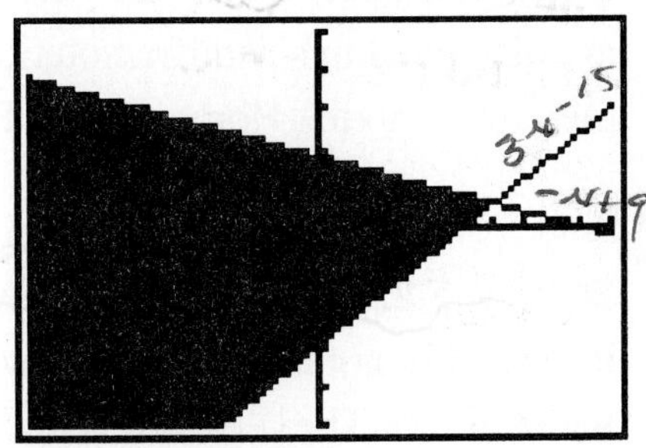

Figure 4.4

Chapter 5 Polynomials

Section 5.1 Exponents and Their Properties

Using [TRACE] to Show Two Graphs that Appear to Be Identical

Use [TRACE] to support rules of exponents by showing that two graphs appear to be identical.

Example 1: Determine graphically if $(2x^3)(3x^2) = 6x^5$ is correct.

Select Connected mode and enter

$Y_1 = (2X \wedge 3)(3X \wedge 2)$ and $Y_2 = 6X \wedge 5$.

Graph in the [–2, 2] by [–10, 10] window. Press [TRACE]. See Fig. 5.1. In Section 3.1 in the second example we described how to move from one curve to the other by pressing [▼] and [▲]. When you press these cursor-movement keys, the cursor does not appear to move to another graph and the coordinates of the Trace cursor don't change. This is graphical evidence that both Y_1 and Y_2 are the same function.

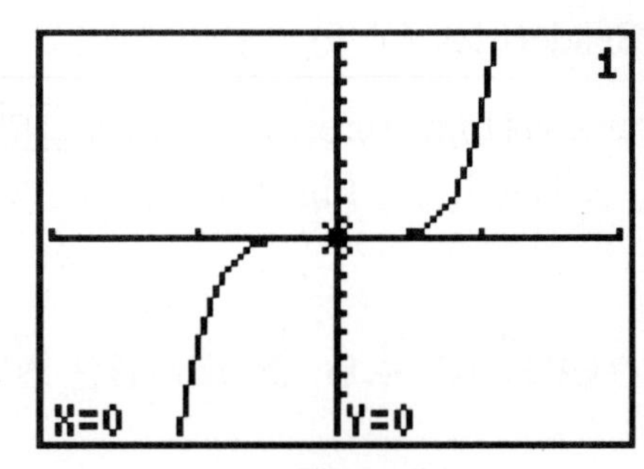

Figure 5.1

Scientific Notation, the [EE] Key, and the Normal Mode

When you select **Sci** on the first line of the Mode screen, numbers are displayed in scientific notation. Press [MODE] and highlight 'Sci,' then return to the Home screen. Press

[CLEAR] **1230** [ENTER].

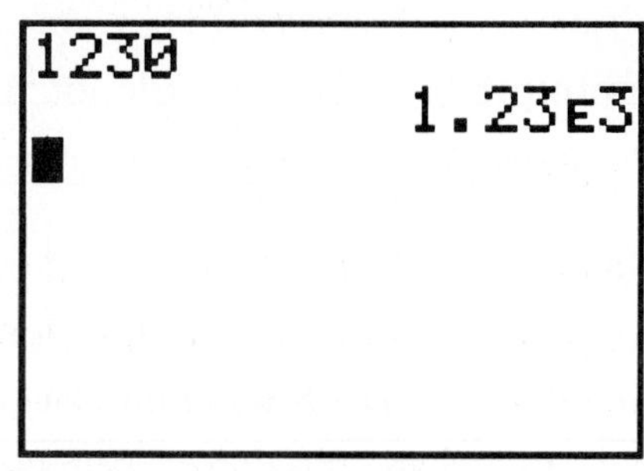

Figure 5.2

See Fig. 5.2. This notation means 1.23×10^3. You can enter numbers in scientific notation with [EE] whether the TI-82 is in Scientific mode or not. The [EE] key is the second function of the key found on row six, column two. Press [MODE] and switch back to **Normal mode** by highlighting 'Normal.' Now return to the Home screen and press

1 [.] **23** [2nd] [EE] **3** [ENTER].

The number was entered in scientific mode but displayed in normal mode. See Fig. 5.3.

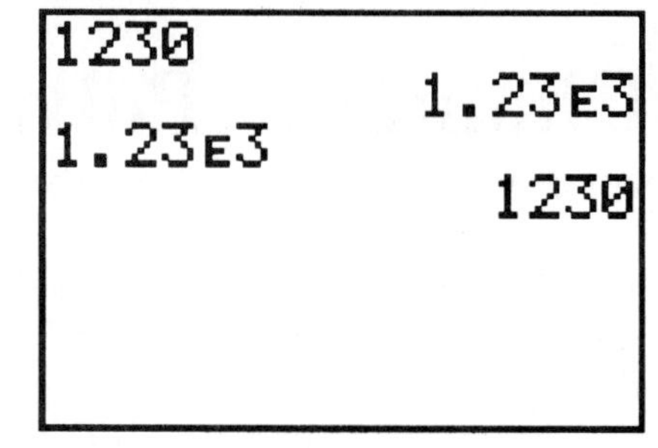

Figure 5.3

Large Numbers on the TI-82

The TI-82 displays numbers with up to 10 digits. If the number has more than 10 digits, the TI-82 converts the number to scientific notation. Press

2 [^] **33** [ENTER].

on the Home screen. The entire number is displayed since it is 10 digits long. Now enter 2 ^ 34 on the Home screen. See Fig. 5.4. This number has 11 digits, so it is converted to scientific notation and only the first 10 digits are displayed.

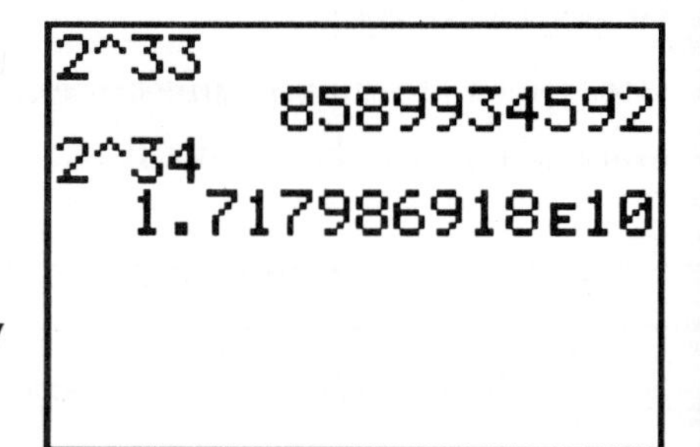

Figure 5.4

Small Numbers on the TI-82

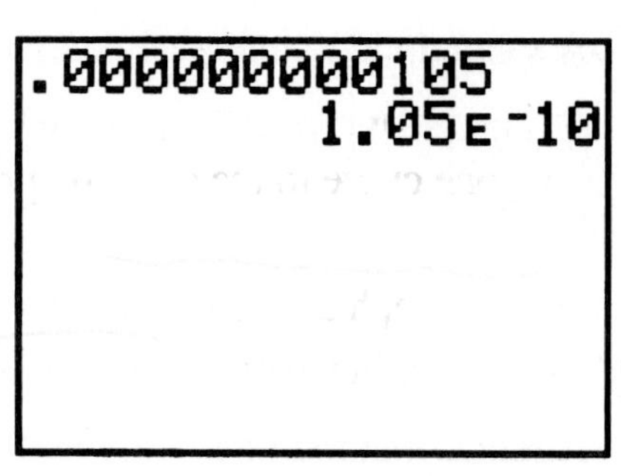

Figure 5.5

Numbers that are close to zero are usually represented in scientific notation on the TI-82. Enter .000000000105 on the Home screen. See Fig. 5.5. The decimal is converted to scientific notation. This notation means 1.05×10^{-10} which is the same as .000000000105. This number is very close to zero.

Section 5.2 Polynomial Expressions

Evaluating a Polynomial Expression on the Grapher

Section 1.3 described one method of evaluating a polynomial expression at a particular value of x. Now we will describe another method to evaluate a polynomial that uses the Y= menu. We will enter the polynomial in Y_1, store the value of x in the TI-82 variable 'X,' and then print the value of Y_1 on the Home screen. We print the value of Y_1 by copying it from the **Y-vars menu** to the Home screen.

The Method of Entering the Expression in the Function-Defining Menu

Example 1: Evaluate $2x^3 - 4x^2 + 5x - 7$ at $x = -1$.

Enter $Y_1 = 2X \wedge 3 - 4X \wedge 2 + 5X - 7$ in the Y= menu, then exit to the Home screen and clear the Home screen. Store –1 in X by pressing

[(-)] **1** [STO▶] [ALPHA] [X].

Now copy the variable Y_1 to the Home screen by pressing

[2nd] [:] [2nd] [Y-VARS] [1:Function] [1:Y_1]

and then print the value of Y_1 by pressing [ENTER]. See Fig. 5.6. The TI-82 evaluates Y_1 at the current value of 'X.'

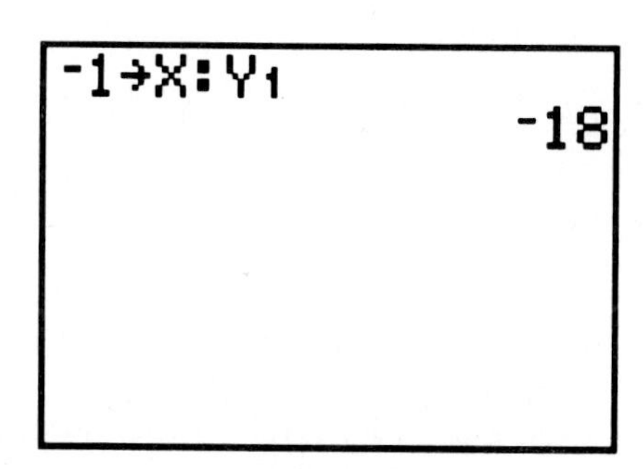

Figure 5.6

A second method to evaluate the polynomial at $x = -1$ is to press

[2nd] [Y-VARS] [1:Function] [1:Y_1] [(] [(-)] **1** [)] [ENTER].

See Fig. 5.7.

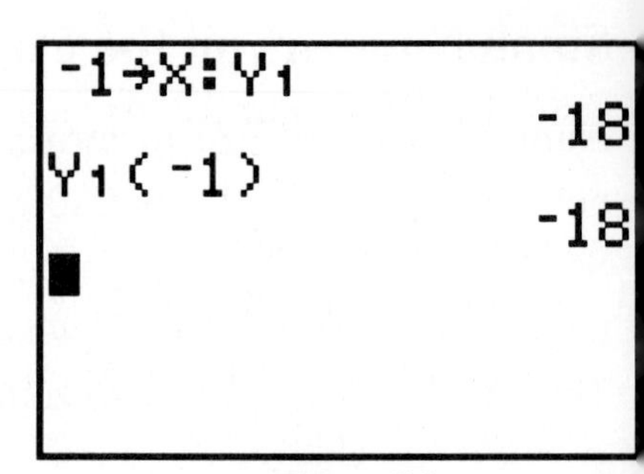

Figure 5.7

Table Building for Evaluating Polynomial Expressions

Another method of evaluating the polynomial is to build a table with the polynomial stored in Y_1. Table building was discussed in Section 1.5.

Entering the Sum of Two Polynomials into a Grapher

In Section 5.1 we used [GRAPH] and [TRACE] to provide visual support for rules of exponents. In this section we will use a similar method to provide support for addition of polynomials. We will enter the two polynomials in Y_1 and Y_2. Then we will enter $Y_3 = Y_1 + Y_2$ and enter the sum we derived algebraically in Y_4. Next we will deselect Y_1 and Y_2 and then graph Y_3 and Y_4 in the same window. If there appears to be only one graph in the window, we have support that Y_3 and Y_4 are the same. We can gather more support by Tracing both curves to see if they are the same.

Example 2: Find the sum of the polynomials $6x^4 + 2x^3 + 2x^2 + 10x + 1$ and $-5x^4 - 7x^3 + 3x^2 - 3x - 8$, and support your answer graphically.

The algebraic answer is $x^4 - 5x^3 + 5x^2 + 7x - 7$. Support this answer by clearing the Y= menu and entering: $Y_1 = 6X \wedge 4 + 2X \wedge 3 + 2X \wedge 2 + 10X + 1$ and $Y_2 = -5X \wedge 4 - 7X \wedge 3 + 3X \wedge 2 - 3X - 8$. Enter $Y_3 = Y_1 + Y_2$ by placing the cursor next to 'Y_3 =' and pressing

[2nd] [Y-VARS] [1:Function] [1:Y_1] [+] [2nd] [Y-VARS] [1:Function] [2:Y_2] [ENTER].

Then enter

$Y_4 = X \wedge 4 - 5X \wedge 3 + 5X \wedge 2 + 7X - 7$.

Deselect Y_1 by moving the cursor onto the '=' symbol beside Y_1 and pressing [ENTER]. This should unhighlight the '=' symbol. Deselect Y_2

the same way. See Fig. 5.8. Graph these polynomials in the [–5, 5] by [–10, 50] window. There appears to be just one graph. Reinforce this by pressing [TRACE] and moving the Trace cursor along the graph with the cursor-movement keys. As you move the Trace cursor along the graph, move from the graph of Y_3 to the graph of Y_4 by pressing [▲]. See Fig. 5.9. Since the coordinates don't change when you move from the graph of Y_3 to the graph of Y_4, we have support that the two polynomials are the same.

```
Y1=6X^4+2X^3+2X^
2+10X+1
Y2=-5X^4-7X^3+3X
^2-3X-8
Y3=Y1+Y2
Y4=X^4-5X^3+5X^2
+7X-7
Y5=
```

Figure 5.8

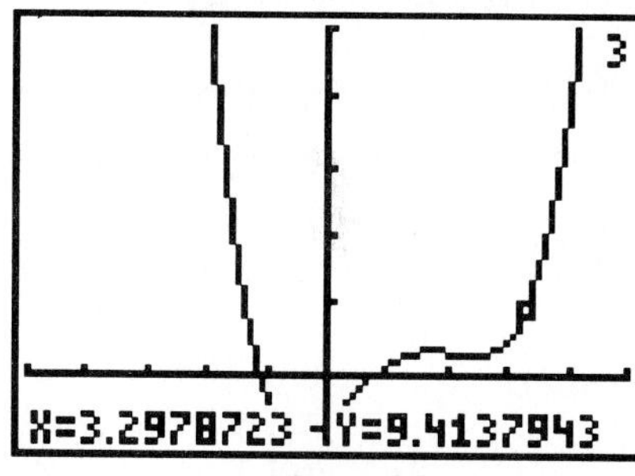

Figure 5.9

Evaluating Polynomials with Two or More Variables

We will use [STO▶] to assign values to the variables in the polynomial and then evaluate the polynomial. This method was introduced in Section 1.3.

Example 3: The volume of a tennis ball container is given as the polynomial $\pi x^2 y + (^2/_3)\pi x^3$, where x and y are dimensions of the container. Find the volume of the container when $x = 3.75$ and $y = 16.25$.

Clear the Home screen and press

3 [.] **75** [STO▶] [ALPHA] [X] [2nd] [:] **16** [.] **25** [STO▶] [ALPHA] [Y] [2nd] [:] [2nd] [π] [X/θ/T] [x²] [ALPHA] [Y] [+] [(] **2** [÷] **3** [)] [2nd] [π] [X/θ/T] [^] **3** [ENTER].

See Fig. 5.10. The volume is about 828.35.

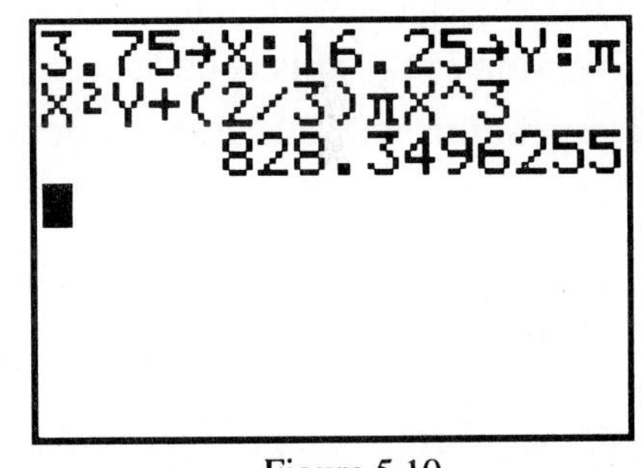

Figure 5.10

Section 5.4 Polynomial Equations and Factoring

See Section 2.1 for a discussion of the 0.1 window.

Chapter 6 Rational Expressions and Equations

Section 6.1 Simplifying Rational Expressions

Using Parentheses in Rational Expressions

When evaluating and graphing rational expressions, you must use parentheses to enter the expression correctly.

Example 1: Graph $y = \frac{x-2}{x}$.

Enter

$Y_1 = (X - 2)/X$ and $Y_2 = X - 2/X$.

Graph in the Standard window. See Fig. 6.1. Since there are two different graphs, we can see that Y_1 and Y_2 are not the same. The correct version is Y_1. This is because of the order of operations. Order of operations was discussed in Section 1.1. The parentheses in Y_1 force the subtraction to take place before the division. In Y_2 the division takes place before the subtraction since division has higher precedence than subtraction.

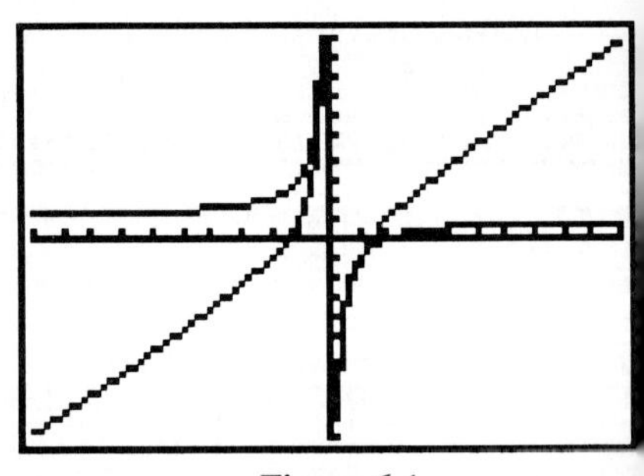

Figure 6.1

Error Messages with Rational Expressions

From the Home screen, press

0 [STO▶] [ALPHA] [X] [ENTER] **1** [÷] [X/θ/T] [ENTER].

You should see the same 'DIVIDE BY 0' error that you saw in the Preliminary Chapter. This is because we have instructed the TI-82 to divide by zero. You may encounter this difficulty when evaluating other rational functions. For example, you cannot evaluate $1/(x - 2)$ at $x = 2$ because division by zero is undefined.

Section 6.2 Multiplying and Dividing Rational Expressions

See Section 3.1 for a discussion of selecting and deselecting graphs.

Section 6.3 Adding and Subtracting Rational Expressions

Graphing Rational Expressions in Different Viewing Windows

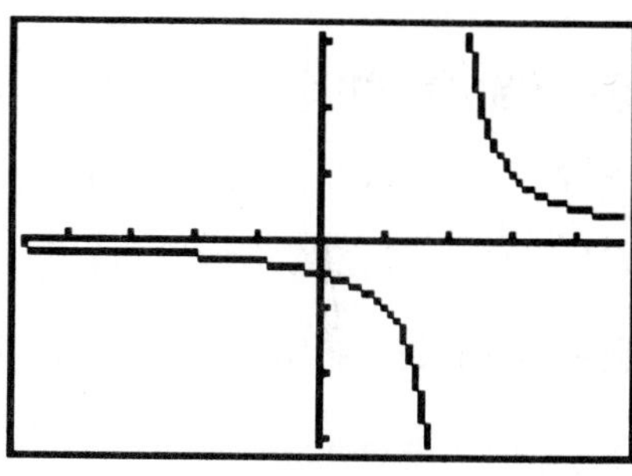

Figure 6.2

Example 1: Graph $y = 1/(x - 2)$.

Enter $Y_1 = 1/(X - 2)$ and graph in the Decimal window. See Fig. 6.2. Now regraph in the [–5, 5] by [–5, 5] window. See Fig. 6.3. The vertical line that you see in Fig. 6.3 is not really a part of the graph of $1/(x - 2)$. In some viewing windows, the bottom branch of the graph of a rational expression may be connected to the top branch by a vertical line. This vertical line is not a part of the correct graph. The occasional connecting of the branches has to do with the way the TI-82 graphs functions. It plots some of the points in the function and then connects those points to produce the final graph. In some windows, when the points are connected, a vertical line appears.

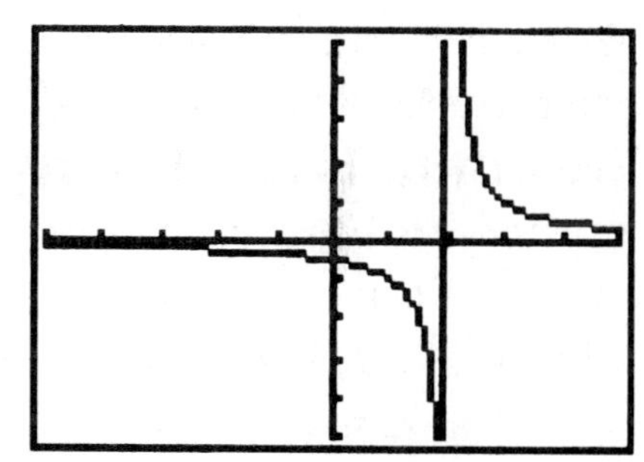

Figure 6.3

Section 6.4 Complex Fractions

Using Parentheses with Complex Fractions

Parentheses are required to enter expressions involving complex fractions. Make sure you have parentheses around the entire numerator and around the entire denominator.

Example 1: Evaluate $\dfrac{\frac{1}{2}+\frac{2}{3}}{\frac{5}{6}-\frac{3}{4}}$.

Clear the Home screen and enter

$({}^1/_2 + {}^2/_3)/({}^5/_6 - {}^3/_4)$.

See Fig. 6.4. The answer is 14. The parentheses force the TI-82 to calculate the numerator and denominator first and then divide.

Figure 6.4

Section 6.6 Solving Equations Containing Rational Expressions

See Section 2.1 for a discussion of the 0.2 and Integer windows.

Chapter 7 Rational Exponents, Radicals, and Complex Numbers

Section 7.1 Roots and Radicals

Error Messages with Grapher Limitations in Evaluating Square Roots of Negative Numbers

From the Home screen press

[CLEAR] [2nd] [√] [(-)] **3** [ENTER].

An ERROR message appears indicating a DOMAIN error. The DOMAIN error occurs because the square root of negative three is not a real number. The expression $\sqrt{-3}$ is not real because no real number squared is –3.

Evaluating Expressions with Fractional Exponents

Example 1: Evaluate $9^{1/2}$ and $125^{1/3}$.

Go to the Home screen and press

[CLEAR] **9** [^] [(] **1** [÷] **2** [)] [ENTER] **125** [^] [(] **1** [÷] **3** [)] [ENTER].

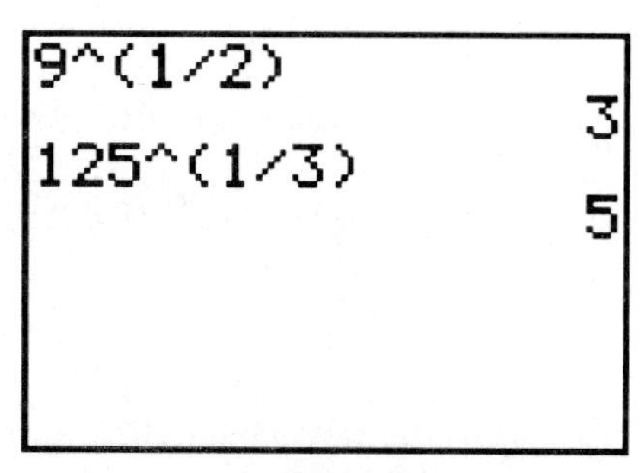

Figure 7.1

See Fig. 7.1. You must place parentheses around the exponent.

Example 2: Compare the graphs of $y = x \wedge ({}^{1}/_{4}) \wedge 2$ and $y = x \wedge 2 \wedge ({}^{1}/_{4})$.

Enter $Y_1 = X \wedge ({}^{1}/_{4}) \wedge 2$ and graph in the Decimal window. See Fig. 7.2. Now erase Y_1 and graph

$Y_1 = X \wedge 2 \wedge ({}^{1}/_{4})$

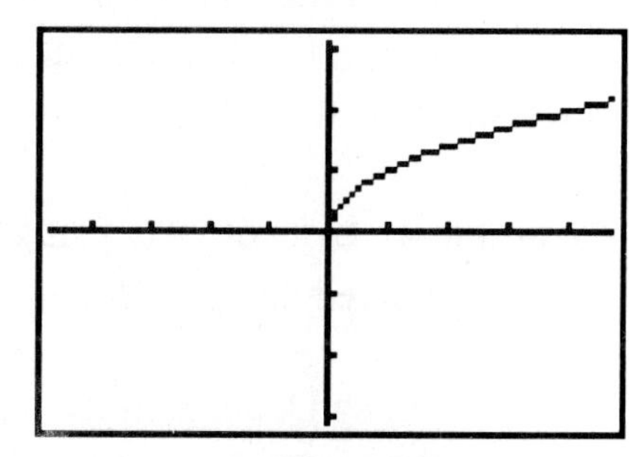
Figure 7.2

in the Decimal window. See Fig. 7.3. Both graphs appear in the first quadrant, but only the second graph also appears in the second quadrant where x is negative. This is because the TI-82 evaluates the expression from left to right. When it evaluates the first expression it takes the fourth root of x and then squares. The fourth root of a negative number is not real, so the first expression is not defined when x is negative. When the TI-82 evaluates the second expression it squares and then takes the fourth root. The square of a negative number is positive so the second expression is defined when x is negative.

Figure 7.3

Evaluating Roots with Your Grapher

Square roots can be evaluated with [√], and cube roots can be evaluated with the fourth option in the MATH menu. Other roots can be evaluated with the fifth option in the MATH menu.

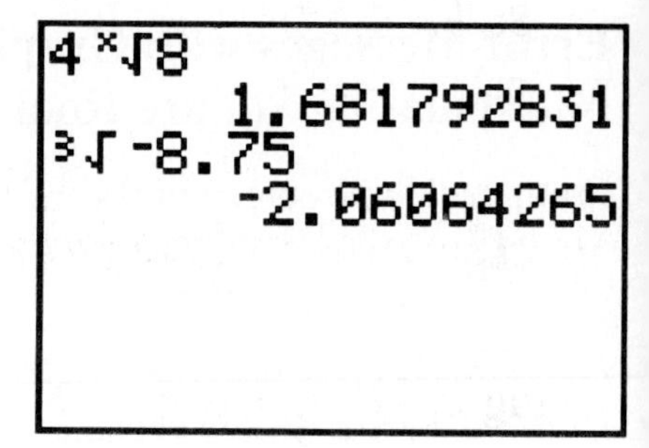

Figure 7.4

Example 3: Evaluate the fourth root of eight by pressing

[CLEAR] **4** [MATH] [5: ˣ√] **8** [ENTER].

Then evaluate the cube root of –8.75 by pressing

[MATH] [4: ∛] [(-)] **8** [.] **75** [ENTER].

See Fig. 7.4.

Adjusting the Number of Decimal Places Displayed on Your Grapher

We will adjust the number of decimal places displayed with the Fixed mode option in the Mode screen. This was discussed in Section 1.1.

Example 4: Evaluate $\sqrt[4]{8}$ using a floating decimal display setting and then a fixed decimal display setting of two digits.

Select Floating Decimal mode and enter 8 ^ ($^1/_4$) on the Home screen. Next, select Fixed Decimal mode by highlighting the '2' on the second line of the Mode screen. Enter 8 ^ ($^1/_4$) on the Home screen. See Fig. 7.5. In Fixed Decimal (2) mode, only two digits to the right of the decimal are displayed.

```
8^(1/4)
     1.681792831
8^(1/4)
            1.68
```

Figure 7.5

Section 7.2 Rational Exponents and Radicals

See Section 7.1 for a discussion of evaluating an expression with rational exponents.

Section 7.6 Complex Numbers

Using the Grapher to Work with Complex Numbers

The TI-82 cannot work with complex numbers.

Chapter 8 Quadratic Functions

Section 8.4 Sketching Graphs of Quadratic Functions

See Section 2.1 to review the 0.2 window.

Section 8.6 Vertex and Symmetry of a Parabola

See Section 1.1 to review the Frac Command.

Chapter 9 Higher Order Systems of Equations and Matrices

Section 9.1 Solving Systems of Equations in Three Variables

Equations in three variables cannot be graphed on the TI-82. The only support the TI-82 can provide for a solution to a system of equations in three variables is numerical.

Section 9.2 Matrix Algebra

Entering a Matrix on the TI-82

The matrix key is on the fourth row, second column. Press [MATRX] [▶] [▶] to see the Matrix edit screen. See Fig. 9.1. You can use up to five matrices on the TI-82. The notation for these matrices is [*A*], [*B*], [*C*], [*D*], and [*E*].

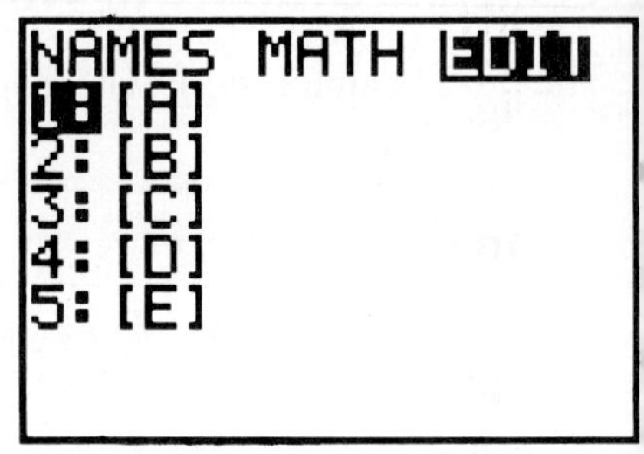

Figure 9.1

Example 1: Enter the 2×2 matrix

$$A = \begin{bmatrix} 4 & -3 \\ 2 & 6 \end{bmatrix}$$

and the 2×2 matrix

$$B = \begin{bmatrix} 2 & 5 \\ 1 & 2 \end{bmatrix}$$

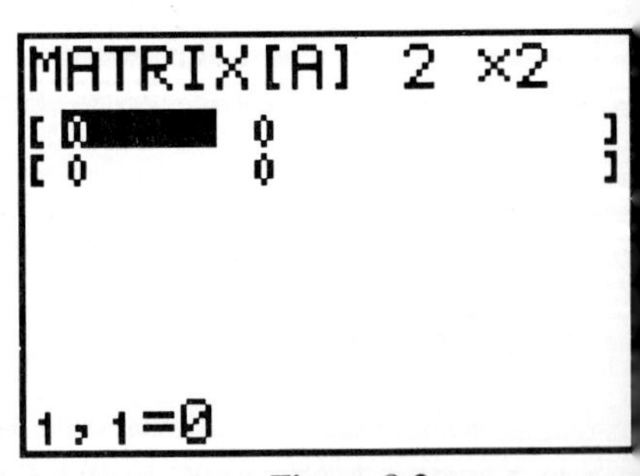

Figure 9.2

We must first change the dimension of matrix [*A*] to two rows and two columns. From the Matrix edit screen press

12 [ENTER] **2** [ENTER].

See Fig. 9.2. Now enter the values in the matrix by pressing

4 [ENTER] [(-)] **3** [ENTER] **2** [ENTER] **6** [ENTER].

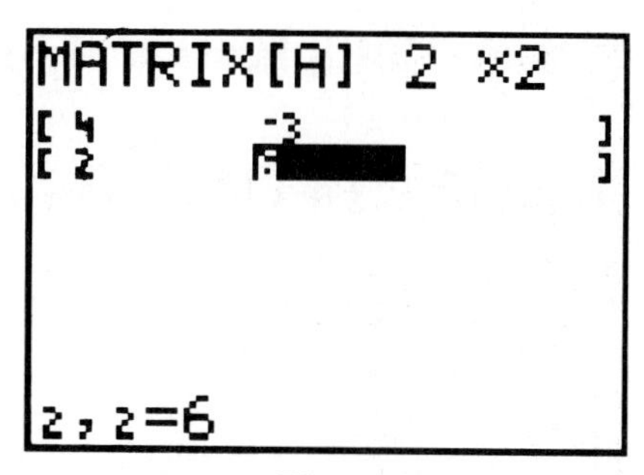

Figure 9.3

See Fig. 9.3. Now edit matrix [*B*] by pressing

[MATRX] [▶] [▶] **22** [ENTER] **2** [ENTER] **2** [ENTER] **5** [ENTER] **1** [ENTER] **2** [ENTER].

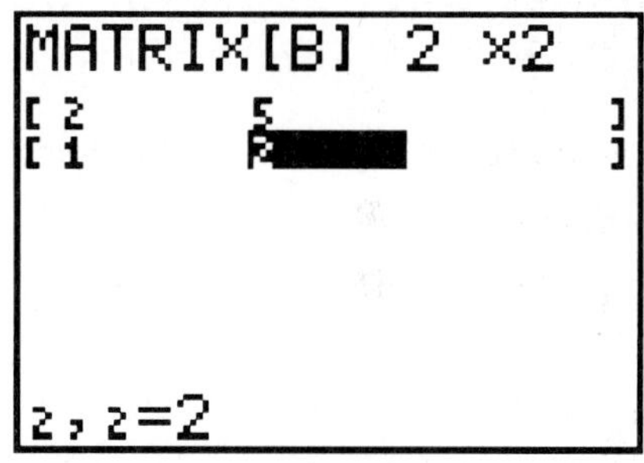

Figure 9.4

See Fig. 9.4. Press

[2nd] [QUIT]

to return to the Home screen. You can print the values in matrix [*A*] and matrix [*B*] by pressing

[MATRX] [1:[A]] [ENTER] [MATRX] [2:[B]] [ENTER].

See Fig. 9.5.

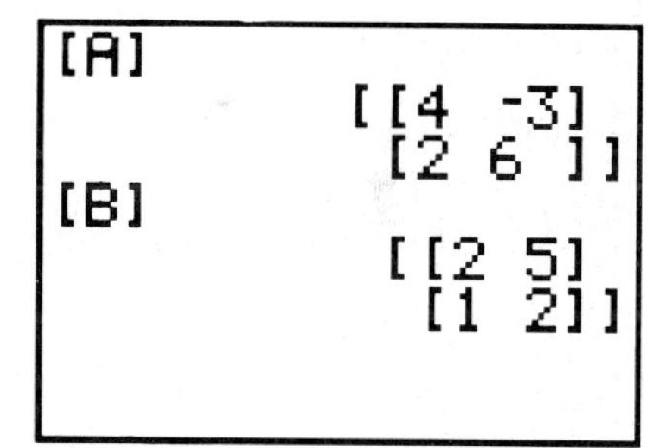

Figure 9.5

Adding and Subtracting Matrices

Example 2: Find [*A*] + [*B*] and [*A*] – [*B*] where [*A*] and [*B*] have the values assigned in Example 1.

Press

[CLEAR] [MATRX] [1:[A]] [+] [MATRX] [2:[B]] [ENTER]

to see the matrix sum. Press

[MATRX] [1:[A]] [–] [MATRX] [2:[B]] [ENTER]

to see the difference. See Fig. 9.6.

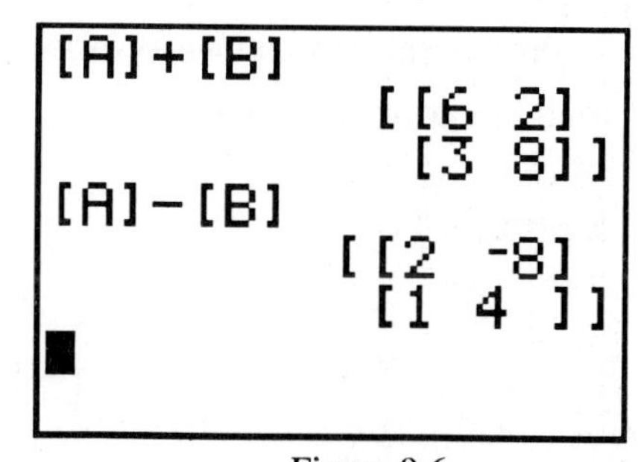

Figure 9.6

Scalar Multiplication of Matrices

Example 3: Find 3[*A*] where

$$[A] = \begin{bmatrix} 2 & 1 & 5 \\ 6 & -1 & 3 \end{bmatrix}.$$

Enter matrix [*A*] by pressing

MATRX ▶ ▶ **12** ENTER **3** ENTER **2** ENTER **1** ENTER **5** ENTER **6** ENTER (–) **1** ENTER **3** ENTER.

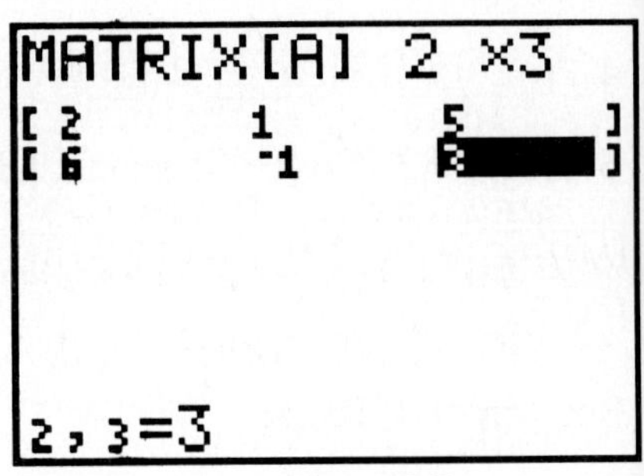

Figure 9.7

See Fig. 9.7. Notice you enter matrices one row at a time on the TI-82. Press

2nd QUIT

to return to the Home screen. Now perform the scalar multiplication by pressing

CLEAR **3** MATRX [1:[A]] ENTER.

See Fig. 9.8.

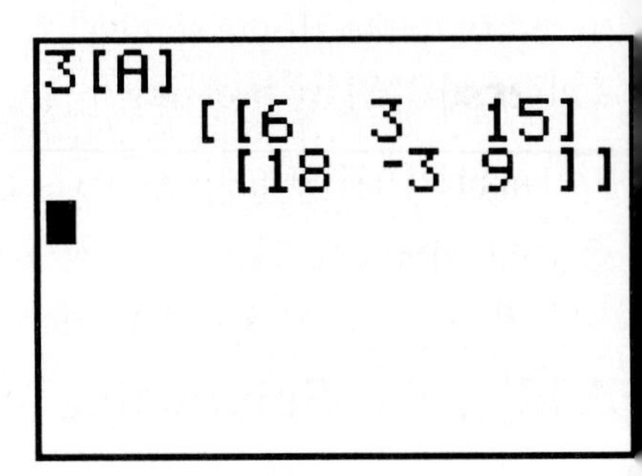

Figure 9.8

Section 9.3 Multiplication of Matrices

Matrix Multiplication

Once matrices have been entered on the TI-82, you can multiply them by simply entering their names side-by-side on the Home screen. A multiplication symbol is not needed between the two matrix variable names.

Example 1: If

$$A = \begin{bmatrix} 3 & -4 \\ 2 & -8 \end{bmatrix} \quad \text{and} \quad B = \begin{bmatrix} 4 & -2 \\ -1 & 9 \end{bmatrix},$$

find [*A*] × [*B*].

Enter the matrices A and B with the Matrix edit screen. (Entering matrices was explained in Section 9.2.) Then from the Home screen press

CLEAR MATRX [1:[A]] MATRX [2:[B]] ENTER.

See Fig. 9.9.

[A][B]
[[16 -42]
[16 -76]]

Figure 9.9

The Identity Matrix on the Grapher

Identity matrices of various dimensions can be easily created with the fifth option in the MATRX [MATH] menu. To create a 3×3 identity matrix, press

MATRX ▶ **53** ENTER.

See Fig. 9.10.

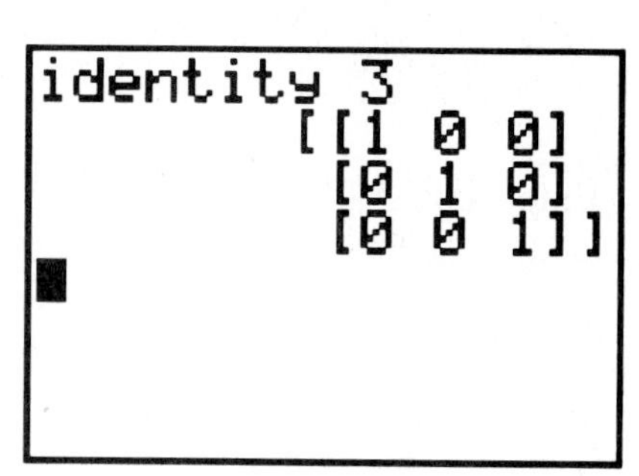

Figure 9.10

Inverse Matrices Using the x^{-1} Key

If a matrix can be inverted, you can usually find its inverse with x^{-1}. If the inverse doesn't exist, you will see an error message or an odd-looking matrix with entries that are very near zero.

Example 2: Find the inverse of $A = \begin{bmatrix} 5 & 1 \\ 4 & 1 \end{bmatrix}$.

Enter matrix A with the Matrix edit screen and then from the Home screen press

CLEAR MATRX [1:[A]] x^{-1} ENTER.

See Fig. 9.11.

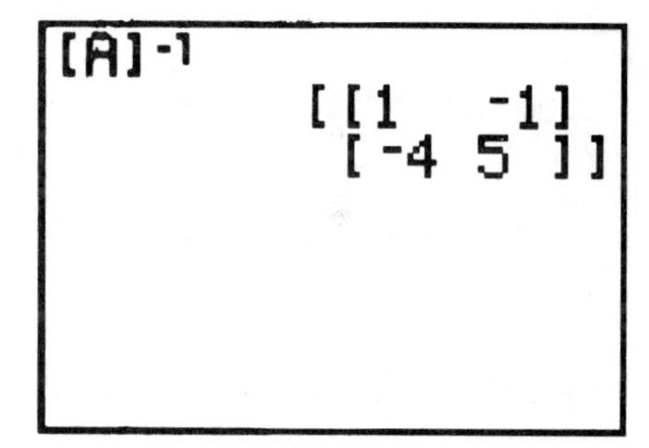

Figure 9.11

Example 3: Find the inverse of $A = \begin{bmatrix} 3 & 1 \\ 3 & 1 \end{bmatrix}$.

Enter matrix [A] with the Matrix edit screen, and from the Home screen press

CLEAR MATRX [1:[A]] x^{-1} ENTER.

See Fig. 9.12. This error message tells you that this matrix cannot be inverted. If you see an error message or if all the entries are very

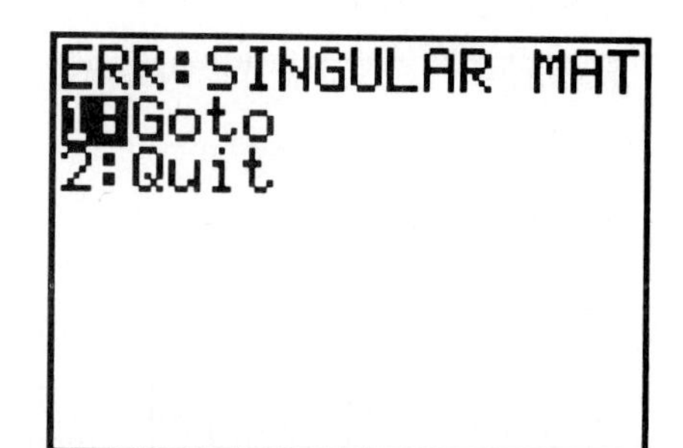

Figure 9.12

near zero when you attempt to find the inverse of a matrix, the inverse likely does not exist.

Determinants of a Square Matrix

[1:det] (determinant) is the first option in the MATRX [MATH] menu. This feature can be used to find determinants of square matrices.

Example 4: Find the determinant of

$$A = \begin{bmatrix} 3 & 1 \\ 3 & 1 \end{bmatrix}.$$

Enter matrix $[A]$ with the Matrix edit screen. Then from the Home screen press

CLEAR MATRX ▶ [1:det] MATRX [1:[A]] ENTER.

See Fig. 9.13. The determinant is zero.

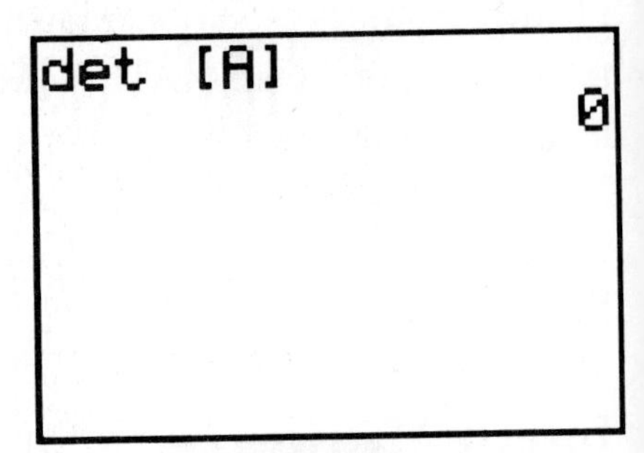

Figure 9.13

Section 9.4 Solving a System of Equations Using Matrices

How to Solve a System of Equations with Matrices on the Grapher

Systems of equations can be expressed with matrices. When this is done, the solution of the system is a matrix. We can solve for this solution matrix with the inverse matrix feature of the TI-82.

Example 1: Solve the system

$$\begin{cases} 4x - 3y = -26 \\ 15x + 11y = 36 \end{cases}.$$

This system can be expressed in matrix form as $AX = B$, where

$$A = \begin{bmatrix} 4 & -3 \\ 15 & 11 \end{bmatrix}, \quad X = \begin{bmatrix} x \\ y \end{bmatrix}, \quad \text{and} \quad B = \begin{bmatrix} -26 \\ 36 \end{bmatrix}.$$

We then solve the matrix equation $AX = B$ for X, obtaining $X = A^{-1}B$. Matrix X contains the solution to the original system of equations. Enter the values just described for the 2×2 matrix A and the 2×1 matrix B with the Matrix edit screen. Don't forget to change the dimension of $[B]$ to two rows, one column. After you have entered A and B, return to the Home screen and find the product $A^{-1}B$ by pressing

CLEAR MATRX [1:[A]] x^{-1} MATRX [2:[B]] ENTER.

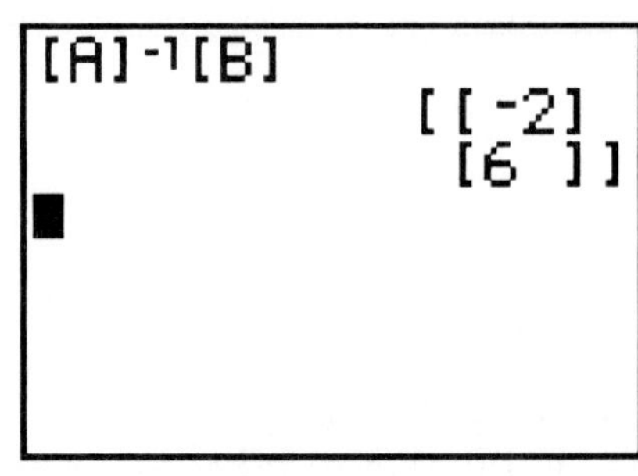

Figure 9.14

See Fig. 9.14. The solution is $x = -2$, $y = 6$. Matrix multiplication is not commutative, so you must enter $A^{-1}B$, not BA^{-1}.

Chapter 10 Conic Sections

Conic Sections

In general, when we graph conic sections on the TI-82, we must rewrite the conic equation as $y = \ldots$ and then graph. While solving for a conic equation for y, we usually obtain two new equations. A lengthy radical will probably be created in both equations. We will enter this radical into the Y_1 slot and enter the two equations in Y_2 and Y_3 with Y_1 used to represent the radical. Then we will deselect Y_1 and graph Y_2 and Y_3. There may be gaps between the graphs of Y_2 and Y_3 because of screen resolution.

Example 1: Graph the ellipse $(x - 1)^2/4 + (y + 2)^2/9 = 1$.

We solve for y and obtain

$$y = -2 \pm \sqrt{(4-(x-1)^2)/2}.$$

Enter

$$Y1 = 3\sqrt{(4-(X-1)^2)/2}, \quad Y_2 .= -2 + Y_1, \quad \text{and} \quad Y_3 = -2 - Y_1$$

Deselect Y_1. See Fig. 10.1. Now graph in the Standard viewing window. See Fig. 10.2.

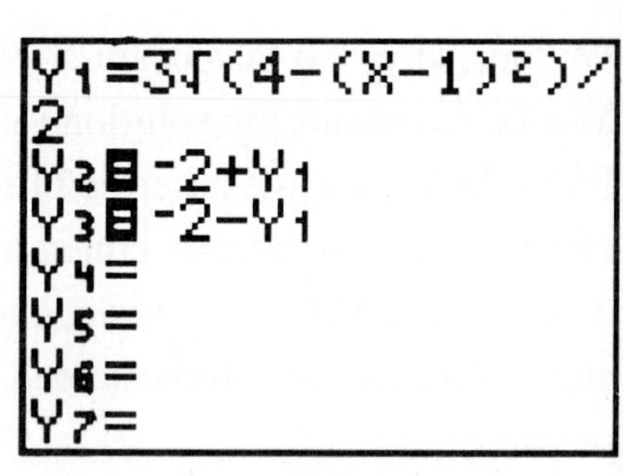

Figure 10.1

Section 10.1 Circles

Graph in the Square Window for a Circle

A circle must be graphed in a square window for the true shape of the circle to appear. In the Standard window, the circle will look distorted.

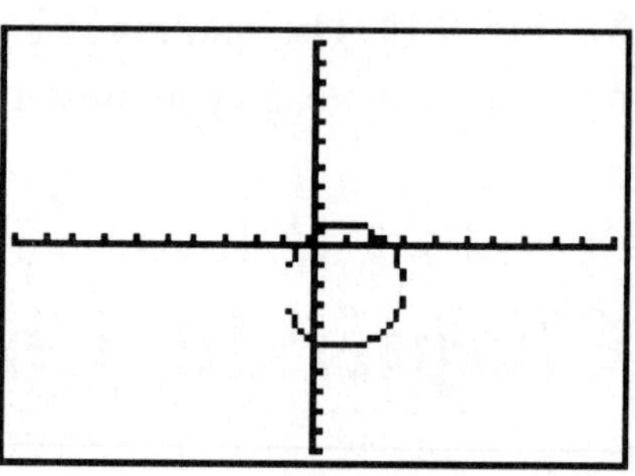

Figure 10.2

Example 1: Graph the circle $x^2 + y^2 = 25$.

Enter

$Y_1 = \sqrt{(25 - X \wedge 2)}$ and $Y_2 = -Y_1$.

Graph in the Standard window. See Fig. 10.3. The circle looks like an ellipse. Now press

[ZOOM] [5:Square].

See Fig. 10.4. The proportions of the circle are correct in a square window.

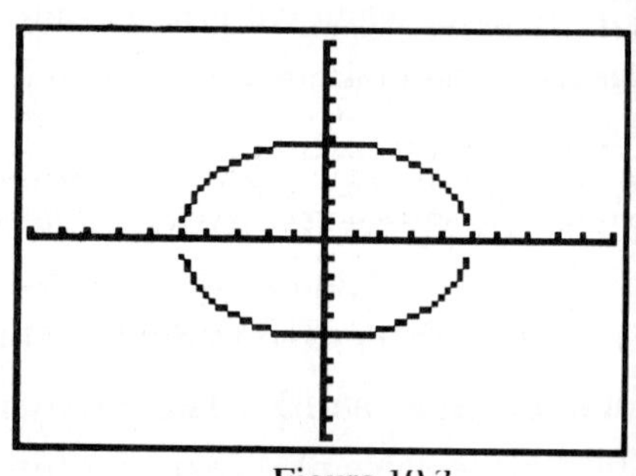

Figure 10.3

Section 10.5 Nonlinear Systems of Equations

A Review on Solving Systems of Equations Graphically

See Chapter 3 for a review of how to solve systems of equations graphically. Recall that we set the scale marks to 0.01 to assist with a solution with at most 0.01 error.

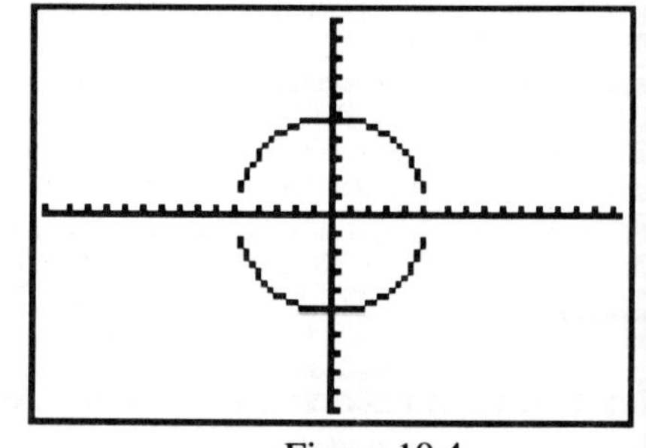

Figure 10.4

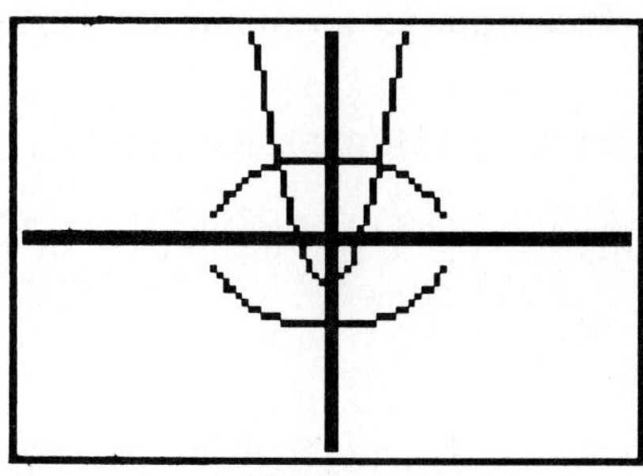
Figure 10.5

Example 1: Solve the system $x^2 + y^2 = 16$ and $y = 2x^2 - 2$.

We must solve the first equation for y before we can solve graphically. This gives

$$y = \pm\sqrt{(16 - x^2)}.$$

Enter

$$Y_1 = \sqrt{(16 - X\wedge 2)} \quad \text{and} \quad Y_2 = -Y_1.$$

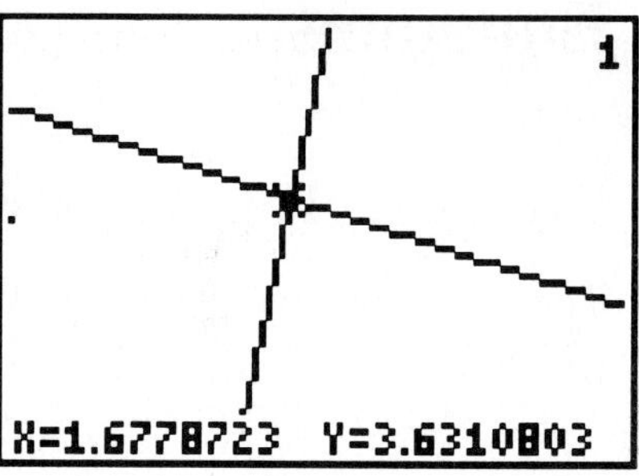

Figure 10.6

Enter $Y_3 = 2X \wedge 2 - 2$ and graph all three in the [−10, 10] by [−10, 10] window with scale values of 0.01. See Fig. 10.5. Before we Zoom In, notice that Y_2 has no intersection points with Y_3, so we will deselect Y_2 for faster graphing. Now we Zoom In several times on the intersection in the first quadrant. See Fig. 10.6. Since the intersection is contained between the scale marks, we know that the error is at most 0.01. The other solution can be found by regraphing in the original window and then Zooming In on the intersection in the second quadrant.

Chapter 11 Exponential and Logarithmic Functions

Section 11.3 Logarithmic Functions

The [LOG] and [LN] Keys

The base 10 log key [LOG] is on the seventh row, first column. The natural log key [LN] is just below the base 10 log key.

Example 1: Evaluate log 125 and ln 12.3.

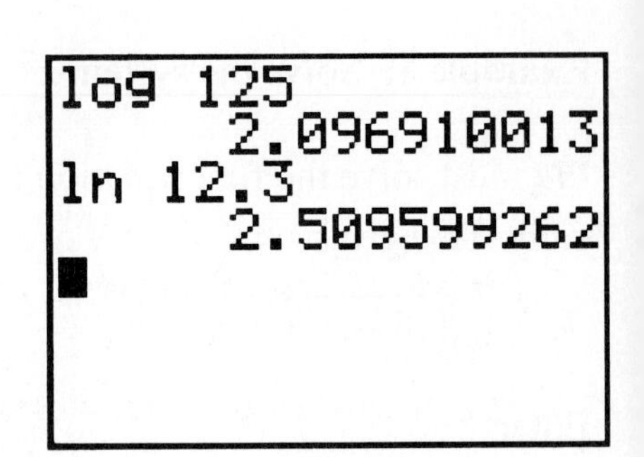

Figure 11.1

From the Home screen press

[CLEAR] [LOG] **125** [ENTER] [LN] **12** [.] **3** [ENTER].

See Fig. 11.1.

Section 11.5 Solving Logarithmic Equations

Evaluating Y_1 at 3 by Using Notation $Y_1(3)$

The TI-82 can evaluate Y_1 at 3 by using the notation $Y_1(3)$. See Section 5.2 to review how to evaluate an expression on the TI-82.

Revisiting the Grapher-Generated Tables on the TI-82

Tables were discussed in Section 1.5.

Example 1: Use a table to find the value of x where $y = 3000$ in the equation $y = P(1.07)^x$ when $P = 1000$. This is the time needed for the investment described by this equation to triple.

Enter $Y_1 = 1000(1.07)$ ^ X. Press

[2nd] [Tblset]

and enter 1 for both TblMin and ΔTbl. Now view the table by pressing

[2nd] [TABLE].

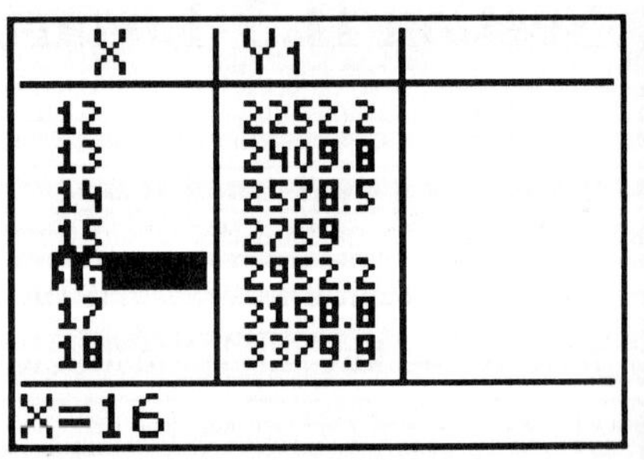

Figure 11.2

Press [▼] to scroll down the table to find the entry with a y value close to 3000. See Fig. 11.2. From this table, we know that the investment triples between 16 and 17 years. We can increase the detail in the table by changing TblMin and ΔTbl. Press

[2nd] [Tblset] **16** [ENTER] **0** [.] **1**

to change these values, and then press

2nd TABLE

to see the new graph. See Fig. 11.3. We can see from the new table that y is 3000 when x is between 16.2 and 16.3. To magnify the table numerically again, change TblMin to 16.2 and ΔTbl to 0.01 in the Table Setup screen, and then view the new table. See Fig. 11.4. Now the answer appears to be about 16.24.

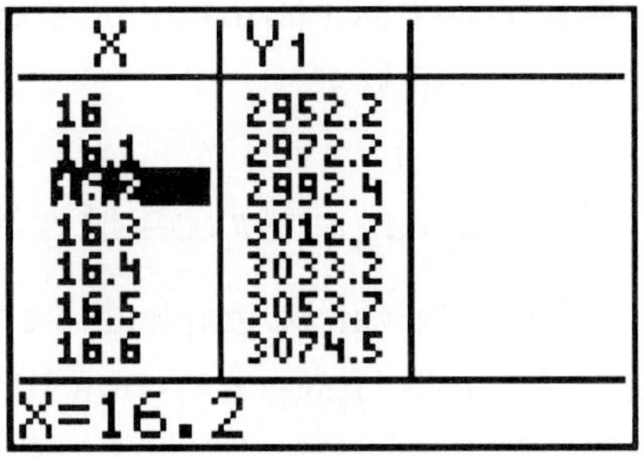

X	Y_1
16	2952.2
16.1	2972.2
16.2	2992.4
16.3	3012.7
16.4	3033.2
16.5	3053.7
16.6	3074.5

X=16.2

Figure 11.3

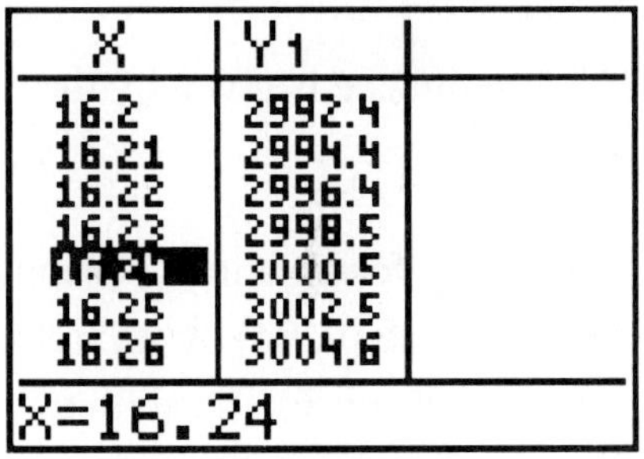

X	Y_1
16.2	2992.4
16.21	2994.4
16.22	2996.4
16.23	2998.5
16.24	3000.5
16.25	3002.5
16.26	3004.6

X=16.24

Figure 11.4

Chapter 12 Sequences and Series and the Binomial Theorem

Section 12.1 Arithmetic Sequences and Series

The TI-82 Generate Sequences

It is possible to generate sequences with a **seq(** command on the TI-82. This command has five parameters. They are (1) the formula for the nth term of the sequence, (2) the index, (3) the initial and (4) final values of the index, and (5) the amount by which the index will increase.

Example 1: Generate the sequence 3,5,7,9,...,($2n$+1),....

From the Home screen press

CLEAR 2nd L [5:seq(] **2** ALPHA N + **1** , ALPHA N , **1** , **10** , **1** ENTER.

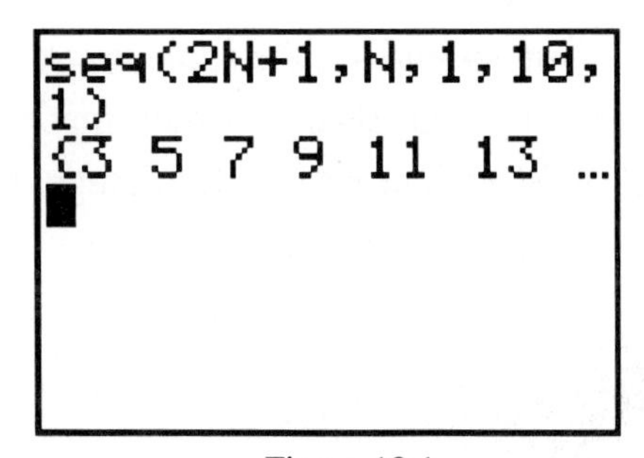

Figure 12.1

See Fig. 12.1. The first 10 terms of the sequence are generated. You can see the last terms in the sequence by pressing the right cursor-movement key.

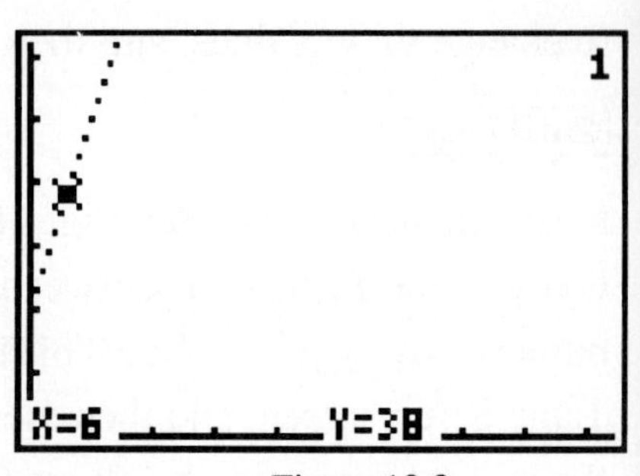

Figure 12.2

Displaying a Graph of a Sequence

Sequences can be graphed like functions if we use the Dot mode and graph in the Integer window [0, 94] by [0, 62].

Example 2: Display the graph of the sequence 23, 26, 29,... and determine the value of the sixth term.

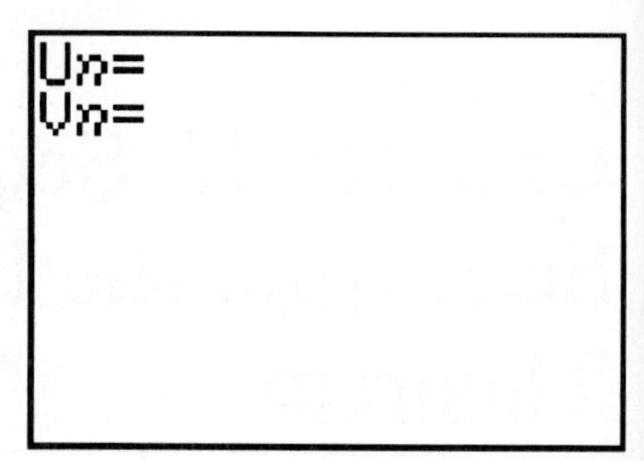

Figure 12.3

This sequence is given by the formula

$23 + (X - 1)3.$

Enter this formula in Y_1, change from Connected to Dot mode, and graph in the [0, 94] by [0, 62] window. Use x and y scales of 10. Now trace until $x = 6$. The sixth term is 38. See Fig. 12.2.

An alternative solution to this Example is to use the Sequence Graphing mode on the TI-82. Press [MODE] and select **Seq** (sequence) and Dot, then press [Y=]. See Fig. 12.3. You can enter two different sequences. Now press

23 [+] [(] [2nd] [*n*] [−] **1** [)] **3**.

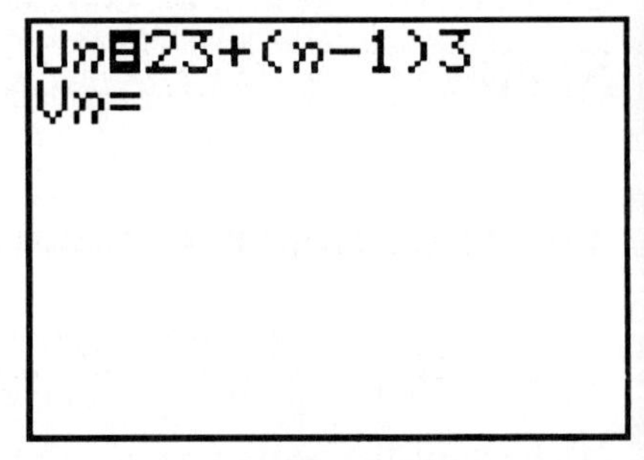

Figure 12.4

[*n*] is the second function of the **9** key. See Fig. 12.4. Set up the viewing window by pressing [WINDOW] [▼], and assign the following values to the WINDOW variables: U*n*Start = 1, V*n*Start = 0, *n*Start = 1, *n*Min = 1, *n*Max = 94, and *X*min = 0, *X*max = 94, *X*scl = 10, *Y*min = 0, *Y*max = 62, and *Y*scl = 10. Now press [GRAPH] and trace until $n = 6$. You should see the same graph as in Fig. 12.2. The advantage of using Sequence Graphing mode is that you can graph the sequence in windows other than the Integer window.

Section 12.2 Geometric Sequences and Series

See Section 12.1 for guidelines on graphing sequences.

Section 12.3 Binomial Expansion

Factorials

The Factorial function is the fourth option in the MATH [PRB] menu.

Example 1: Evaluate 5! and 8!5!/(4!3!).

From the Home screen, press

[CLEAR] **5** [MATH] [▶] [▶] [▶] [4:!] [ENTER]

to evaluate the first expression and

8 [MATH] [▶] [▶] [▶] [4:!] **5** [MATH] [▶] [▶] [▶] [4:!] [÷] [(] **4** [MATH] [▶] [▶] [▶] [4:!] **3** [MATH] [▶] [▶] [▶] [4:!] [)] [ENTER]

to evaluate the second expression. See Fig. 12.5.

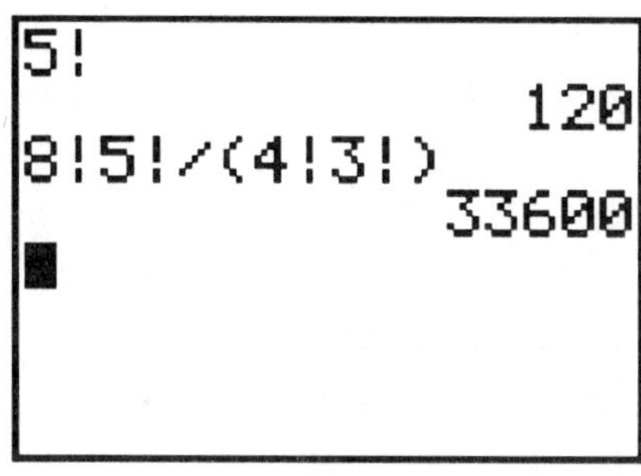

Figure 12.5

Evaluating a Binomial Coefficient $\binom{n}{r}$ on the Grapher

$\binom{n}{r}$ means the same thing as ${}_nC_r$. The ${}_nC_r$ notation is the notation for a combination of *n* things taken *r* at a time. This option is found in the MATH [PRB] (probability) menu.

Example 2: Determine the value of the binomial coefficient $\binom{5}{3}$. $\binom{5}{3} = {}_5C_3$.

From the Home screen press

[CLEAR] **5** [MATH] [▶] [▶] [▶] [3: ${}_nC_r$] **3** [ENTER].

See Fig. 12.6. The result is 10.

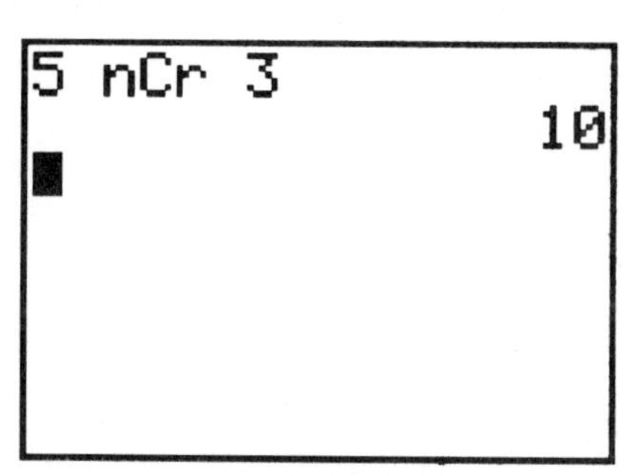

Figure 12.6

Section 12.4 Counting Principle, Permutations, and Combinations

Evaluating a Permutation $_nP_r$

This option is also found in the MATH [2:Prb] menu. The notation '$_nP_r$' means a permutation of n things taken r at a time.

Example 1: Evaluate $_{26}P_6$.

From the Home screen, press

[CLEAR] **26** [MATH] [▶] [▶] [▶] [2:$_nP_r$] **6** [ENTER].

See Fig. 12.7.

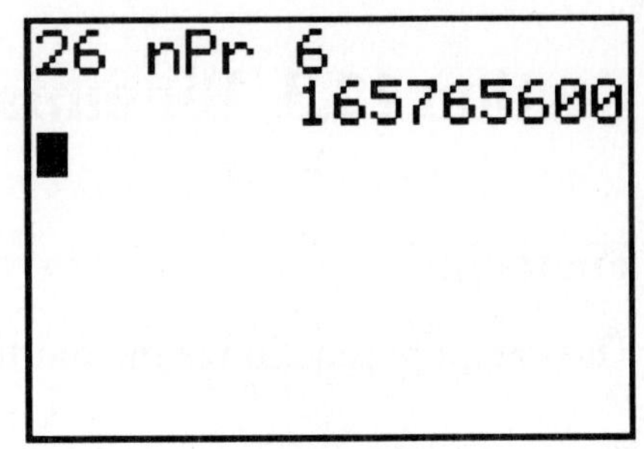

Figure 12.7

Evaluating a Combination $_nC_r$

This feature was described in Section 12.3.

Example 2: Evaluate $_4C_3$.

Press

4 [MATH] [▶] [▶] [▶] [3:$_nC_r$] **3** [ENTER].

The answer is 4.

Preliminary Chapter Getting Started Using the TI-85

Section 1 The Keyboard

The keys on the TI-85 are organized by color and position. See Fig. 1. The grey keys on the lower part of the keyboard are used to enter numbers. The black keys on the lower right-hand side of the keyboard are used to perform arithmetic. The grey keys with arrows in the upper right-hand corner of the keyboard are used to move the cursor on the screen. They are called the cursor-movement keys. The black keys on the top row just under the screen are used to draw graphs of equations. The other black keys are used for scientific functions and menus.

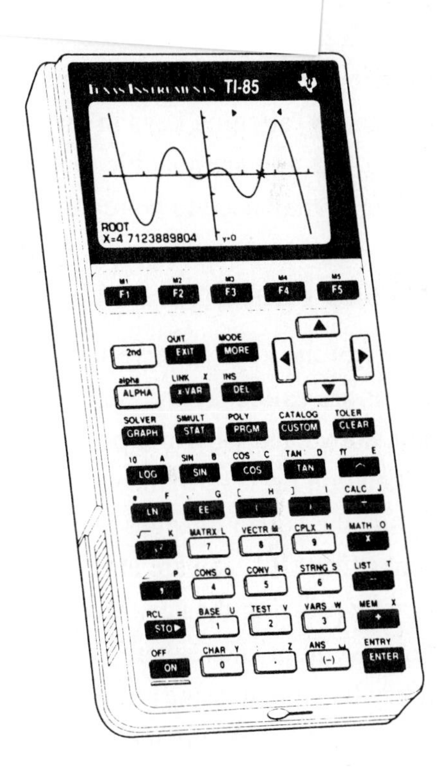

Figure 1

The [ON] and [OFF] Keys and the Automatic Off

The black key in the lower left-hand corner of the keyboard is used to turn the calculator on. Notice the key is labeled 'ON' and the word 'OFF' appears in yellow lettering above the key. Press [ON] to turn the calculator on. If you press [2nd] (the yellow key in the upper left-hand corner of the keyboard) and then press [ON], you will turn the calculator off. Pressing [2nd] lets you use the feature printed in yellow above the key. If you leave the TI-85 on for about five minutes without pressing any keys, it will turn itself off. This automatic off feature saves the batteries.

Section 2 The Cursor and Adjusting the Contrast

After you turn the calculator on, you should see a blinking rectangle displayed on the screen. This rectangle is called the cursor. The cursor indicates where the first number or letter you enter will be placed on the screen. If the screen seems too light, you can darken the display by pressing [2nd] and then holding [▲] until the contrast looks right. If the screen seems too dark, you can lighten the display by pressing [2nd] and holding [▼] until you have the contrast you want. ([▲] and [▼] are grey keys in the upper right-hand corner of the keyboard.) If you are not able to make the display dark enough, your batteries may need to be replaced.

Second Functions and The Alpha Key

Many of the keys have yellow labels above them. These yellow labels are called second functions. You can access the second function above a key by first pressing [2nd] and then pressing the key. When you press [2nd] the cursor changes to ' ⬛ '. If you press [2nd] again, the cursor changes back to its original shape. While the cursor looks like ' ⬛ ', you are able to access the second function above a key by pressing the key. Many of the keys also have a light blue letter or symbol above them. You can access the light blue letter above a key by first pressing [ALPHA] and then pressing the key. When you press [ALPHA] the cursor changes to ' ⬛ '. We say the cursor is in the Alpha mode. Pressing [ALPHA] twice locks the cursor in the Alpha mode. If you press [ALPHA] a third time, the cursor changes back to a blinking rectangle. While the cursor looks like ' ⬛ ', you can access the letter above a key by pressing the key. If you press

[2nd] [ALPHA],

the cursor contains a lowercase 'a'. This means when you press a key, you will see the lowercase version of the letter above the key.

You can exit the lowercase Alpha mode by pressing [ALPHA] twice. The role of [2nd] and [ALPHA] is similar to the shift key on a typewriter or computer keyboard.

Section 3 Keyboarding Instructions

When you see instructions to press a certain key, the instructions will include a box containing the label on the key. If the key is a number key, however, the label will not be enclosed in a box. For example, if you follow directions to turn on the calculator and press

[CLEAR] **2** [+] **4** [ENTER],

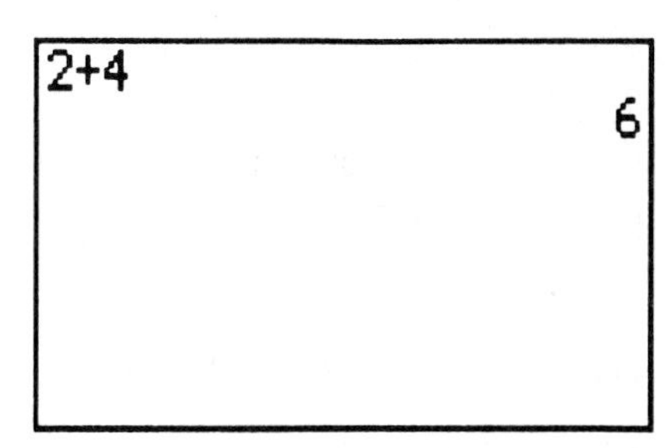

Figure 2

your screen should look like Fig. 2. When you are instructed to access a second function, the directions will include [2nd] followed by a box containing the second function label above the key rather than the label on the key. For example, notice the second function of [x^2] is $\sqrt{\ }$ ([x^2] is midway down the first column). The directions to calculate the square root of 2 would be to press

[CLEAR] [2nd] [$\sqrt{\ }$] **2** [ENTER].

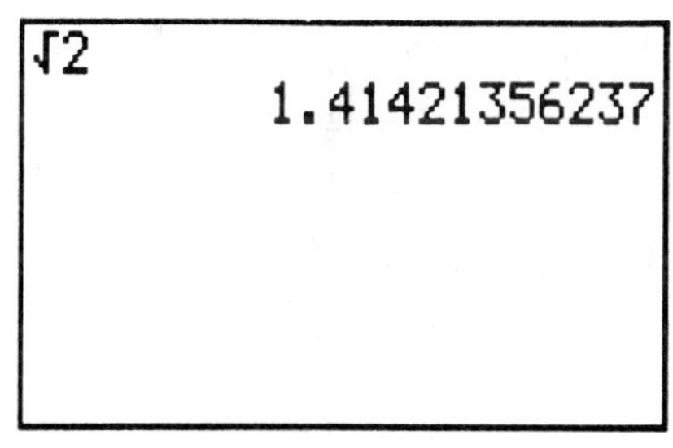

Figure 3

See Fig. 3. The reference to [$\sqrt{\ }$] in these directions means to press the key with x^2 on the key and $\sqrt{\ }$ above the key. The instructions to access a letter will include [ALPHA] followed by a box containing the letter above the key.

Resetting Your Grapher

You should reset the TI-85 to its factory settings. Turn on the calculator, press [2nd], and then press [+]. ([+] is a black key in the lower right-hand corner of the keyboard. Notice the word "MEM" in yellow lettering above [+].) After you press [2nd] and [+] you should see the **MEMORY** menu on the screen. The MEMORY menu should look like Fig. 4. Now press [F3:RESET] to see the **MEMORY [F3:RESET]** menu. [F3] is on the top row. This menu should look

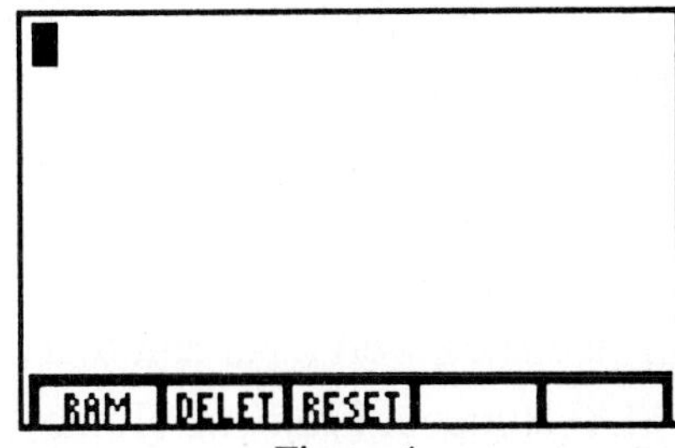

Figure 4

like Fig. 5. Press [F3:DFLTS] to select **Defaults,** the third item in the MEMORY [F3:RESET] menu. The TI-85 will ask you, "Are you sure?" Press [F4:YES] to select 'yes'. The display should show the message **Defaults set**. Press [CLEAR] to clear the screen. ([CLEAR] is a black key on the fourth row, fifth column of the keyboard.) You may need to adjust the contrast after you reset your TI-85.

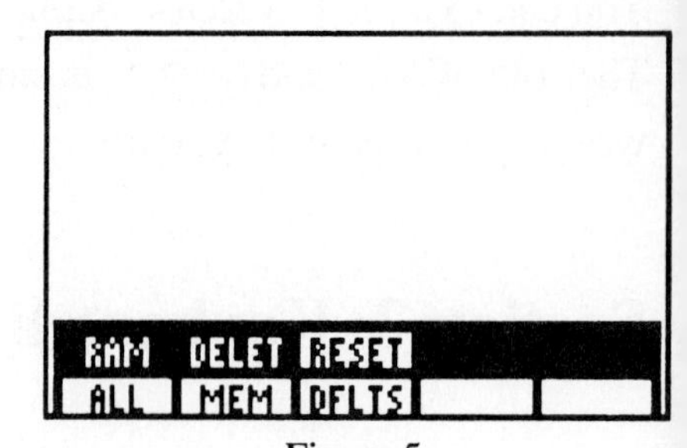

Figure 5

Section 4 The Home Screen, CLEAR, Menus, and QUIT

When you first turn on the TI-85, you see the **Home screen**. The calculations you performed in the keyboarding instructions were done on the Home screen. The CLEAR key you used before each calculation clears the Home screen of previous entries. If you make a mistake in a calculation, you can start over by pressing [CLEAR]. You can access **menus** from the Home screen. If you Reset your TI-85 you used the MEMORY [F3:RESET] menu. You can exit a menu by pressing [EXIT]. [EXIT] is next to [2nd]. Press

[2nd] [MATH]

to see the **MATH** menu. The [MATH] key is the second function of the key on the seventh row, fifth column of the keyboard. See Fig. 6. Each of the items in this menu is a heading for a submenu. Notice that [F5:MISC] is directly beneath the 'MISC' option in the MATH menu. To access the Miscellaneous submenu, press [F5:MISC]. See Fig. 7. There are five options visible on this menu, and the arrow by the fifth feature indicates there are more options available. To see the other options, press [MORE]. [MORE] is next to [EXIT]. Press [EXIT] to return to the MATH menu, and then press [EXIT] again to exit the MATH menu and return to the Home screen.

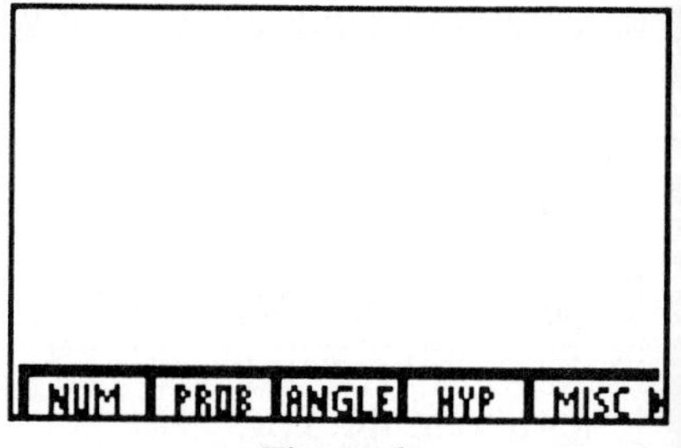

Figure 6

You can convert a percent to a decimal by using a feature in the MATH menu. From the Home screen press

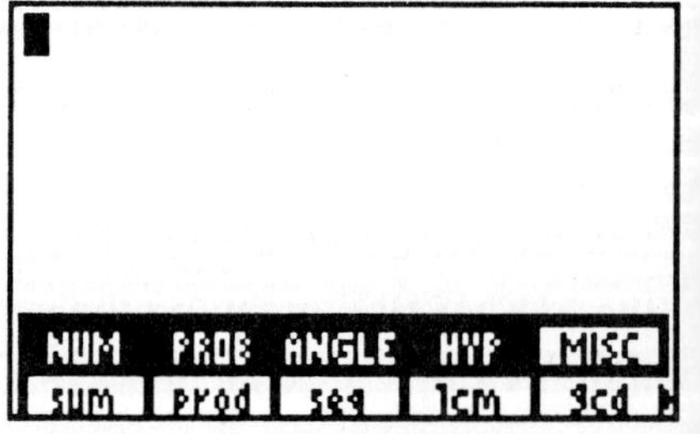

Figure 7

[CLEAR] **7** [2nd] [MATH] [F5:MISC] [MORE] [F2:%] [ENTER].

See Fig. 8. Seven percent equals .07. The notation [F5:MISC] in the keystroke instructions means to press [F5] after you are in the MATH menu. The notation [F2:%] means to press [F2] when you are in the MATH [MISC] menu. Did you notice pressing [F2:%] caused the '%' symbol to be printed on the Home screen? Press [EXIT] [EXIT] to leave these menus.

Figure 8

The MODE or Set Up Menu

From the Home screen press

[2nd] [MODE] (second row, third column).

Your screen should look like Fig. 9. Each of the left-hand entries should be highlighted. The left-hand entries are called the Default modes. They were selected when you reset the TI-85 earlier in this chapter. We will discuss the meaning of these entries in later sections. If any of the left-hand entries are not highlighted on your calculator, use the cursor-movement keys to move the cursor to the unhighlighted item and then press [ENTER] to highlight that item. Press [CLEAR] to return to the Home screen.

Figure 9

Section 5 Basic Operations Keys

The four black keys [+] [−] [×] [÷] on the lower right-hand side of the keyboard are used for addition, subtraction, multiplication, and division. Press the following keys to perform the calculation 19 minus 27 plus 23:

[CLEAR] **19** [−] **27** [+] **23** [ENTER].

See Fig. 10. Notice you don't see the result of the calculation until you press [ENTER]. Then you see the expression on the left-hand side

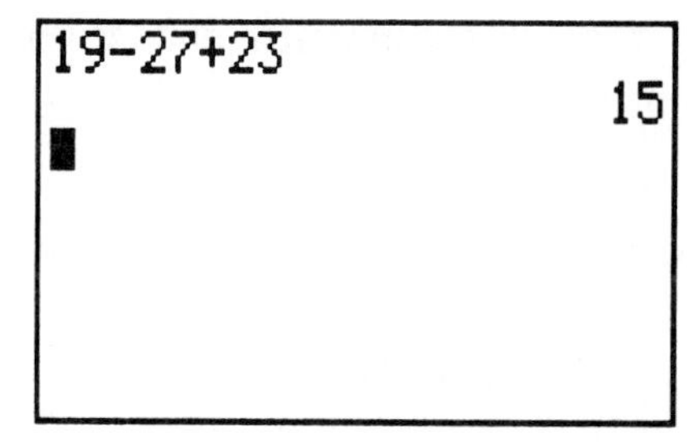

Figure 10

of the Home screen and the result of the calculation on the right-hand side of the next line. Press the following keys to multiply 18 by 145:

[CLEAR] **18** [×] **145** [ENTER].

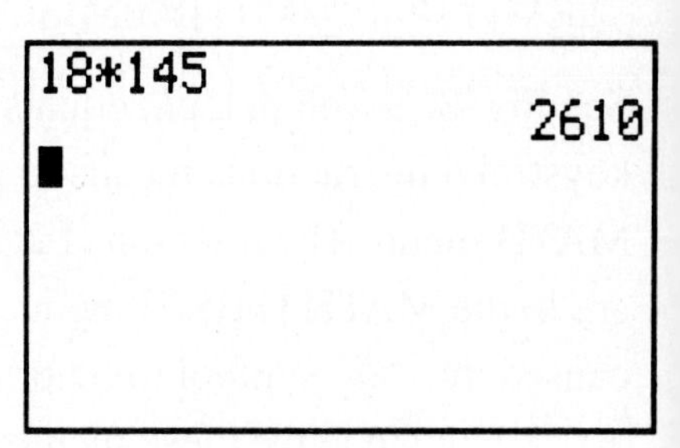

Figure 11

The result is 2610. See Fig. 11. Notice pressing [×] produces the '*' symbol on the Home screen. Both '*' and '×' mean multiplication. Now press

8 [÷] **156** [ENTER]

to divide 8 by 156. See Fig. 12. Notice pressing [÷] produces the '/' symbol on the Home screen. Both '÷' and '/' mean division. Also notice the previous calculation '18*145' remains on the screen. This is because you did not clear the screen by pressing [CLEAR] before entering the division problem.

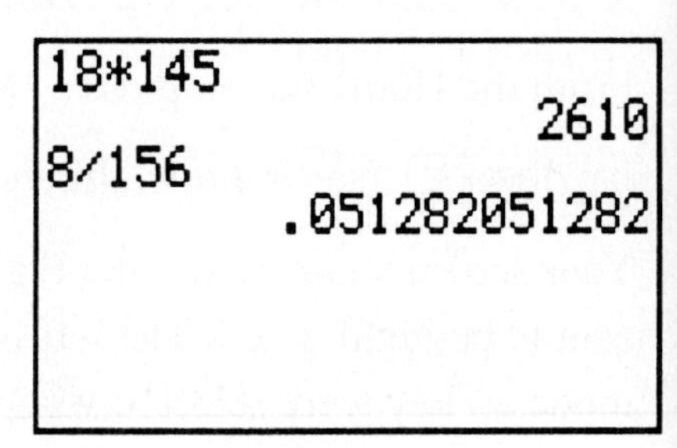

Figure 12

The Subtraction Key [−] versus the Opposite Key [(−)]

There are two keys which appear to be minus signs on the TI-85. One is the black subtraction key [−] mentioned in the previous section. The other is the grey key [(−)] on the bottom row. This grey key is the "opposite of" or negative key. For example, –3 is denoted on the Home screen by pressing

[CLEAR] [(−)] **3** [ENTER].

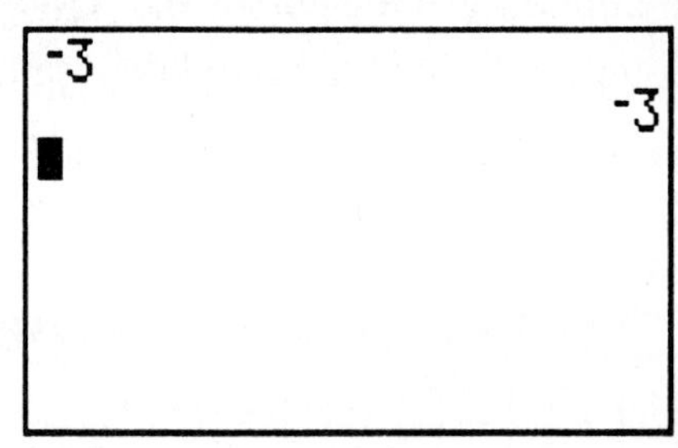

Figure 13

See Fig. 13. When the number –3 is added to 3, the result is zero. To see this on your calculator, press

3 [+] [(−)] **3** [ENTER] or [(−)] **3** [+] **3** [ENTER].

See Fig. 14. The negative key is different from the subtraction key. You can see the difference if you press

[CLEAR] [(−)] [−].

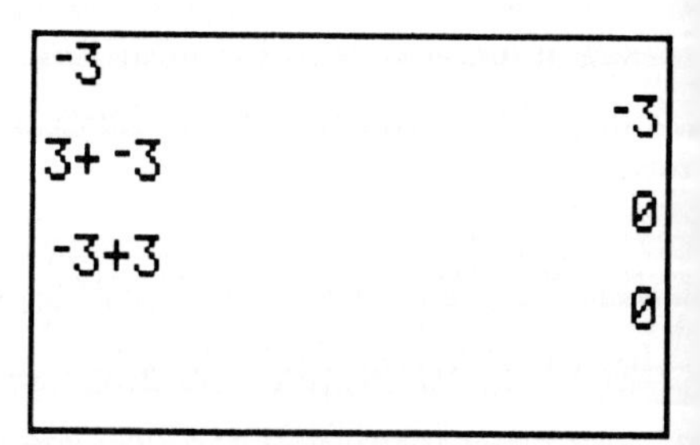

Figure 14

Observe that the negative key screen symbol is smaller in length and its position is higher than the subtraction symbol. See Fig. 15. Press CLEAR to clear the Home screen for the next example.

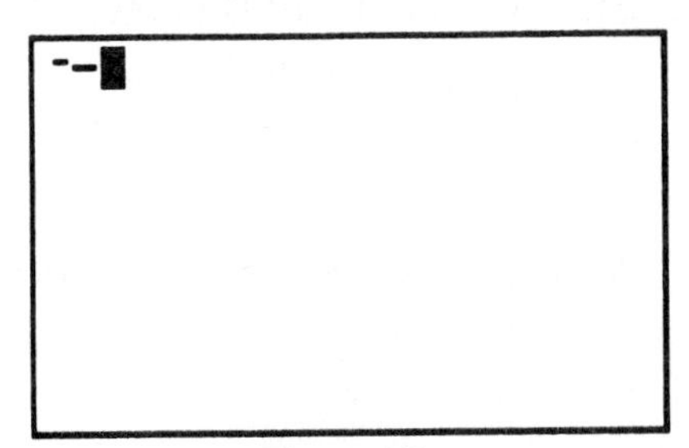

Figure 15

Section 6 Error Messages

If you want to see the result of –3 subtracted from 7 on the calculator, you must press

7 [–] [(–)] **3** [ENTER].

If you press

7 [(–)] [–] **3** [ENTER],

the TI-85 will give you a syntax error message. See Figs. 16 and 17. If you press [F1:GOTO] from the error message screen, the TI-85 will return you to the expression on the Home screen. The cursor blinking on the minus symbol tells you this is the mistake. You should have used the subtract key before the negative key. If you want to calculate 2(–3), then press

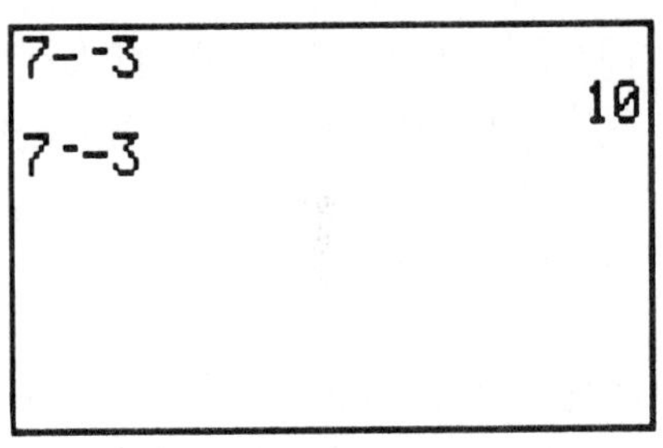

Figure 16

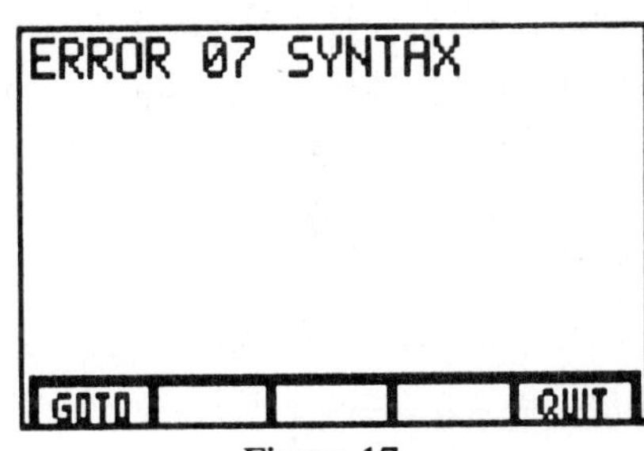

Figure 17

[CLEAR] **2** [x] [(–)] **3** [ENTER] or 2 [(] [(–)] **3** [)] [ENTER].

If you press

2 [(–)] **3** [ENTER]

the TI-81 and the TI-85 will return a syntax error message; however, this syntax works on the TI-85. Why do you see an error when you press

2 [÷] **0** [ENTER]?

Press

[F5:QUIT]

to return to the Home screen.

Section 7 How to Input with the x^2 and x^{-1} Keys

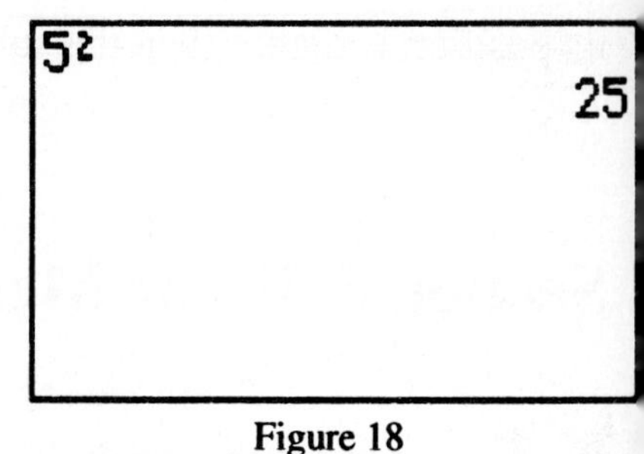

Figure 18

Figure 19

You can calculate squares with [x^2]. This key is in the first column. For example, the square of 5 can be calculated by pressing

[CLEAR] **5** [x^2] [ENTER].

See Fig. 18. Notice when you pressed [x^2], the ^{2} symbol was printed on the Home screen next to the 5, but the answer did not appear until you pressed [ENTER]. You can calculate reciprocals with [x^{-1}]. This feature is the second function of the key on the sixth row, second column. The reciprocal of 4 can be determined by pressing

4 [2nd] [x^{-1}] [ENTER].

See Fig. 19.

Section 8 The Insert and Delete Keys

You can edit expressions with the **Insert** key [INS] and the **Delete** key [DEL]. The [DEL] key is on the third row, third column, and [INS] is the second function of [DEL]. Press

[CLEAR] **36** [+] **10**,

but don't press [ENTER]. The cursor should be at the end of the expression. If you would like to change the expression to '316 + 10' by inserting a 1 between the '3' and the '6', press the left cursor-movement key [◄] until the cursor is blinking on the '6' in '36'. Now press

[2nd] [INS].

Notice the cursor changes from a blinking rectangle to a blinking underline. Press **1**. The expression changes to '316 + 10'.

Press [ENTER] to evaluate this new expression. The result is 326. See Fig. 20. Now we will use [DEL] to delete a digit. Press

123 [−] **10**

but don't press [ENTER]. The cursor should be blinking at the end of this new expression. Press the left cursor-movement key until the cursor is blinking on the '2' in '123'. Now press [DEL]. The expression has been changed to '13 – 10'. Press [ENTER] to evaluate this new expression. See Fig. 20.

```
316+10
                326
13-10
                  3
```

Figure 20

Chapter 1 Numerical Mathematics and the Graphing Calculator

Section 1.1 Real Numbers and the Graphing Calculator

Many of the computational skills you will need in this chapter are explained in the Preliminary Chapter. Be sure to read that chapter before beginning Chapter 1.

Changing Decimals to Fractions with the FRAC Command

Some decimals can be changed to equivalent fractions with the **►Frac** option in the MATH [F5:MISC] menu.

Example 1: Convert 0.44 to an equivalent simplified fraction.

From the Home screen press

[CLEAR] **0** [.] **44** [2nd] [MATH] [F5:MISC] [MORE] [F1:►Frac] [ENTER].

See Fig. 1.1. The equivalent fraction is 11/25. Exit the menu by pressing

[EXIT] [EXIT].

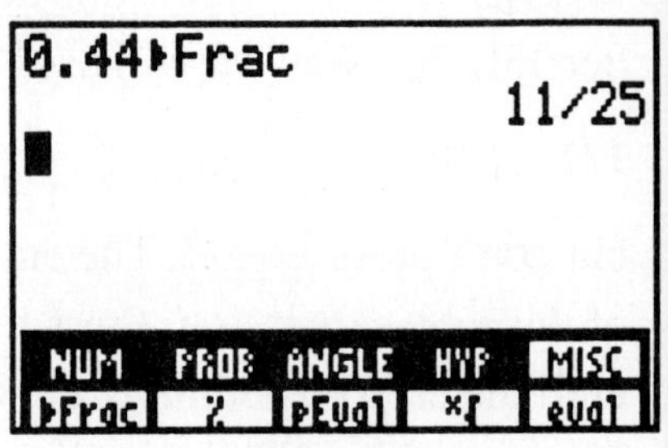

Figure 1.1

Decimal Representation of Fractions and Repeating and Terminating Decimals

Example 2: Convert the following fractions to decimals: 11/25, 143/999, 1/17.

Press

[CLEAR] **11** [÷] **25** [ENTER] **143** [÷] **999** [ENTER] **1** [÷] **17** [ENTER].

See Fig. 1.2. The decimal representation of 11/25 is exact because it is a terminating decimal. The TI-85 decimal representations of 143/999 and 1/17 are approximations because the correct representations of these fractions are repeating decimals. Although the screen shows an approximation to 143/999, you can probably tell what the repeating pattern is from looking at your calculator result. 1/17 is a repeating decimal but you cannot see the repeating block in the first 12 digits. The TI-85 can show only 12 digits in a number. If the decimal representation repeats or terminates after the first 12 digits, you may not be able to determine the exact form from the TI-85 representation.

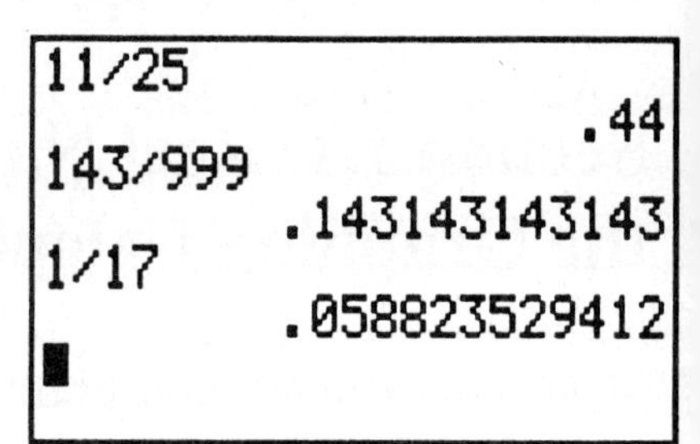

Figure 1.2

Floating Decimal Display versus Fixed-Point Display

When your calculator displayed the first 12 digits of the decimal representation of 143/999 in the previous example, it was in Floating Decimal mode. If you only want to see the first two digits, you can switch to Fixed Decimal mode. Do this by pressing

[2nd] [MODE] [▼] [▶] [▶] [▶] [ENTER].

See Fig. 1.3. Now press

CLEAR **143** ÷ **999** ENTER.

Figure 1.3

You should see the first two decimals in the answer: .14. See Fig. 1.4. Press

2nd MODE

and change the number of decimals in Fixed mode to four by using the cursor-movement keys to highlight the '4' on the second line of the Mode screen, then press ENTER. Press CLEAR to return to the Home screen and again divide 143 by 999. You should see the first four digits of the decimal representation, .1431. See Fig. 1.4. Now switch back to Floating Decimal mode by pressing

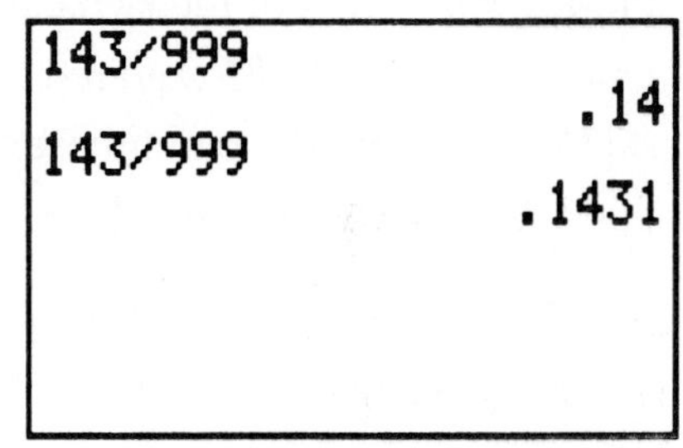

Figure 1.4

2nd MODE,

highlighting 'Float', and pressing ENTER and CLEAR.

Square Root Key and Use of 2nd Keys

The use of these keys was described in the Preliminary Chapter. The decimal representation of $\sqrt{2}$ shown in Fig. 4 is an approximation because $\sqrt{2}$ is a nonterminating, nonrepeating decimal. However, it is a very accurate approximation.

Powers and Exponentiation

The Preliminary Chapter described how to square numbers. Now we will see how to raise numbers to other powers. You can calculate powers with the power key ^ on the fifth row, fifth column of the keyboard.

Example 3: Compute 3^5.

To see 3^5 press

CLEAR **3** ^ **5** ENTER.

See Fig. 1.5.

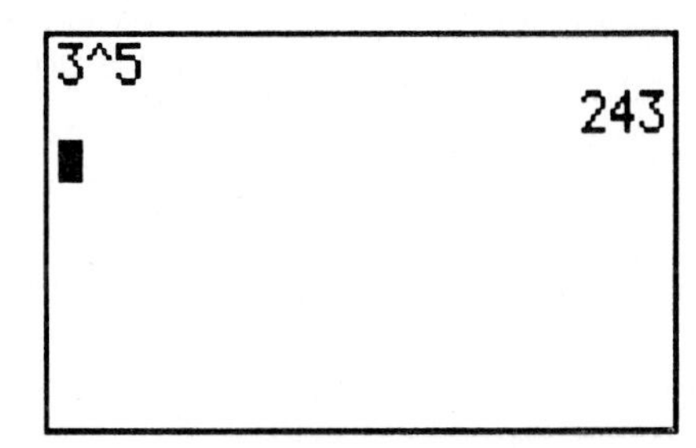

Figure 1.5

Example 4: Calculate -3^2 and $(-3)^2$.

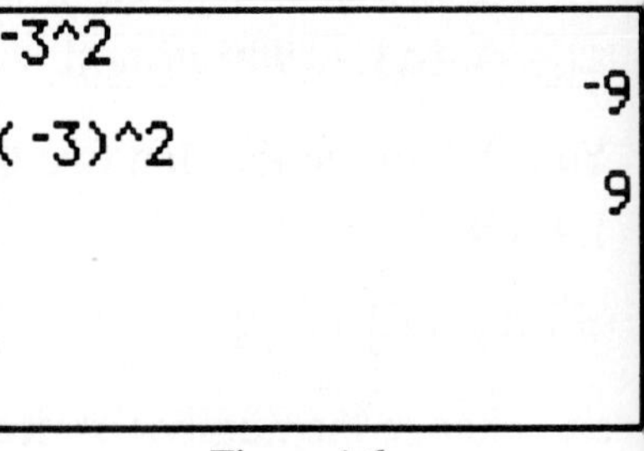

Figure 1.6

Press

CLEAR (-) **3** ^ **2** ENTER

to see -3^2, and press

((-) **3**) ^ **2** ENTER

to see $(-3)^2$. See Fig. 1.6.

Notice the results of the two calculations are different. In the first calculation, the TI-85 squared three first and then negated. In the second calculation, the parentheses forced the TI-85 to negate three first and then square negative three. You can square numbers with the x^2 or with the ^ keys.

Example 5: $\left(\frac{2}{5}\right)^3$.

You must use parentheses to raise a fraction to a power. Press

CLEAR (**2** ÷ **5**) ^ **3** ENTER.

The answer is .064.

Order of Operations

Example 6: Calculate $2 + 3 \times 4^2 - 6 \div 2$.

Press

CLEAR **2** + **3** × **4** ^ **2** − **6** ÷ **2** ENTER.

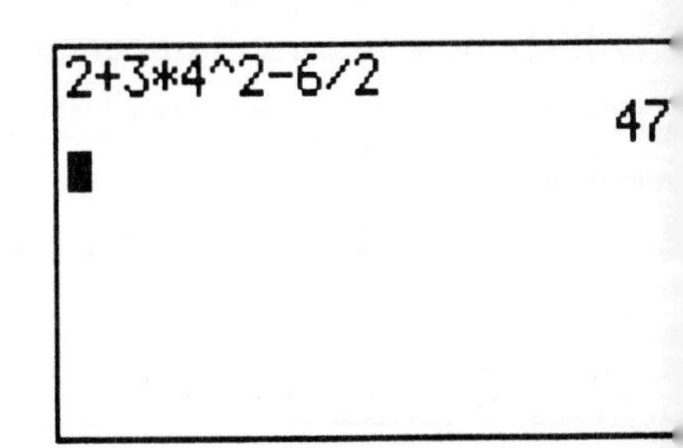

Figure 1.7

See Fig. 1.7. The TI-85 does powers first, which reduces the expression to $2 + 3 \times 16 - 6 \div 2$. Next the calculator does multiplication and division, which simplifies the expression further to $2 + 48 - 3$. Last it does addition and subtraction to yield a result of 47.

Comparing with the TEST Menu

To see how the TI-85 interprets 3 < 5, we will need to use the TEST menu. TEST is the second function of the **2** key. Before we evaluate the expression 3 < 5, let's look at the TEST menu by pressing

CLEAR 2nd TEST.

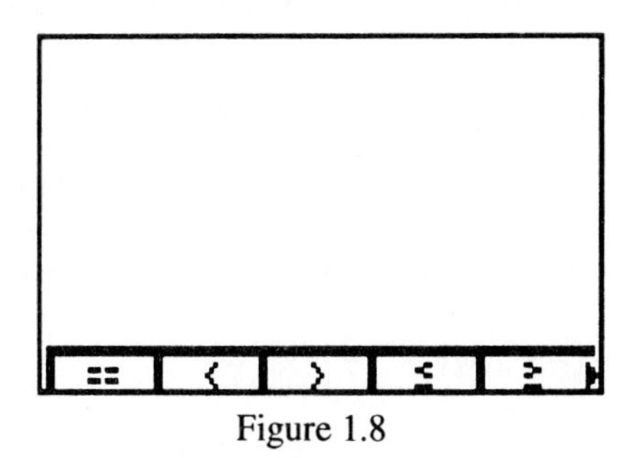

Figure 1.8

See Fig. 1.8. This menu contains equality and inequality symbols. Press EXIT to return to the Home screen, and press CLEAR again to clear the Home screen. Now evaluate the expression
3 < 5 by pressing

3 2nd TEST [F2:<] **5** ENTER.

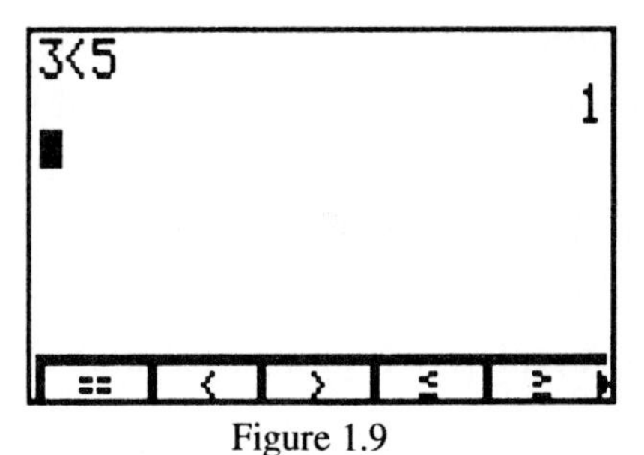

Figure 1.9

See Fig. 1.9. The TI-85 returns the value 1 because the expression is true. To see the result of 3 > 5 press

3 [F3:>] **5** ENTER.

The calculator shows a result of 0 because this expression is false. You should also see a result of 0 when you press

(-) **1** [F5:≥] (-) **0** . **5** ENTER

because –1 is not greater than or equal to –0.5. Press EXIT to leave the TEST menu.

Section 1.2 Properties of Real Numbers and the Basic Rules of Operations

Properties of Opposites and the Subtraction Key

These keys were discussed in the Preliminary Chapter.

The Absolute Value Key

Absolute value is in the MATH N (Number) menu and in the Catalog.

Example 1: Calculate |7|, |−7|, and −|−7|.

From the Home screen press

CLEAR 2nd MATH [F1:NUM] [F5:abs] 7 ENTER

to see the absolute value of 7, then press

[F5:abs] (−) **7** ENTER

to see the absolute value of −7. We will use a different method to access absolute value for the third calculation in Example 1. All the commands on the TI-85 are not only organized in various menus but also appear in an alphabetical list called the **CATALOG.** The CATALOG key is the second function of the key on the fourth row, fourth column. The Catalog is useful when you cannot remember which menu contains the option you want. Press (−) and then use the Catalog to find the absolute value command by pressing

2nd CATALOG.

Figure 1.10

See Fig. 1.10. Since the arrow points to 'abs', when you press ENTER, the 'abs' command will be copied to the Home screen. Finish the problem by pressing

(−) **7** ENTER

to see −|−7|. See Fig. 1.11.

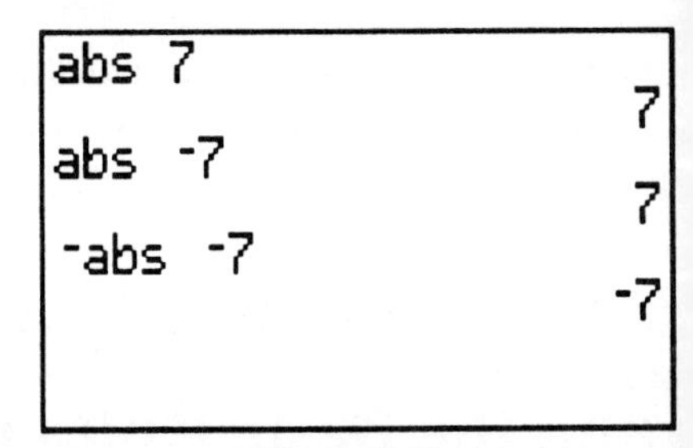

Figure 1.11

Example 2: Calculate |5.2 − 17.8|.

Press

CLEAR 2nd MATH [F1:NUM] [F5:abs] (**5** . **2** − **17** . **8**)
ENTER.

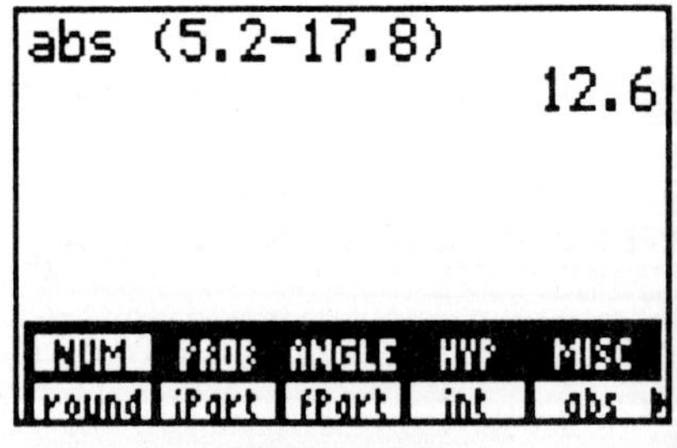

Figure 1.12

The result is 12.6. If you repeat the same keystroke sequence but leave out the parentheses, you will obtain an incorrect result. See Fig. 1.12. With the parentheses, the TI-85 subtracts 17.8 from 5.2 first and then finds the absolute value of the difference. Without the parentheses, the TI-85 takes the absolute value of 5.2 and then subtracts 17.8.

Section 1.3 Algebraic Expressions an Problem Situations

Using a Grapher to Evaluate an Expression and [STO▶] and REPLAY

You can assign values to variables on the TI-85 and then use variables in calculations. TI-85 variables are the letters of the bet found above many of the keys. You assign values to varia with [STO▶]. This key is just above [ON]. After you press [STO TI-85 automatically changes the cursor to the Alpha mode so can access the letters on the keyboard. When the TI-85 is in Alpha mode and you press a key, the letter above the key will be printed. You can exit the Alpha mode by pressing [ALPHA].

Example 1: Evaluate $x^3 + 17$ at $x = 25$, 39, and 52.

Store the value of 25 in X by pressing

[CLEAR] **25** [STO▶] [X] [ENTER].

The reference to [X] in these instructions means to press the key with 'X' above it. This key happens to be [+]. Now press

[ALPHA] [X] [^] **3** [+] **17** [ENTER].

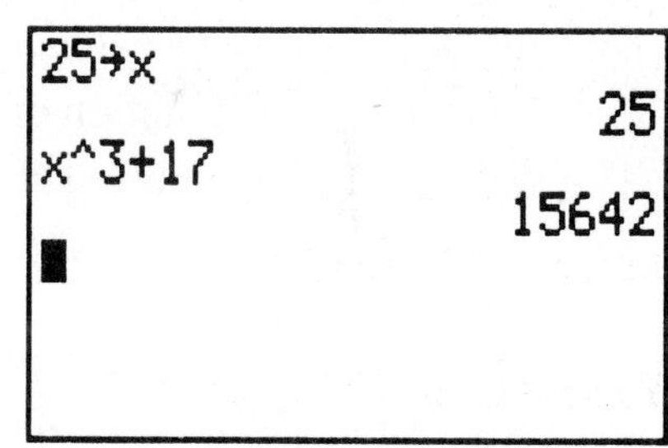

Figure 1.13

The screen should show you that $x^3 + 17 = 15{,}642$ when x is 25. See Fig. 1.13.

We will evaluate $x^3 + 17$ at $x = 39$ and 52 by a different approach. Instead of storing 39 in X and then evaluating the expression, we will enter 39 directly into the expression. Press

[CLEAR] **39** [^] **3** [+] **17** [ENTER].

39^3+17
59336
39^3+17

Figure 1.14

The answer is 59,336. See Fig. 1.14. Now we will use the **REPLAY** feature to edit the expression and replace 39 with 52. You can REPLAY an expression by using [ENTRY], which is the second function of [ENTER]. Recall the expression by pressing

[2nd] [ENTRY].

The previous expression is copied on the next line and the cursor appears at the end of the line. See Fig. 1.14. Now press the left cursor-movement key [◀] until the cursor is on the '3' in '39'. Replace '39' with '52' by simply typing 52. See Fig. 1.15. Finish the calculation by pressing [ENTER]. The answer is $52^3 + 17 = 140{,}625$.

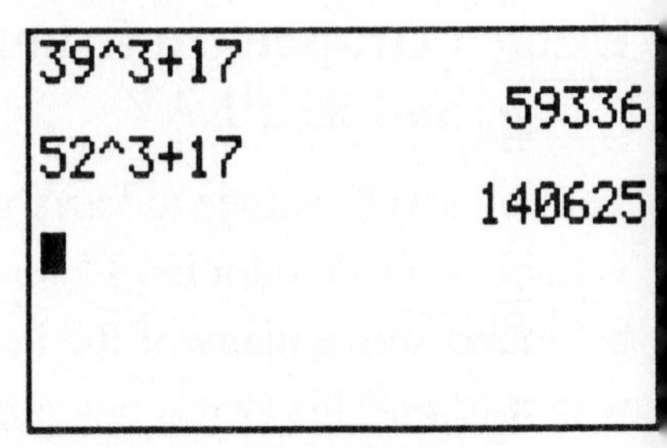

Figure 1.15

Example 2: Evaluate $4T - \dfrac{3S^3}{5R}$ at $R = 13.5$, $S = -23.7$, and $T = 3.28$.

Assign values to the variables by pressing

[CLEAR] **13** [.] **5** [STO▶] [R] [ENTER] [(-)] **23** [.] **7** [STO▶] [S] [ENTER] **3** [.] **28** [STO▶] [T] [ENTER].

Now evaluate the expression by pressing

4 [ALPHA] [T] [−] **3** [ALPHA] [S] [^] **3** [÷] [(] **5** [ALPHA] [R] [)] [ENTER].

See Fig. 1.16.

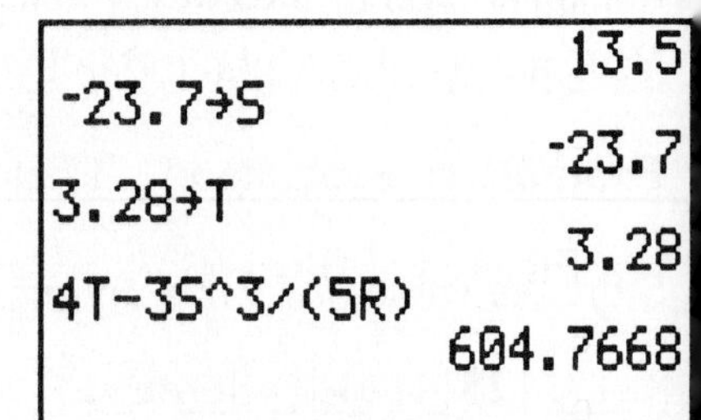

Figure 1.16

Concatenation and the Enter Key

Several calculator steps can be combined with the concatenation key [:]. [:] is the second function of [.]. Concatenate means to combine. The advantage of combining steps is that when you replay the last expression, it really recalls several expressions. This feature is illustrated in the next problem.

Example 3: Evaluate $x^3 + 17$ at $x = 25$ and 39.

In Example 1 we evaluated the polynomial at $x = 25$ in two steps. The first step was to store the value of 25 in X, and the second step was to enter the polynomial. Now we concatenate these steps by pressing

[CLEAR] **25** [STO▶] [X] [ALPHA] [2nd] [:] [ALPHA] [X] [^] **3** [+] **17** [ENTER].

25→X:X^3+17
15642

Figure 1.17

See Fig. 1.17. Now recall the expression by pressing

[2nd] [ENTRY].

See Fig. 1.18. Move the cursor onto the '2' in '25', and replace the '25' with '39'. Then press [ENTER]. The polynomial is reeval-uated at $x = 39$. This is a quicker method to reevaluate the polynomial at various values than the method used in Example 1.

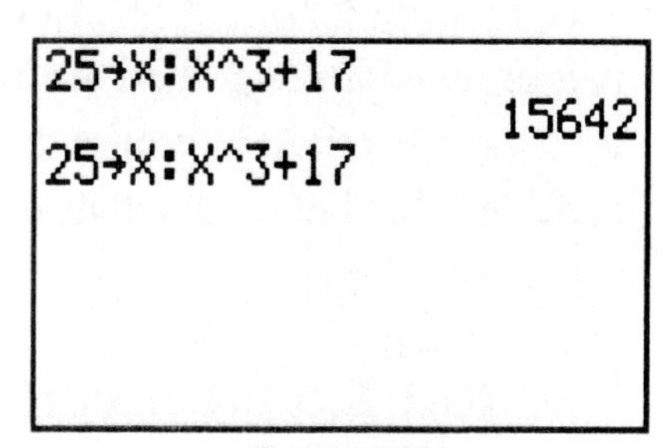

Figure 1.18

Implied Multiplication

Press

[CLEAR] **3** [STO▶] [X] [ENTER] **2** [ALPHA] [X] [ENTER].

See Fig. 1.19. The TI-85 multiplies the 2 and the value of X even though there is no multiplication sign. This is called an implied multiplication because their position next to one an-other implies multiplication. The expression 2 * X is an explicit multiplication because there is a multiplication sign between the two factors. Now press

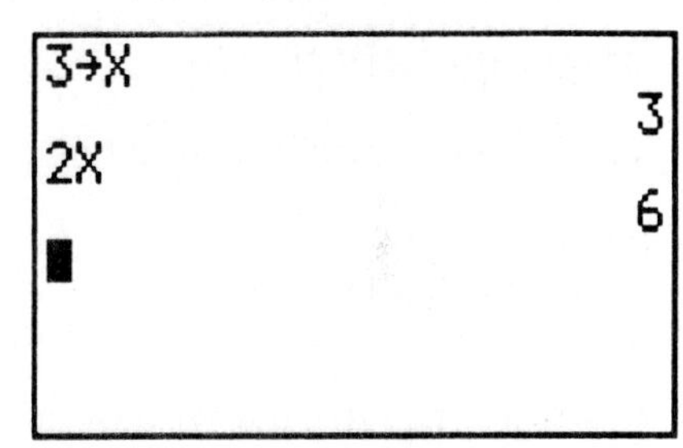

Figure 1.19

12 [÷] **2** [ALPHA] [X] [ENTER].

See Fig. 1.20. The result is 2 because the implied multiplication is done before the division. The TI-85 carries out an implied multiplication before an explicit multiplication or division.

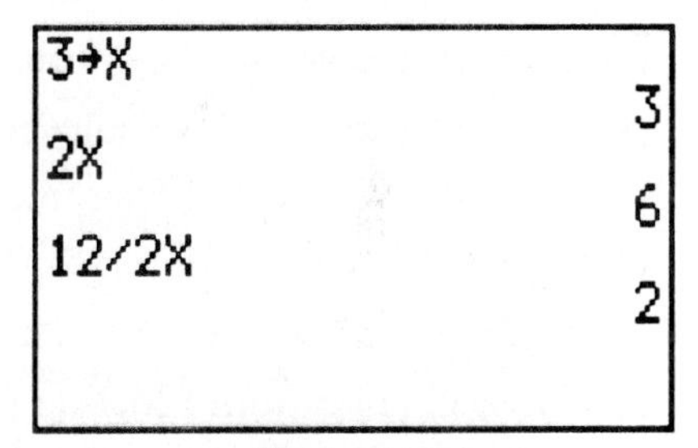

Figure 1.20

Example 4: Evaluate $\frac{5}{9}(41 - 32)$.

You will need to use parentheses to enter this expression correctly in the TI-85. Press

[CLEAR] [(] **5** [÷] **9** [)] [(] **41** [−] **32** [)] [ENTER].

The answer is 5. See Fig. 1.21. The TI-85 performs the calculations inside both pairs of parentheses first and then multiplies the results. Now see what happens if you leave off the first set of parentheses. Press

5 [÷] **9** [(] **41** [−] **32** [)] [ENTER].

The result is different. See Fig. 1.21. This is because the TI-85 performs the implied multiplication between the 9 and the value in the parentheses before it does the division. The expression 5/9(41 – 32)

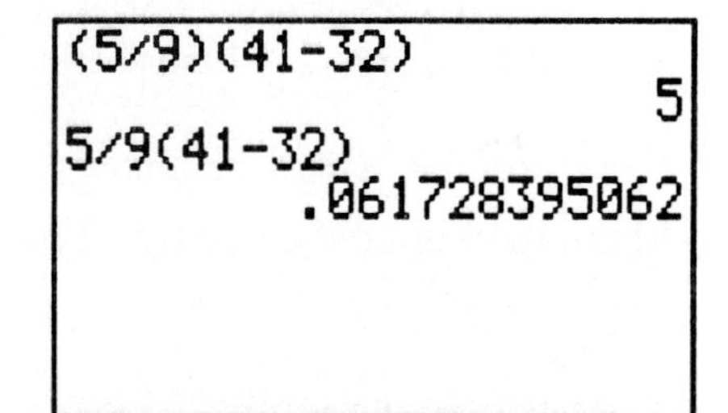

Figure 1.21

is equivalent to 5/[9(41 – 32)]. Now redo Example 2 and omit the pair of parentheses around the $5R$. The result should be the same as with the parentheses because the implied multiplication between 5 and R is done before the division.

Section 1.5 Numerical Representations of Problem Situations

Making Tables

The TI-85 does not have a table building feature.

Finding Percents

Example 1: Find the wholesale price for a book that sells retail for $17.10 if the markup is 20%.

We combine a formula with the method of *guess and correct*. The retail price is equal to the wholesale price plus 20% of the wholesale price. If we let x be the wholesale price, then the retail price is equal to $x + .20x$. We can make an initial guess for x and then check its correctness with the calculator. Since the numbers involved represent money, we are only interested in two decimal places. Use the Mode screen to change from Floating Decimal mode to Fixed Decimal mode by highlighting the '2' on the second line of the Mode screen. Then press [ENTER] and then [CLEAR] to return to the Home screen. To check an initial guess of $15.00, press

[CLEAR] **15** [.] **00** [+] [.] **20** [×] **15** [.] **00** [ENTER].

See Fig. 1.22. Since the answer is $18.00, our initial guess is too high. To change the guess to $14.00, press

[2nd] [ENTRY]

to Replay the expression and use the cursor-movement keys to replace

```
                   18.00
14.00+.20*14.00
                   16.80
14.50+.20*14.50
                   17.40
14.25+.20*14.25
                   17.10
▮
```

Figure 1.22

each '5' with a '4'. This is done by placing the cursor on each '5' and then pressing **4**. After you have changed the expression, press [ENTER]. This guess is too low. Use the Replay feature to change the guess to \$14.50. Since the answer is \$17.40, we are still high. Now Replay and change the expression to guess \$14.25. This value works. See Fig. 1.22. Change to Floating Decimal mode when you are finished with this problem.

Section 1.6 Algebraic Representations of Problem Situations

The Parametric Mode and the Simultaneous Mode

Example 1: Suppose that at the instant Juan is running 30 feet per second, he is 10 feet behind Carlos, who is running at 28 feet per second. How long will it take Juan to catch Carlos? The runners in this problem can be simulated by the TI-85 with the **Parametric Graphing mode** and the **Simultaneous Graphing mode**. Press

[2nd] [MODE]

and highlight 'Param'. Then press

[GRAPH] [MORE] [F3:FORMT]

and highlight 'SimulG'. Next call up the **Range edit screen** by pressing

[GRAPH] [F2:RANGE].

See Fig. 1.23. This screen is used to set up the viewing window for the simulation. The values on this screen can be changed by using the cursor-movement keys to move the cursor to the line you want to change and then typing in the new value. Change the values of the Range variables to the following: tMin = 0, tMax = 5.5, tStep = 0.01, xMin = 0, xMax = 150, xScl = 10, yMin = 0, yMax = 8,

```
RANGE
 tMin=█
 tMax=6.28318530718
 tStep=.130899693899...
 xMin=-10
 xMax=10
↓xScl=1
F(t)= RANGE ZOOM TRACE GRAPH▸
```

Figure 1.23

and *y*Scl = 1. See Fig. 1.24. Now press

[F1:E(t)=].

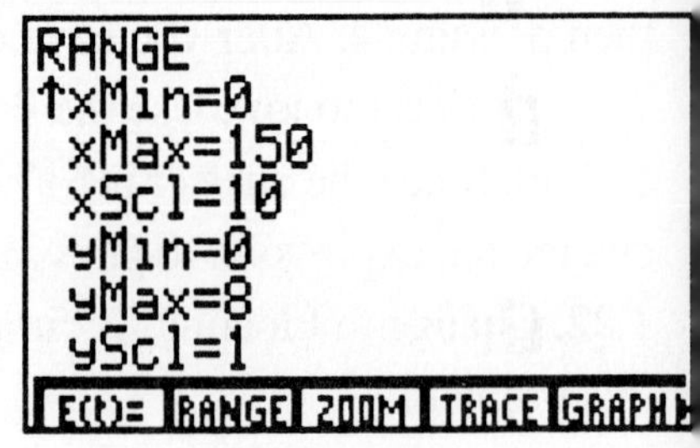

Figure 1.24

We will enter the distance equations on this screen. Remember that distance = rate × time. Press

[CLEAR] **30** [F1:t] [ENTER] [CLEAR] **5** [ENTER] [CLEAR] **28** [F1:t] [+] **10** [ENTER] [CLEAR] **2** [ENTER].

See Fig. 1.25. To see the simulation, press

[EXIT] [F5:GRAPH].

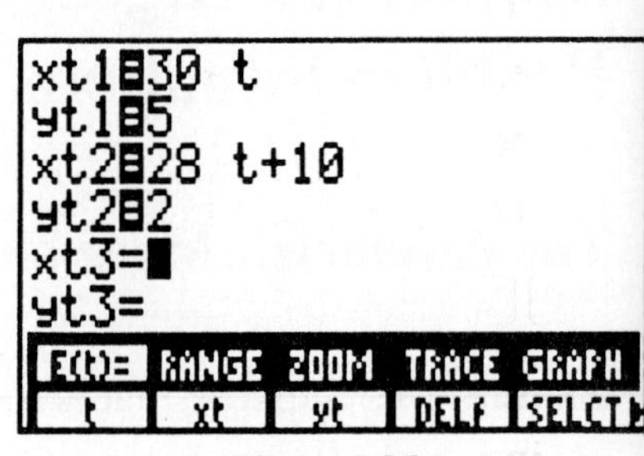

Figure 1.25

The two moving lines represent the runners. You can trace their progress by pressing

[F4:TRACE].

The Trace cursor appears at the beginning of the top line. The value of *t* printed at the bottom of the screen is the elapsed time, and the value of *x* is the distance traveled by Juan. Press [▶] repeatedly to see Juan's progress each 0.01 second along the way. To see the distance traveled by Carlos at the same time, press [▼]. Press [▲] to move the cursor back to the top graph. When *t* = 5, both runners have traveled a distance of 150 feet. See Fig. 1.26. Juan will catch Carlos after five seconds. Reselect Sequential mode for future examples by pressing

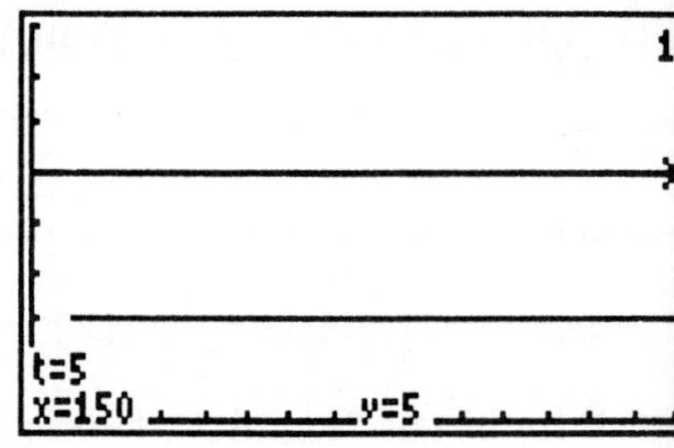

Figure 1.26

[GRAPH] [MORE] [F3:FORMT]

and highlighting 'SeqG'. Pressing [ENTER] highlights 'SeqG' and pressing [CLEAR] returns you to the Home screen.

Chapter 2 Graphing Equations and Functions

Section 2.1 Rectangular Coordinate Plane

Representing a Coordinate Plane on a Grapher

The TI-85 can use the screen to display a portion of the coordinate plane. This portion is called a **viewing window**. Graphs of functions can be displayed in the viewing window. The black keys on the top row are used to define the viewing window and create graphs. Press

[2nd] [MODE]

and select the Function (Func) mode before continuing with the following examples.

The RANGE and Range Adjustments

The use of [GRAPH] [F2:RANGE] determines which portion of the coordinate plane will be displayed in the viewing window. Press

[GRAPH] [F2:RANGE]

to see the Range edit screen. See Fig. 2.1. Your screen probably doesn't look like Fig. 2.1. The variables on the Range edit screen are called Range variables. The *x*Min is the coordinate of the left side of the viewing window, and the *x*Max is the coordinate of the right side of the viewing window. The *x*Scl is the distance between the tick marks which are drawn on the *x*-axis. The *y*Min is the coordinate of the bottom of the viewing window. The *y*Max is the coordinate of the top, and the *y*Scl is the distance between the tick marks on the *y*-axis. You can change the values of your Range variables by moving the cursor to the value you would like to change and typing in the new value. Modify your Range edit screen as needed until it looks like Fig. 2.1.

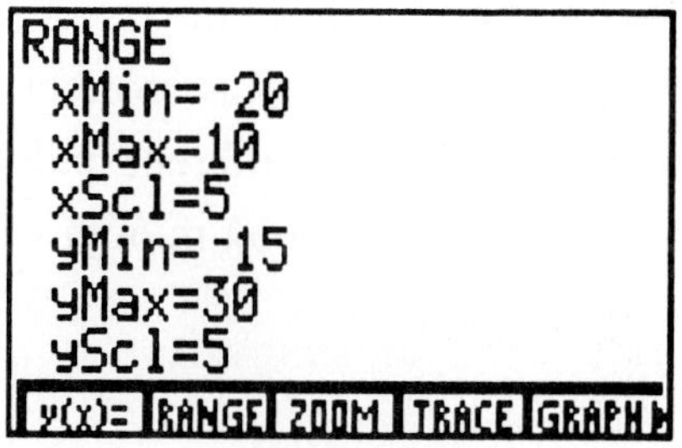

```
RANGE
 xMin=-20
 xMax=10
 xScl=5
 yMin=-15
 yMax=30
 yScl=5
y(x)= RANGE ZOOM TRACE GRAPH
```

Figure 2.1

GRAPH, Screen Coordinates, and Pixels

Clear the equations in the y(x)= menu by pressing

[GRAPH] [F1:y(x)=]

and then moving the cursor to each line with an equation and pressing [CLEAR]. Press

[EXIT] [F5:GRAPH]

to see the viewing window. Press [CLEAR] to remove the menu from the window. See Fig. 2.2. This viewing window represents points whose x-coordinates are between –20 and 10 and whose y-coordinates are between –15 and 30. The tick marks are each five units apart. We say that this is the window [–20, 10] by [–15, 30]. This notation means xMin = –20, xMax = 10, yMin = –15, and yMax = 30.

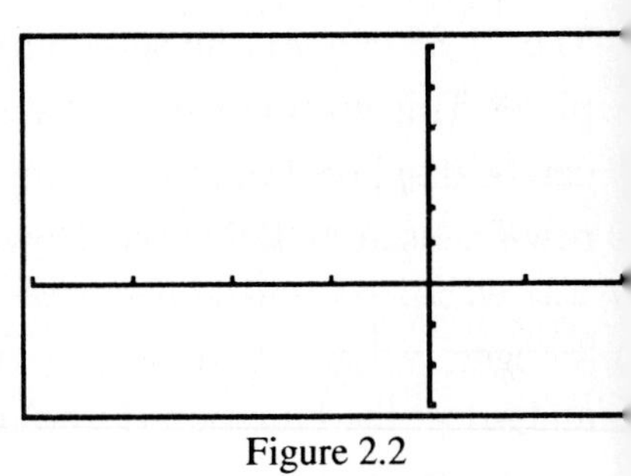
Figure 2.2

Now press [▶]. A cursor in the shape of a '+' with a blinking center appears on the screen. This is called the **free-moving cursor**. The blinking rectangle in the center is a pixel. The x- and y-coordinates of this pixel are printed at the bottom of the screen. See Fig. 2.3. You can move this cursor around the viewing window with the cursor-movement keys. The coordinates printed at the bottom of the screen are approximations to the actual coordinates of the free-moving cursor. Move the cursor to the origin. The coordinates should be (0, 0), but the y-coordinate is slightly off. Press

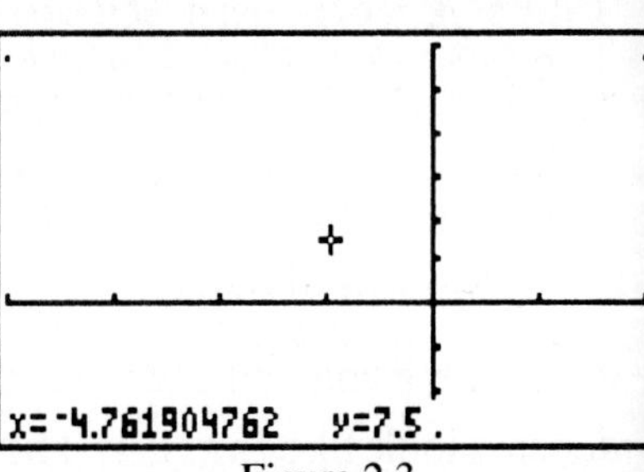

Figure 2.3

[GRAPH] [F2:RANGE]

and change the Range values to [–10, 10] by [–10, 10] with xScl and yScl values of 1. Then press [F5:GRAPH] and use the cursor-movement keys to move the free-moving cursor to the origin. Notice the cursor coordinates are correct this time. The coordinates of the cursor may be slightly different in different windows even though the cursor is on the same point in both windows.

Integer Windows

With the cursor at the origin of the [–10, 10] by [–10, 10] window, press

GRAPH [F3:ZOOM] MORE MORE [F4:ZINT] ENTER.

Now use the cursor-movement keys to move the free-moving cursor. The *x*- and *y*-coordinates of the cursor are always integers. This is because the viewing window has been changed. Press

GRAPH [F2:RANGE]

to see how the values of the Range variables have changed. They are [–63, 63] by [–31, 31]. This is called the **Integer window**. There are other windows where the cursor coordinates are integers, but since this window has the origin at the center, your textbook calls it the Integer window. You can find other Integer windows by moving the screen cursor away from the origin after you press

GRAPH [F3:ZOOM] MORE MORE [F4:ZINT]

but before you press ENTER. The Range variables [0, 126] by [0, 62] also form an Integer window, as do [0, 252] by [0, 124]. In an Integer window, *x*Max – *x*Min is a multiple of 126 and *y*Max – *y*Min is a multiple of 62.

The Decimal or 0.1 Window

Now press

GRAPH [F3:ZOOM] MORE [F4:ZDECM]

and use the cursor-movement keys. The *x*- and *y*-coordinates are always tenths. See Fig. 2.4. This is called the **Decimal Window** or **0.1 Window**. Press

EXIT [F2:RANGE]

to see the values of the Range variables. They should be [–6.3, 6.3] by [–3.1, 3.1]. [–6.3, 6.3] by [–10, 10] is a window where the *x*-coordinates are always tenths, but not the *y*-coordinates.

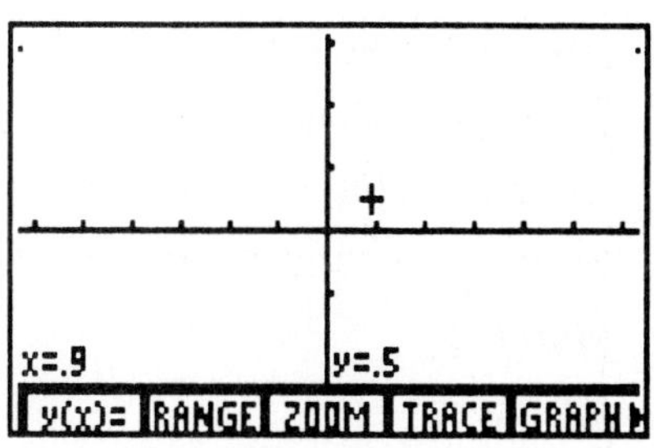

Figure 2.4

The 0.2 Window

The **0.2 Window** is [–12.6, 12.6] by [–6.2, 6.2]. Why is this called the 0.2 window?

Section 2.2 Equations in Two Variables and Their Graphs

Plotting Points with PtOn(and STAT

Example 1: Plot the following solution pairs to $14x + 28y = 168$ with the points (–15, 13.5), (–10, 11), (0, 6), and (15, –1.5).

Use

[GRAPH] [F2:RANGE]

to create a [–20, 20] by [–15, 15] viewing window. Press

[2nd] [QUIT]

to move from the Range edit screen to the Home screen. We will use the **point-on** option to plot these points. We will place the point-on option in a **custom menu** so it is easy to use.

To customize our menu we must first find the PtOn(command in the catalog. Press

[2nd] [CATALOG] [,].

Since [,] has a 'P' above it, the Catalog cursor jumps to the commands that begin with 'P'. This is quicker than using the cursor-movement keys to scroll down the Catalog. Now press [F1:page ↓] [F1:page ↓] to scroll down the Catalog one page at a time until the PtOn(command comes into view. Move the cursor to 'PtOn(' by pressing

.

See Fig. 2.5. Now place 'PtOn(' in the custom menu by pressing [F3:CUSTM] [F1]. See Fig. 2.6. Press

[EXIT] [EXIT]

to return to the Home screen.

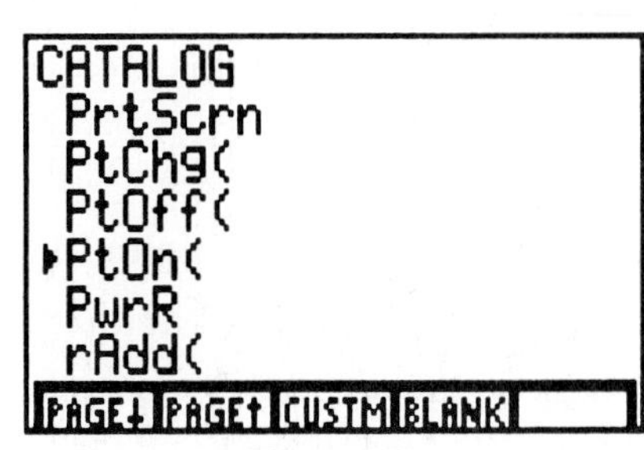

Figure 2.5

We can use our new custom menu to plot points. Press

[CLEAR] [CUSTOM] [F1:PtOn(].

The PtOn(command is copied to the Home screen. Now press

[(–)] **15** [,] **13** [.] **5** [)].

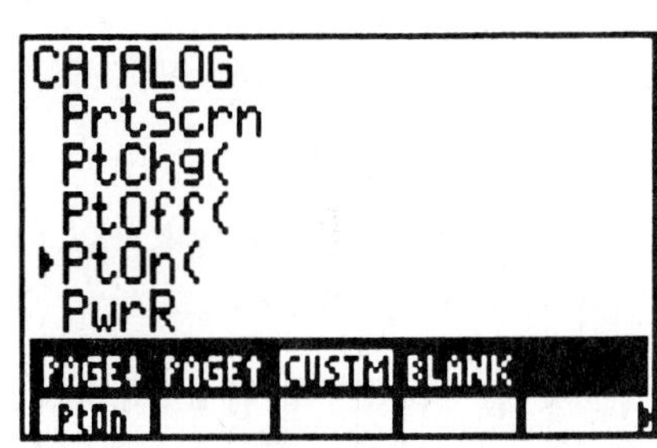

Figure 2.6

See Fig. 2.7. This command tells the TI-85 to plot the point (–15, 13.5). Press [ENTER] to see the point plotted on the viewing window. The point should be plotted in the upper left-hand corner of the viewing window. Now press [EXIT] to return to the Home screen. Then press

[CUSTOM] [F1:PTOn(] [(–)] **10** [,] **11** [)] [ENTER]

to plot the second point. To plot the third point press

[EXIT] [CUSTOM] [F1:PtOn(] **0** [,] **6** [)] [ENTER].

Figure 2.7

You don't see this point because it is on the *y*-axis. Plot the last point by pressing

[EXIT] [CUSTOM] [F1:PtOn(] **15** [,] [(–)] **1** [.] **5** [)] [ENTER].

See Fig. 2.8. Press

[2nd] [CATALOG] [COS] [▼] [ENTER]

to copy the ClDrw (Clear Draw) command from the catalog to the Home screen. Then press [ENTER] to execute the ClDrw command and [CLEAR] to clear the viewing window.

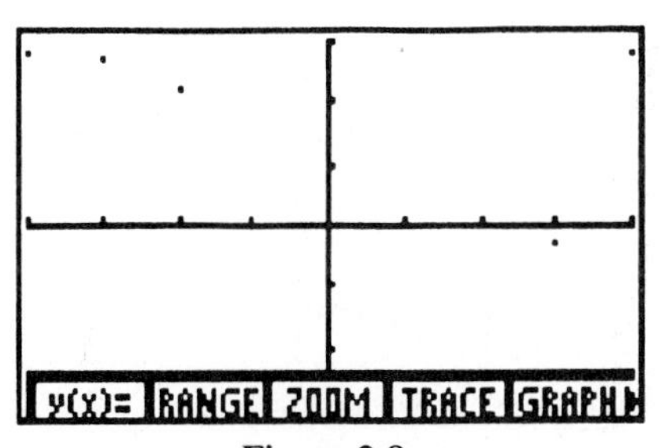

Figure 2.8

Another way to plot the points is with the Statistical features of the TI-85. You should still have the same values for the Range variables that were just used in the first solution to Example 1. The Statistical menu is accessed with the key on the fourth row, second column.

The first feature of the Statistical menu we will access is the **Data Edit screen**. To do this press

STAT [F2:EDIT] ENTER ENTER.

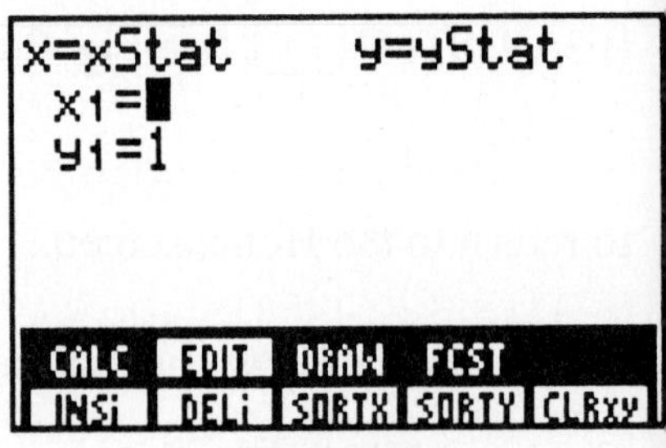

Figure 2.9

See Fig. 2.9. We will enter the ordered pairs in this screen. If your Data Edit screen does not look like Fig. 2.9, you can erase the numbers on the screen by pressing

[F5:CLRxy].

Enter the ordered pairs by pressing

(-) **15** ENTER **13** . **5** ENTER (-) **10** ENTER **11** ENTER **0** ENTER **6** ENTER **15** ENTER (-) **1** . **5**.

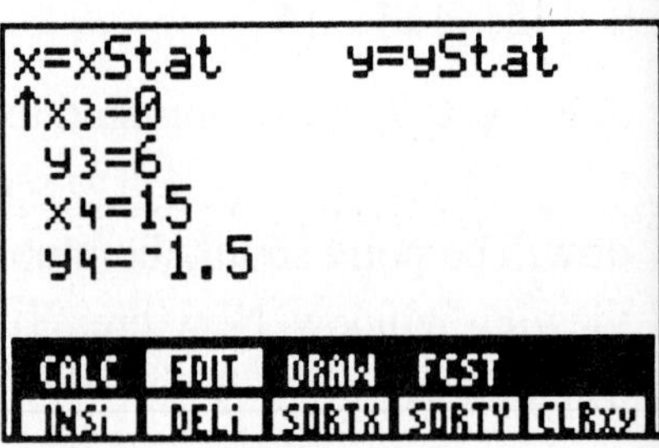

Figure 2.10

Your screen should look like Fig. 2.10. Now plot the points by pressing

EXIT [F3:DRAW] [F2:SCAT].

This picture is sometimes called a **scatter plot**. The points in this picture are the same ones you plotted previously with the PtOn(option. Press

EXIT EXIT

to return to the Home screen.

How to Graph an Equation Using Trace and the Function Defining Menu

Example 2: Graph $y = 2x - 8$.

Press

GRAPH [F2:RANGE]

and enter a [–63, 63] by [–31, 31] viewing window. This is the Integer window discussed in Section 2.1. Pressing

GRAPH [F1:y(x)=]

is used to enter the equation to be graphed. Press

GRAPH [F1:y(x)=]

to see the **y(x)=** menu. You can erase an old equation by pressing CLEAR when the cursor is on the equation. We will enter the equation in $y1$. Press

2 X-VAR − 8.

See Fig. 2.11. Pressing the X-VAR key produces a lowercase 'x'. The independent variable in the entered equation must be a lowercase 'x'. Press

EXIT [F5:GRAPH]

to see the graph of $2x - 8$. Press [F4:TRACE]. The Trace cursor appears on the graph. See Fig. 2.12. Notice the Trace cursor looks like a blinking 'X' with a box in the center. The coordinates at the bottom of the screen are the coordinates of the Trace cursor. Use the left and right cursor-movement keys to see the coordinates of points on the line. Each pair of coordinates you trace on the screen should be a solution to the equation $y = 2x - 8$.

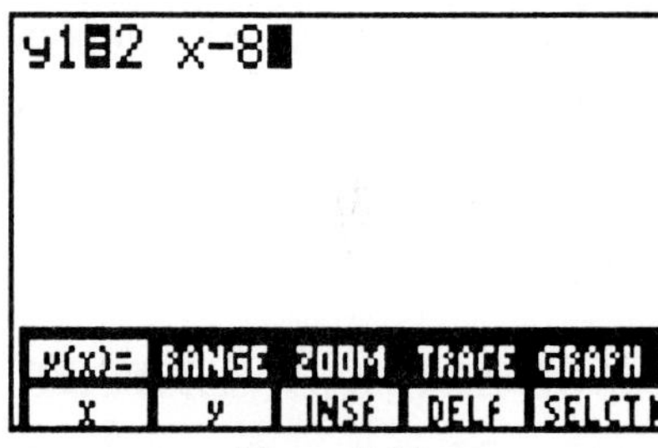

Figure 2.11

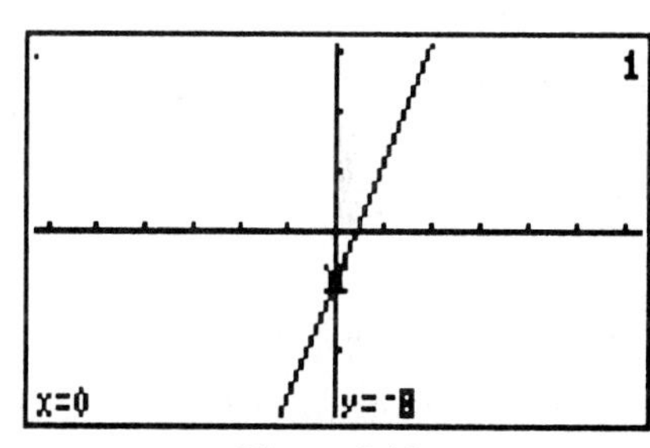

Figure 2.12

Example 3: Graph $y = \frac{3}{4}x - 6$.

Press

GRAPH [F1:y(x)=]

and press CLEAR to erase the previous equation. Press

(**3** ÷ **4**) X-VAR − **6**.

The parentheses are needed here because of implied multiplication (see Section 1.3). The viewing window from Example 2 will work here so there is no need to use the Range edit screen. Press

2nd [M5:GRAPH]

to see the graph. The M5 key is the second function of F5. This key allows you to access the top row of menu options when there are two rows on the screen. Trace to the point where $x = -9$ and $y = -12.75$. See Fig. 2.13. This ordered pair is a solution to the equation $y = (3/4)x - 6$. The y-coordinate of the Trace cursor is calculated by replacing 'x' from the equation in the y(x)= menu with the x-coordinate of the Trace cursor. Now we will move the free-moving cursor to the same point currently occupied by the Trace cursor. Press GRAPH to clear the Trace cursor from the viewing window, and then use the cursor-movement keys to move the free-moving cursor until its x-coordinate is –9 and the cursor appears to be on the line. See Fig. 2.14. The y-coordinate of the free-moving cursor is not the same as the y-coordinate of the Trace cursor when it was on this point. This ordered pair (–9, –13) is not a solution to the equation. This is because the y-coordinate of the free-moving cursor is not calculated from the equation in the y(x)= menu.

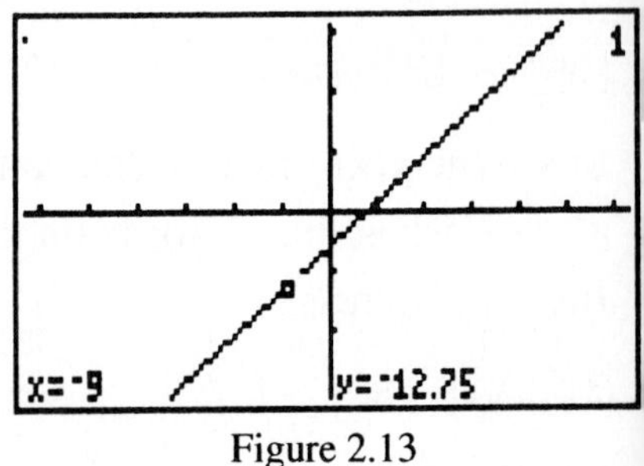

Figure 2.13

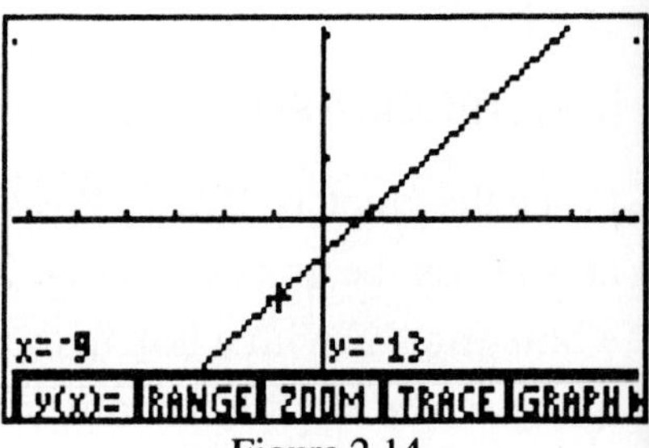

Figure 2.14

Section 2.3 Linear Equations and the Slope of a Line

Graphing a Vertical Line

Example 1: Graph $x = 3$.

Select the viewing window by pressing GRAPH [F2:RANGE] and entering the following values for the Range variables: xMin = –6.3, xMax = 6.3, xScl = 1, yMin = –5, yMax = 5, and yScl = 1. Erase any functions in the y(x)= menu and return to the Home screen. The graph of $x = 3$ is a vertical line. This is not a function so we can't enter an equation in the y(x)= menu. Instead we will use an option in the GRAPH [F2:DRAW] menu. Press

GRAPH MORE [F2:DRAW] [F3:VERT].

The free-moving cursor appears in the viewing window. Press ▶ to move a vertical line to the right until the x-coordinate of the line is 3, then press [ENTER]. See Fig. 2.15. Press

[EXIT] [EXIT]

to return to the [GRAPH] menu.

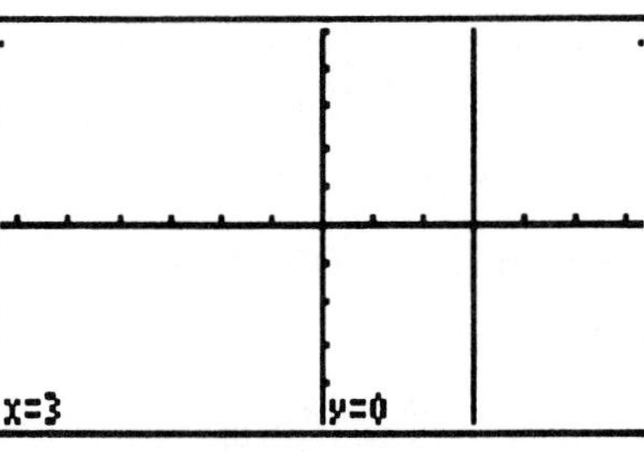

Figure 2.15

Section 2.4 The Slope-Intercept Form and Point-Slope Form of Linear Equations

Squaring a Viewing Window

Example 1: Graph $y = 0.5x + 2$ and $y = -2x - 3$ in a square viewing window.

Press

[GRAPH] [F1:y(x)=].

Erase any existing equations by pressing [CLEAR] on each line that has an equation. With the cursor on the top line, enter the first equation in $y1$ by pressing

0 [.] **5** [X-VAR] [+] **2** [ENTER],

and then enter the second equation in $y2$ by pressing

[(-)] **2** [X-VAR] [−] **3**.

Press [2nd] [M2:RANGE] and select a [–10, 10] by [–10, 10] window. Then press [F5:GRAPH]. The lines should be perpendicular, but they don't look like they are. See Fig. 2.16. The viewing window needs to be adjusted to make the lines look perpendicular. Press

[F3:ZOOM] [MORE] [F2:ZSQR]

to make this adjustment. The ZSQR option in the ZOOM menu adjusts the values of the Range variables so the slopes and proportions of the picture are correct. Press

[2nd] [M2:RANGE]

to see how these values have changed.

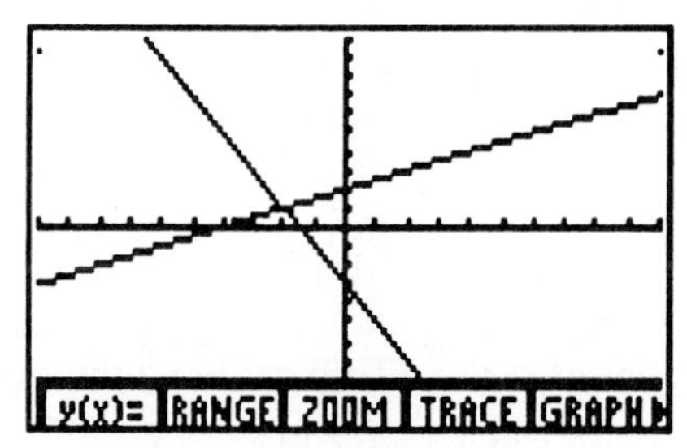

Figure 2.16

Section 2.5 Functions and Graphs of Functions

Functions on the TI-85 and the Standard Viewing Window

The method of graphing functions such as $y = x^2$, $y = \sqrt{x}$, and $y = x^{-1}$ is identical to the one you used in Section 2.2 to find the graph of a line. You enter the equation in the y(x)= menu, select an appropriate window, and graph.

The Standard Window

Example 1: Graph $y = -x^2 - 3x$.

Press

GRAPH [F1:y(x)=],

and use CLEAR and the cursor-movement keys to erase any equations that may be in the menu. Move the cursor next to $y1$ and press

(-) X-VAR x^2 − **3** X-VAR.

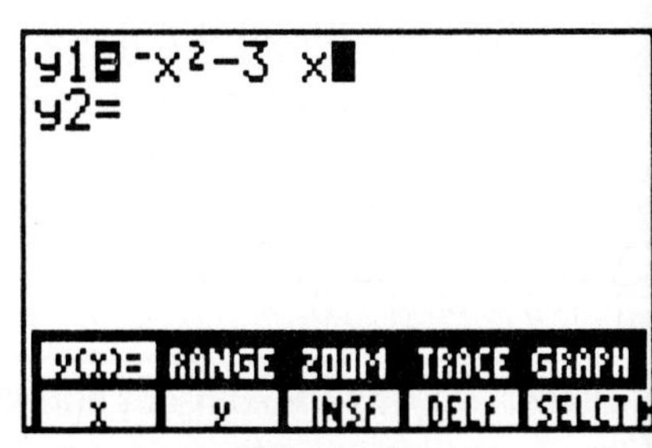

Figure 2.17

See Fig. 2.17. Most graphers use notation like $y1 = -x^2 - 3x$ to represent the function $f(x) = -x^2 - 3x$. Now press

2nd [M2:RANGE],

and enter a [–5, 5] by [–10, 10] window, then press [F5:GRAPH]. See Fig. 2.18. The graph is a parabola. Next we redraw the graph in the [–10, 10] by [–10, 10] window. This window is called the **Standard window**. There is a shortcut to enter the Standard window. Instead of using [F2:RANGE], press

[F3:ZOOM] [F4:ZSTD].

Figure 2.18

Notice the TI-85 redraws the graph in the new window as soon as you press [F4:ZTD]. Press CLEAR to remove the menus from the viewing window. See Fig. 2.19. Press

[GRAPH] [F2:RANGE]

to verify this is the Standard window.

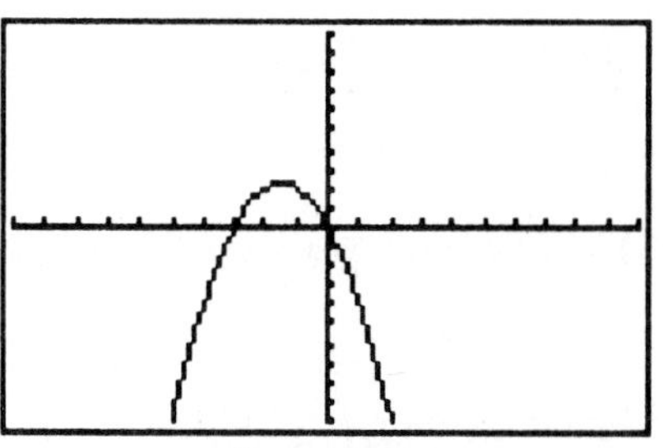

Figure 2.19

Tracing to Find the Domain and Range of a Function and Scrolling

Example 2: Find the domain and range of $f(x) = \sqrt{(x-3)}$.

We will enter this equation in the y(x)= menu and graph in the Integer window. We are going to use a shortcut to enter the Integer window. Press

[GRAPH] [F1:y(x)=],

and erase existing functions from the menu. Move the cursor to $y1$ and press

[2nd] [√‾] [(] [X-VAR] [−] **3** [)].

Now press

[GRAPH] [F3:ZOOM] [MORE] [MORE] [F4:ZINT].

As soon as you press [F4:ZINT] you will see the graph drawn in the current window and the free-moving cursor blinking at the origin. Press [ENTER] to redraw the graph in the Integer window. This option in the ZOOM menu creates an Integer window centered on the free-moving cursor. Now press [F4:TRACE]. See Fig. 2.20. The x-coordinate is 0, and the y-coordinate is blank. This is because the function is not defined for $x = 0$. Press [▶] three times. Now $x = 3$ and $y = 0$. Press and hold [▶] to scroll to the right of the graph. The results of [TRACE] indicate the domain is $[3, \infty)$ and the range is $[0, \infty)$.

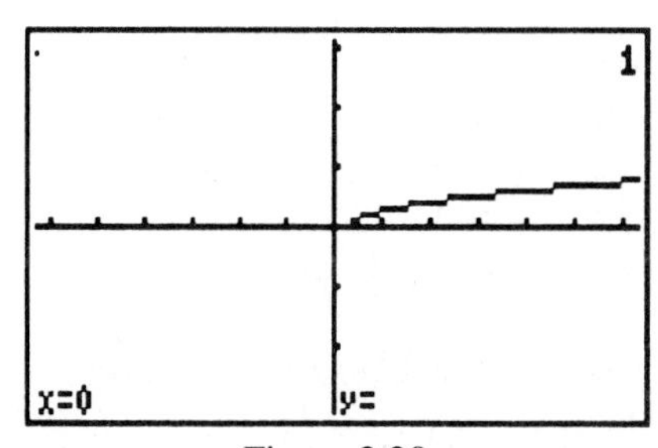

Figure 2.20

Example 3: Find a complete graph of $f(x) = x^4 - 3x^2 + 5x - 6$, and estimate its range.

Press

[GRAPH] [F1:y(x)=]

and erase existing functions. Move the cursor next to $y1$ and press

[X-VAR] [^] **4** [−] **3** [X-VAR] [x^2] [+] **5** [X-VAR] [−] **6.**

Press

[GRAPH] [F2:RANGE]

and enter a [–5, 5] by [–10, 10] window. Then press

[F5:GRAPH] [CLEAR].

See Fig. 2.21. This is not a complete graph because you can't see the bottom portion. Press

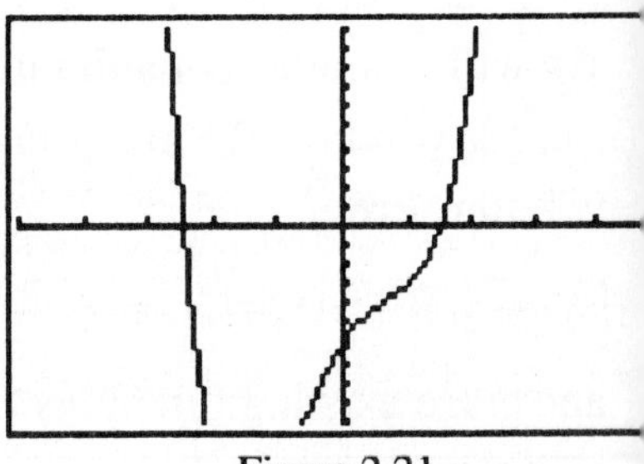
Figure 2.21

[GRAPH] [F4:TRACE],

and repeatedly press [◄]. Notice that even when the Trace cursor moves off the screen, you can still see its coordinates. The smallest y-coordinate that is printed as the Trace cursor moves to the left is about –15.19. If you continue to press [◄], the Trace cursor will reappear on the graph. The results of [TRACE] give us an idea of a better viewing window. Press

[GRAPH] [F2:RANGE],

and enter the [–5, 5] by [–20, 20] window with xScl = 1 and yScl = 2. Then press [F5:GRAPH]. See Fig. 2.22. This is a complete graph. The results of [TRACE] indicate that the range is about [–15.2, ∞).

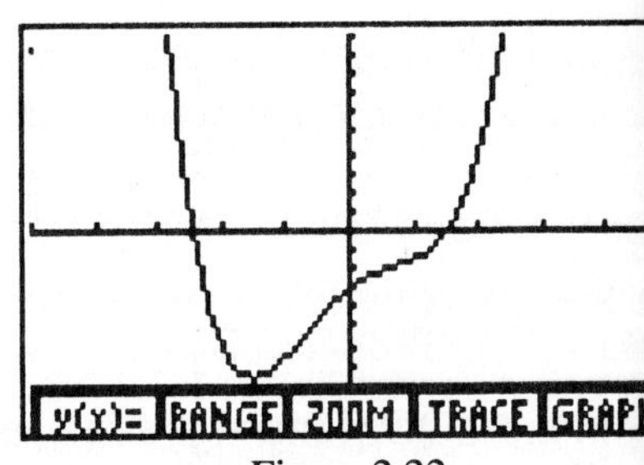

Figure 2.22

How to Graph a Relation that Is Not in Function Form

Example 4: Graph $y^2 = x$.

This relation is not a function, and the TI-85 is in Function mode. However, the relation can be split into two functions by solving for y. If you do this you get $y = -\sqrt{x}$ and $y = \sqrt{x}$. Enter these functions in $y1$ and $y2$ by pressing

[GRAPH] [F1:y(x)=] [CLEAR] [(−)] [2nd] [√] [X-VAR] [ENTER] [CLEAR] [2nd] [√] [X-VAR].

See Fig. 2.23. Graph in the Standard window by pressing

(GRAPH) [F3:ZOOM] [F4:ZSTD].

The graph is a parabola. See Fig. 2.24.

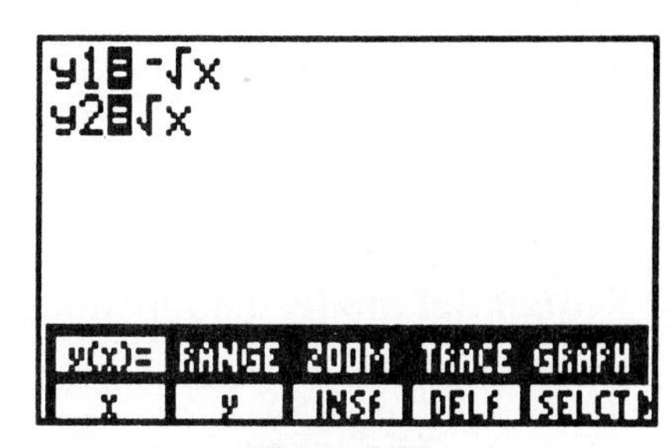

Figure 2.23

Chapter 3 Solving Equations and Systems of Equations

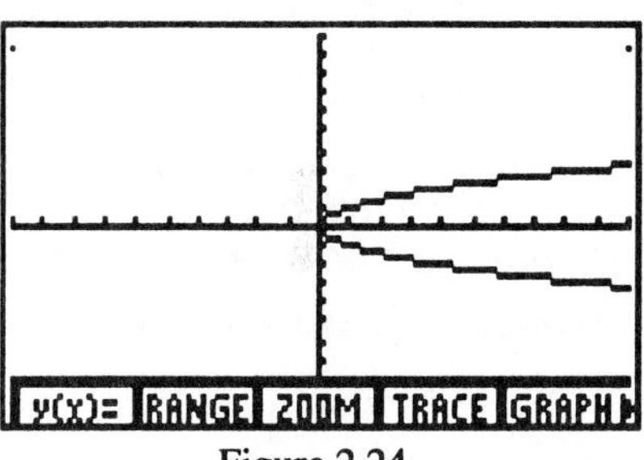

Figure 2.24

Section 3.1 Finding Graphical Solutions

Selecting and Deselecting Graphs and Sequential versus Simultaneous Graphing

When several functions have been entered in the y(x)= menu, you can select some of them for graphing and deselect others so that they won't be graphed. If you select several functions to be graphed in the same window, you can graph them sequentially (one at a time) or simultaneously.

Example 1: Graph $y = 2x + 5$ and $y = 21$.

Clear the y(x)= menu and enter $y1 = 2x + 5$ and $y2 = 21$. Notice the '=' signs next to both $y1$ and $y2$ are highlighted. This means both functions have been selected for graphing. Deselect $y2$ by pressing [F5:SELECT] while the cursor is next to 'y2 = 21'. Notice the '=' is no longer highlighted. This means $y2$ is de-selected. See Fig. 3.1. Now graph in the Integer window. You should see the graph of $y1$ but not $y2$. Now reselect $y2$ by pressing

(GRAPH) [F1:f(x)=] (▼) [F5:SELECT].

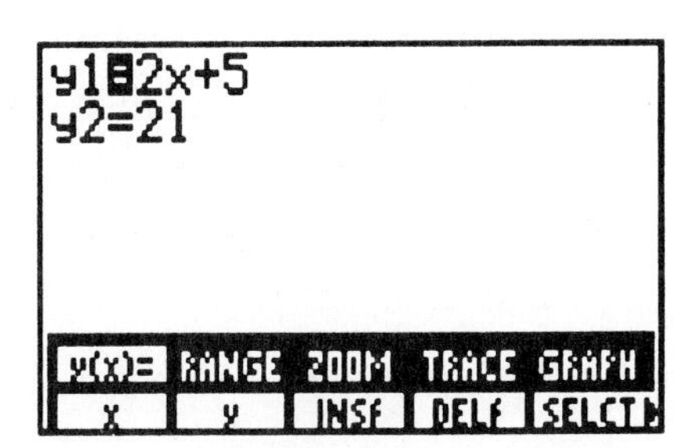

Figure 3.1

Notice that if the '=' is highlighted, pressing [F5:SELECT] unhighlights it. If the '=' is not highlighted, pressing [F5:SELECT] highlights it. Press

[2nd] [M5:GRAPH].

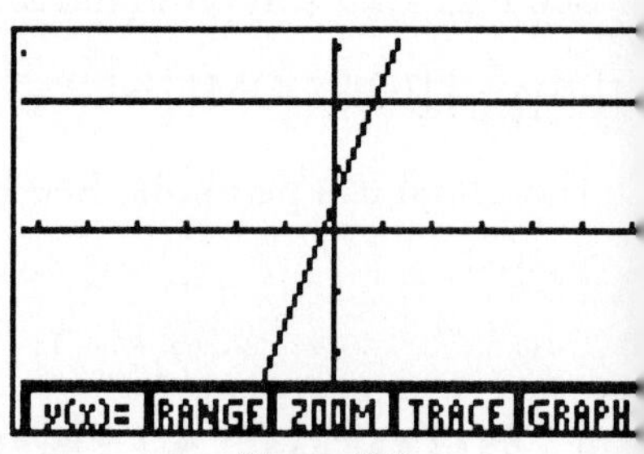

Figure 3.2

Since both functions are highlighted, you see the graph of $y(x) = 2x + 5$, followed by the graph of $y = 21$. See Fig. 3.2. When the TI-85 is in **Sequential mode**, the functions are graphed one at a time. Now press

[GRAPH] [MORE] [F3:FORMT].

Figure 3.3

This is the Format screen. Select **Simultaneous mode** by highlighting 'SimulG' on the Format screen. See Fig. 3.3. Press [F5:GRAPH]. The graphs are drawn at the same time.

Solving Linear Equations Graphically

Example 2: Solve $17 - 1.5x = -10$.

Press

[GRAPH] [F1:y(x)=],

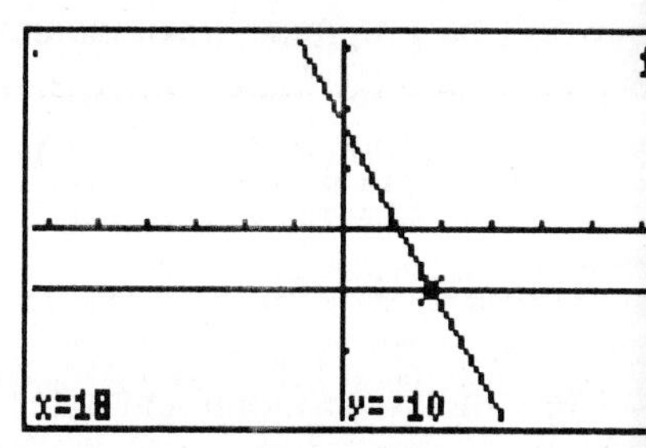

Figure 3.4

and use [CLEAR] and the cursor-movement keys to clear the y(x)= menu. Enter $y1 = 17 - 1.5x$ and $y2 = -10$. Graph in the Integer window. Now press [F4:TRACE] and repeatedly press [▶] until the Trace cursor is on the intersection of the two lines. See Fig. 3.4. The '1' in the upper right-hand corner of the screen indicates that you are tracing on $y1$. If you press [▼] while tracing, the Trace cursor will move to the next graph on the screen. If you press [▼] when the Trace cursor is at the point of intersection, it won't appear to move because the Trace cursor is on both graphs, but the number in the upper right-hand corner changes. If you move the Trace cursor away from the point of intersection and then press [▼], you will be able to see the cursor move from one graph to the other.

Section 3.2 Solving Equations Using Zoom In

Finding a Graphical Solution Using the 0.1 (Decimal) Window Settings

These window settings were described in Section 2.1. They are xMin = –6.3, xMax = 6.3, xScl = 1, yMin = –3.1, yMax = 3.1, and yScl = 1.

Using the Grapher Zoom In Menu

The **ZOOM feature** can be used to quickly magnify a portion of the screen. This is helpful when you need to find an accurate approximation to the solution of an equation. In this section we will discuss the **Set Factors** and **Zoom In** options in the ZOOM menu. Set Factors is used to set the Zoom factors. These factors determine by how much the graph is magnified when you use Zoom In. After you have pressed

[GRAPH] [F3:ZOOM] [MORE] [MORE] [F1:ZFACT],

set the factors to 10 by pressing

10 [ENTER] 10.

See Fig. 3.5. Return to the Home screen by pressing

[2nd] [QUIT].

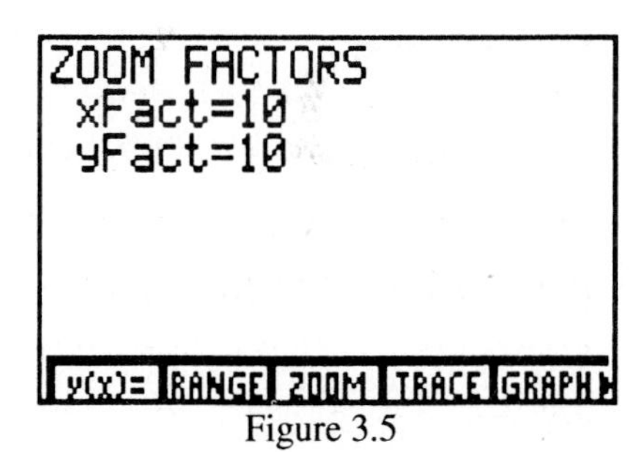

Figure 3.5

Problems Using Zoom In

Accuracy of a Solution with the Grapher

Example 1: Find a solution to $2.7x + 3.28 = 0$ with an error of at most 0.01.

The solution can be found graphically by tracing to the point where the graph of $y = 2.7x + 3.28$ crosses the x-axis. The difficulty with this approach is that if we graph in the Standard window, the Trace cursor won't be accurate enough. We will need to magnify the portion of the graph that crosses the x-axis. We can magnify with Zoom In.

Clear the y(x)= menu and enter $y1 = 2.7x + 3.28$. Enter the following values for the Range variables: xMin = –10, xMax = 10, xScl = 0.01, yMin = –10, yMax = 10, and yScl = 0.01. We use a value of 0.01 for xScl and yScl because the example required an error of at most 0.01. Press

[F5:GRAPH].

The axes look double thick because the tick marks are so close together. The tick marks will spread out as we Zoom In. Now press

[F3:ZOOM] [F2:ZIN].

You should see the graph with the free-moving cursor at the origin. Use ◄ to move the cursor to the x-intercept of the graph. See Fig. 3.6. Press ENTER. The graph is magnified. Use the cursor-movement keys to move the cursor back to the x-intercept. See Fig. 3.7. Press ENTER. The graph is magnified again. Move the cursor to the x-intercept one more time. See Fig. 3.8. Magnify the graph again by pressing ENTER. Now press

GRAPH [F4:TRACE],

and move the Trace cursor to the x-intercept. See Fig. 3.9. The solu-

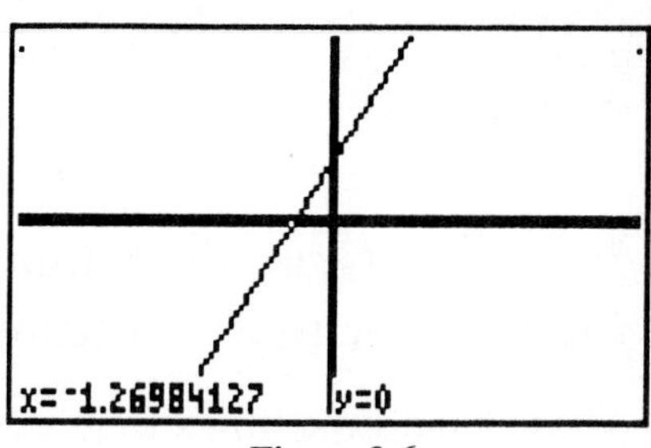

Figure 3.6

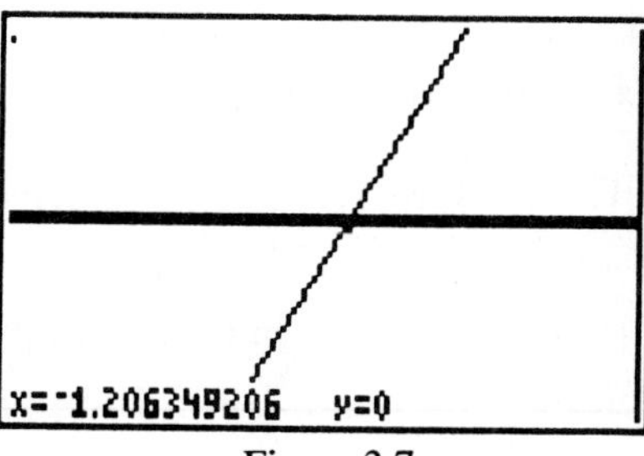

Figure 3.7

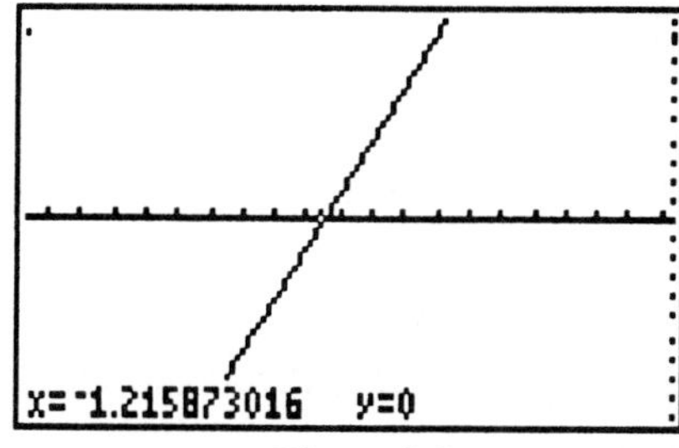

Figure 3.8

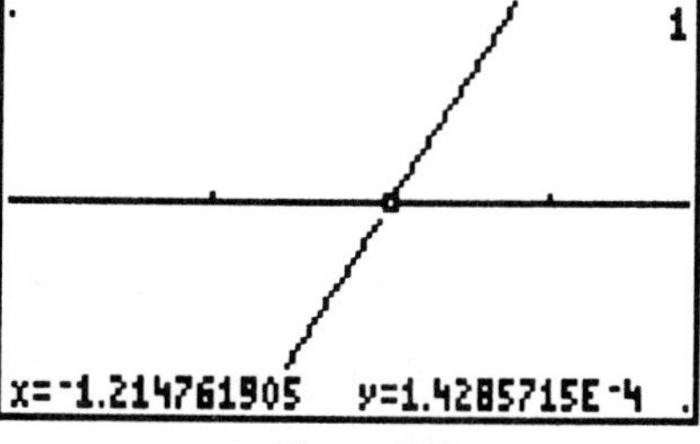

Figure 3.9

tion is approximately $x = -1.215$. We know the error is at most 0.01 because the graph crosses the x-axis between two tick marks that are 0.01 apart.

Example 2: Solve $3x - 2 = 0$.

Clear the y(x)= menu and enter $y1 = 3x - 2$. Enter the following values for the Range variables: xMin = 0.6, xMax = 0.7, xScl = .01, yMin = –0.2, yMax = 0.2, and yScl = .1. Graph and trace to the intersection of the x-axis and the graph. See Fig. 3.10. (Note the tick marks for y appear on the edge of the screen since the y-axis is not included in the window.) The calculator solution is $x = .66666666667$. We could Zoom In to get a better approximation, but no matter how many times we Zoom In, we will still only have an approximation. This is because the exact answer is $2/3$, which has a repeating decimal representation. The grapher supports a solution of $2/3$, but it does not confirm it.

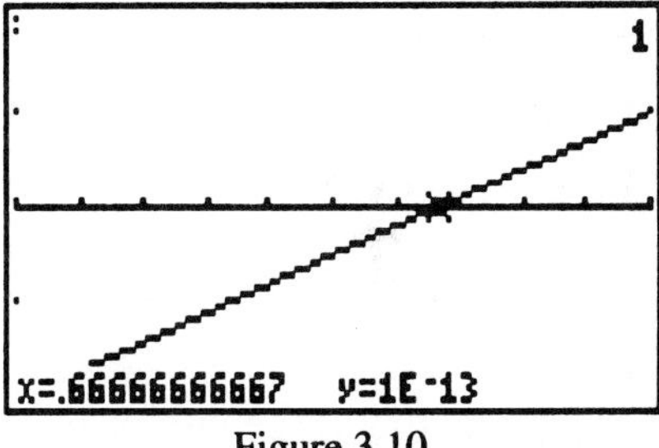

Figure 3.10

How to Determine an Appropriate Window for a Graph

Example 3: Find a complete graph of $y = 32000 + 960x$.

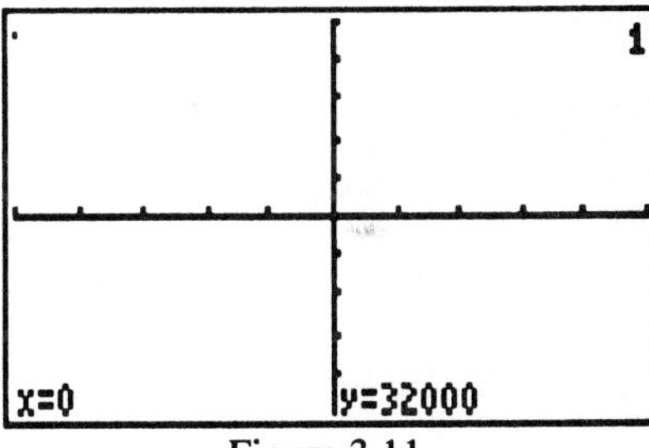

Figure 3.11

Clear the y(x)= menu and enter $y1 = 32000 + 960x$. Graph in the [–5, 5] by [–5, 5] window. The Graph screen is blank because this window does not contain any portion of the graph. We can use Trace to get a better idea of an appropriate window. Press

[F4:TRACE].

Even though you can't see the Trace cursor, you can see its coordinates. See Fig. 3.11. These coordinates tell you that the point (0, 32000) is either on or close to the graph. Based on this information, try the following values for the Range variables: xMin = –5, xMax = 5, xScl = 1, yMin = –1000, yMax = 40000, and yScl = 10000. Press [F5:GRAPH]. This window indicates the complete graph is a line with y-intercept = 32,000. See Fig. 3.12.

Figure 3.12

Section 3.3 Some Nonlinear Equations

Using Zoom with the Multigraph Method

Example 1: Solve $0.7\sqrt{(x-3)} = 0.3$ with the multigraph method.

In the multigraph method, we graph each side of the equation as a separate function. Then we trace to the point where the two functions intersect. This point is a solution to the original equation. In our first attempt, we graph $y1 = 0.7\sqrt{(x-3)}$ and $y2 = 0.3$ in the Standard window. The graph of $y2$ is too close to the x-axis to be seen clearly in this window. Even though you can't see the graph of $y2$ very well, you can still trace it. Press

[F4:TRACE]

and move the Trace cursor to the graph of $y2$ by pressing [▼]. See Fig. 3.13. A better viewing window would be [–10, 10] by [–1, 1]. You can see both graphs as well as their point of intersection in this window. See Fig. 3.14. Now you could Zoom In on the intersection to obtain a solution of $x = 3.18$.

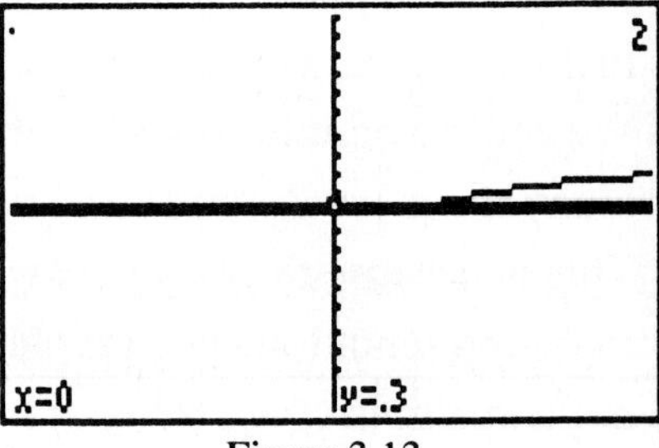

Figure 3.13

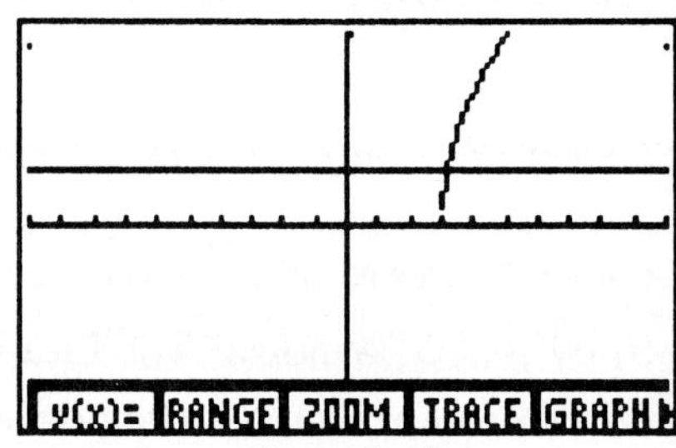

Figure 3.14

Example 2: Solve $x^3 = x - 0.5$ with the multigraph method.

Graph $y1 = x \wedge 3$ and $y_2 = x - 0.5$ in the Standard window. There appear to be two points of intersection. First we Zoom In on the point in the first quadrant. Press

[F3:ZOOM] [F2:ZIN],

and move the free-moving cursor to this point. See Fig. 3.15. When you press [ENTER] to complete the Zoom In, you will see that the two graphs do not intersect after all, so there is no solution in the first quadrant. You could redraw the graphs in the Standard window and then Zoom In on the point of intersection in the third quadrant to obtain a solution of $x = -1.19$. Initially, it looked like there were two solutions to the problem, but Zoom In helped us determine there was only one.

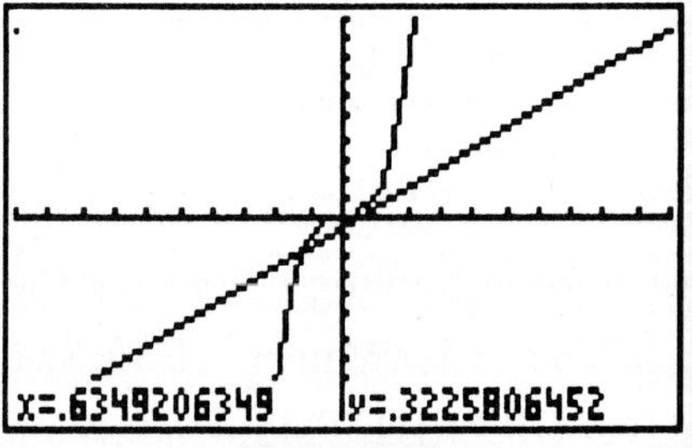

Figure 3.15

Section 3.4 Graphical Representations of Problem Situations

Motion Simulation

Parametric equations are useful in simulating moving objects. A simulation of two runners was described in Section 1.6. In Parametric mode, both the x- and y-coordinates of points on a graph are determined by a third variable, t. In motion simulations, t represents time.

Example 1: Simulate a ball that is thrown straight up with an initial velocity of 88 ft/sec.

The height of the ball is given by the equation $h = -16t^2 + 88t$. Select Parametric (Param) mode on the Mode screen, and then select **DrawDot** mode in the GRAPH [F3:FORMT] menu. Now press

GRAPH [F1:E(t)=],

and erase any functions that may be in the menu. Then enter $xt1 = 3$ and $yt1 = -16t^2 + 88t$. Press

GRAPH [F2:RANGE],

and enter the following values for the Range variables: tMin = 0, tMax = 5.5, tStep = 0.1, xMin = 0, xMax = 5, xScl = 1, yMin = 0, yMax = 150, and yScl = 10. Press

[F5:GRAPH]

to see the simulation. You can review the motion of the ball by pressing TRACE. The Trace cursor appears at the bottom of the screen. Press ▶ repeatedly to see the ball as time moves forward. See Fig. 3.16. This screen tells you that when $t = 2.7$ seconds, the ball has risen to a height of 120.96 feet. Press ◀ to move backward in time through the simulation.

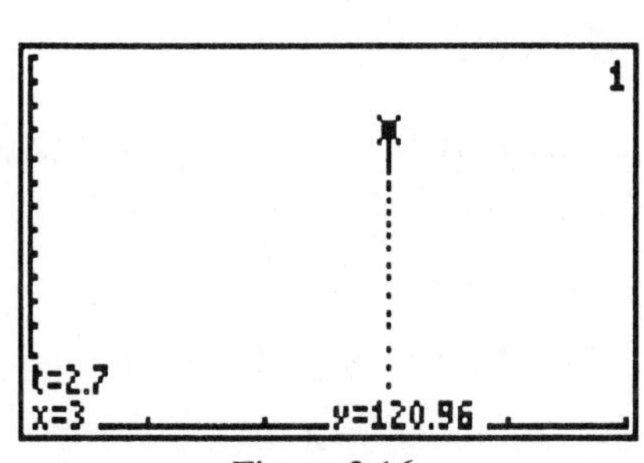

Figure 3.16

The DRAW Menu with Plotting Points and Overlaying a Graph on the Points

Example 2: Plot the points (10, 50), (15, 33.33), (20, 25), and (30, 16.67), and overlay the graph of $y = 500/x$ on the same screen.

Select Function mode from the Mode screen and **DrawLine** mode from the GRAPH [F3:FORMT] menu. In DrawLine mode, the points that are calculated and plotted on the graph are connected with small line segments. This gives the appearance of a smooth graph. Now use

GRAPH [F2:RANGE]

to enter the [0, 60] by [0, 60] viewing window. Clear the y(x)= menu and return to the Home screen. Plot the points given in the example statement with PtOn(or STAT. The PtOn(and STAT options were described in Section 2.2. After you have plotted these points your screen should look like Fig. 3.17. Now we will use the DrawF option from the GRAPH [F2:DRAW] menu to overlay the graph of $500/x$. Press

GRAPH MORE [F2:DRAW] [F5:DrawF]

to copy the DrawF command on the Home screen, then press

500 ÷ X-VAR ENTER.

See Fig. 3.18.

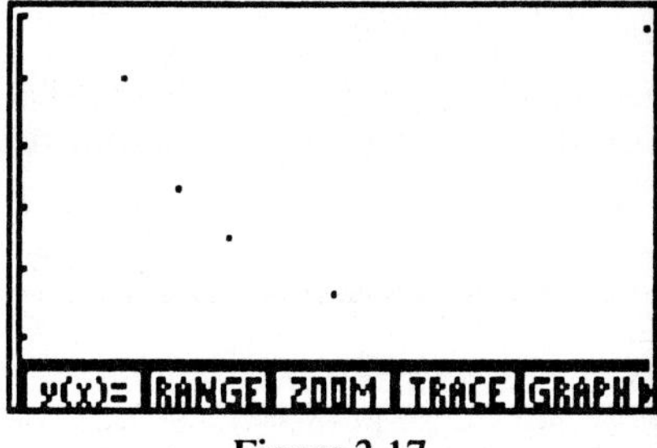

Figure 3.17

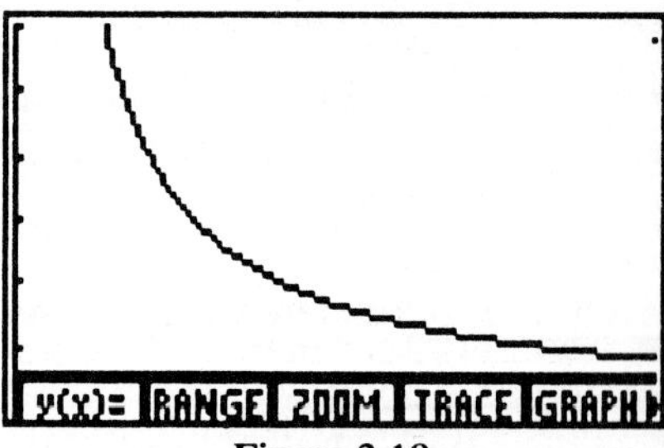

Figure 3.18

Finding an Integer Window that Satisfies [0, *x*Max], [0, *y*Max]

To find an integer window that satisfies [0, *x*Max] by [0, *y*Max], *x*Max must be a multiple of 126, and *y*Max must be a multiple of 62. See Section 2.1.

Section 3.5 Solving Systems of Equations by Graphing

Using Trace on Two Functions

Example 2 in Section 3.1 describes how to Trace on two different graphs in the same viewing window.

Solving Systems of Equations with both Zoom In and Zoom Out and Why 5 and 10 are Convenient Zoom Factors

We will be using Zoom factors of xFact = 5 and yFact = 5 in the next problem, so use the [F1:ZFACT] option in the ZOOM menu to set both factors to 5. This option is described in Section 3.2. Zoom factors of 5 and 10 are convenient since they alter the Range dramatically.

Example 1: Solve the systems $y = -2x + 175$ and $y = 3x - 320$.

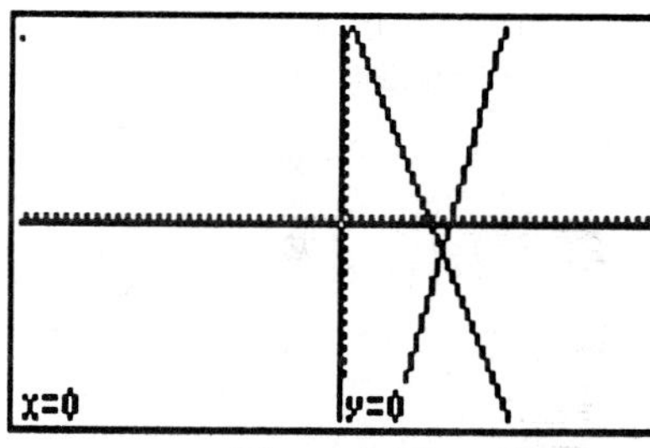

Figure 3.19

Graph $y1 = -2x + 175$ and $y2 = 3x - 320$ in the Integer window [–63, 63] by [–31, 31]. Since you don't see the graphs, we need a different window. We will change the window with **Zoom Out**. Zoom Out provides a more global view by enlarging the viewing window. Press

GRAPH [F3:ZOOM] [F3:ZOUT].

You should see the Graph screen with the free-moving cursor blinking at the origin. When you press ENTER, Zoom Out will enlarge the window centered on the free-moving cursor by a factor of 5. See Fig. 3.19. Now you can see the intersection of the two graphs. Next use the Zoom In option to find an accurate approximation of the point of intersection. See Fig. 3.20. Zoom In was also described in Section 3.2. The coordinates of the point of intersection are $x = 99$, $y = -23$. This is the solution to the system of equations.

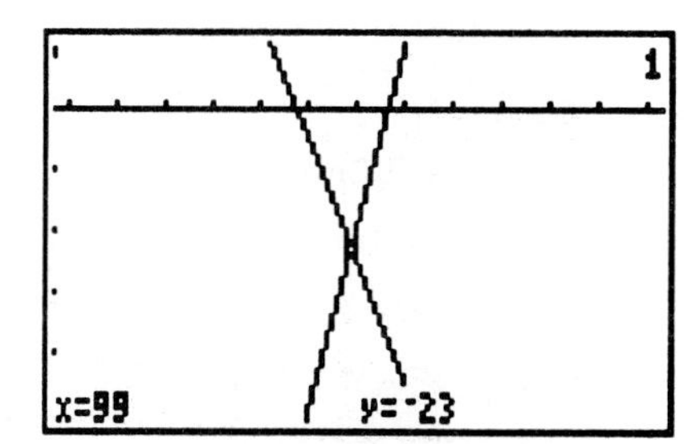

Figure 3.20

Chapter 4 Linear Inequalities and Systems of Linear Inequalities

Section 4.1 Solving Linear Inequalities Algebraically

See Section 1.3 for a discussion of Recall (Replay) to evaluate an expression.

Section 4.4 Solving Absolute Value Equations and Inequalities

Tables cannot be generated on the TI-85.

Section 4.5 Linear Inequalities in Two Variables

Shading Solutions to Inequalities in Two Variables

The **Shade** command is followed by three arguments that are separated by commas. The grapher shades all the points that are above the first argument and below the second. (The third argument refers to the x resolution.) If there is no third argument given, every pixel is turned on in the shaded portion. If the resolution is 2, every other pixel is turned on, 3 every third pixel and so on.

The following example employs two arguments.

Example 1: Graph the solution to $y > 4x - 5$.

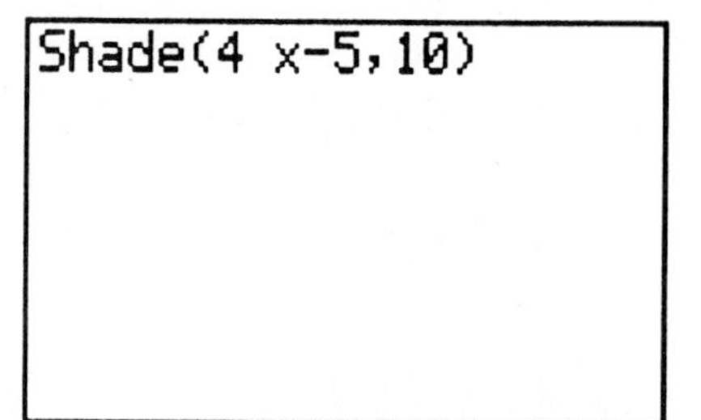

Figure 4.1

Clear the y(x)= menu and select the Standard viewing window. From the Home screen enter the following keystroke sequence:

GRAPH MORE [F2:DRAW] [F1:Shade] **4** X-VAR (−) **5** , **10**).

See Fig. 4.1. Press ENTER and then CLEAR to see the solution. See Fig. 4.2. The solution is all points above the line $y(x) = 4x - 5$. There is no upper bound for the solution, but the TI-85 requires us to provide an upper bound. We provided an upper bound of 10 so that the TI-85 would shade everything between the line and the top of the screen.

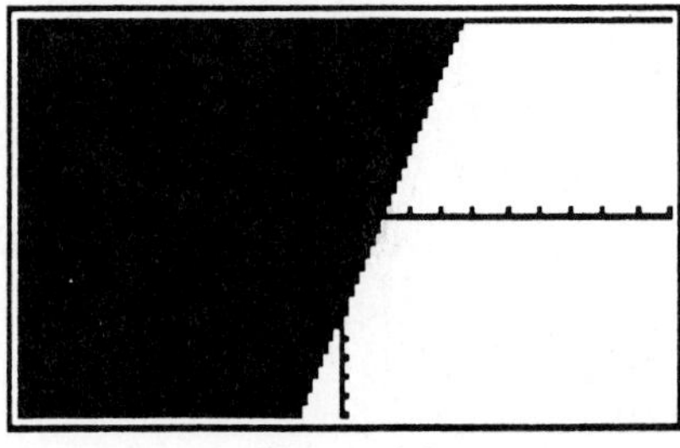
Figure 4.2

From the Home screen press

GRAPH MORE [F2:DRAW] MORE [F5:CLDRW]

to clear the Graphics screen. Now use the GRAPH [F2:DRAW] menu to enter the command

Shade(–10, $4x - 5$)

on the Home screen, and press ENTER to see the solution to $y < 4x - 5$. See Fig. 4.3.

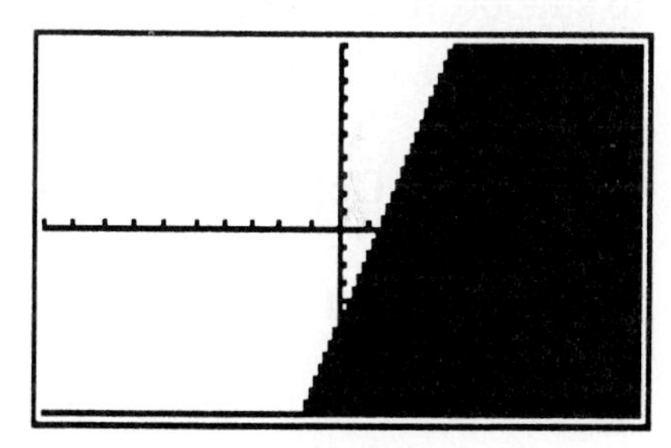
Figure 4.3

Section 4.6 Solving Systems of Inequalities

Graphing Systems of Inequalities Using Shading

Example 1: Graph the solution to the system $y < -x + 9$ and $y > 3x - 15$.

Enter the [–10, 10] by [–25, 25] viewing window. Then enter the following command from the Home screen:

Shade($3x - 15$, $-x + 9$).

See Fig. 4.4. After you press ENTER you will see the lower boundary for the shaded area is the line $y = 3x - 15$ and the upper boundary is the line $y = -x + 9$. See Fig. 4.5.

```
Shade(3 x-15, -x+9)
```

Figure 4.4

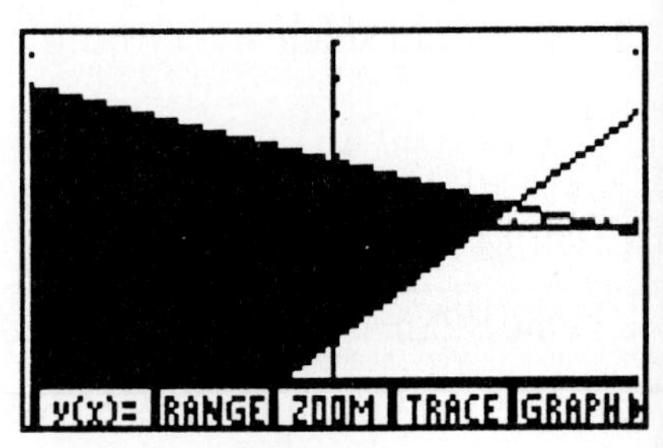

Figure 4.5

Chapter 5 Polynomials

Section 5.1 Exponents and Their Properties

Using TRACE to Show Two Graphs that Appear to Be Identical

Use

GRAPH [F4:TRACE]

to support rules of exponents by showing that two graphs appear to be identical.

Example 1: Determine graphically if $(2x^3)(3x^2) = 6x^5$ is correct.

Select the DrawLine mode from the GRAPH [F3:FORMT] menu, and enter

$y1 = (2x \wedge 3)(3x^2)$ and $y2 = 6x \wedge 5$.

Graph in the [–2, 2] by [–10, 10] viewing window. Press [F4:TRACE]. See Fig. 5.1. In Section 3.1, Example 2 we described how to move from one curve to the other by pressing ▼ and ▲. When you press these cursor-movement keys, the cursor does not appear to move to another graph, and the coordinates of the Trace cursor don't change. This is graphical evidence that both $y1$ and $y2$ are the same function.

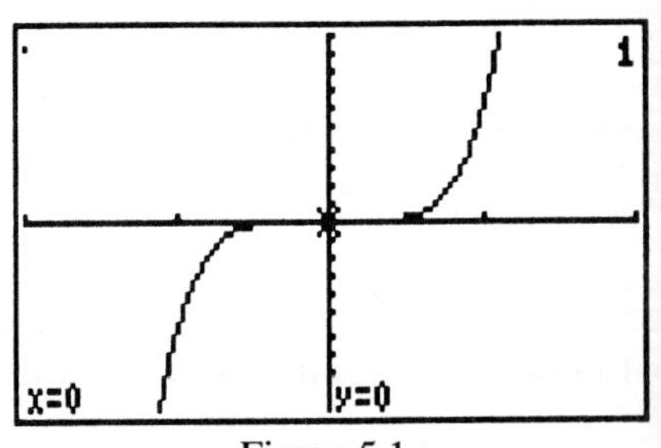

Figure 5.1

Scientific Notation and the [EE] Key, Normal and Scientific Modes

When you select **Sci** on the first line of the Mode screen, numbers are displayed in scientific notation. Press

[2nd] [MODE]

and highlight 'Sci'. Then return to the Home screen. Press

[CLEAR] **1230** [ENTER].

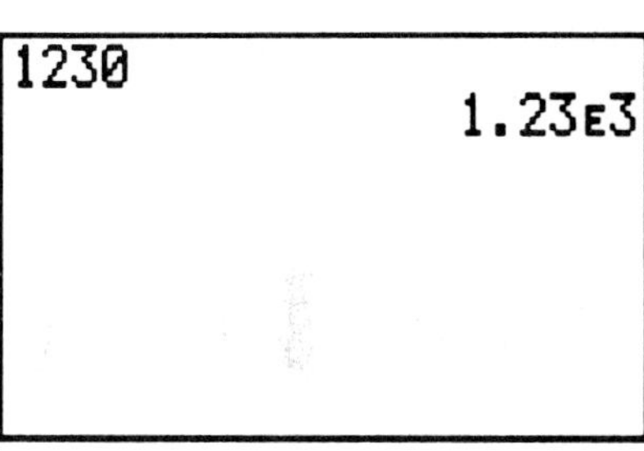

Figure 5.2

See Fig. 5.2. This notation means 1.23×10^3. You can enter numbers in scientific notation with [EE] whether the TI-85 is in Scientific mode or not. [EE] is on row six, column two. Press

[2nd] [MODE],

and switch back to **Normal mode** by highlighting 'Normal'. Now return to the Home screen and press

1 [.] **23** [EE] **3** [ENTER].

The number was entered in Scientific mode but was displayed in Normal mode. See Fig. 5.3.

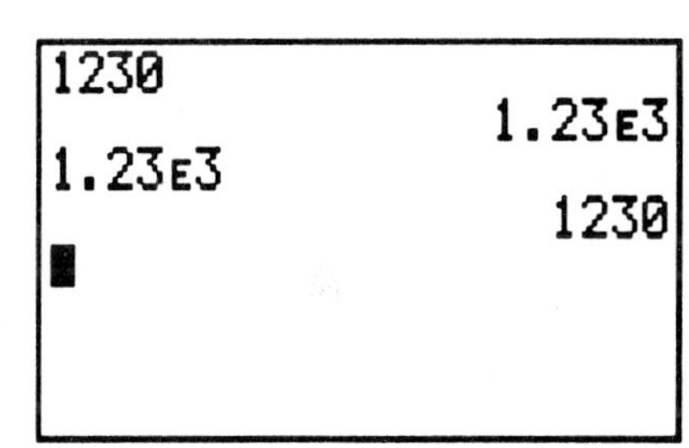

Figure 5.3

Large Numbers on the TI-85

The TI-85 displays numbers up to 12 digits long. If the number has more than 12 digits, the TI-85 converts the number to scientific notation. Enter

2 ^ 39

on the Home screen. The entire number is displayed since it is 12 digits long. Now enter

2 ^ 40

on the Home screen. See Fig. 5.4. This number has 13 digits so it is converted to scientific notation and only the first 12 digits are displayed.

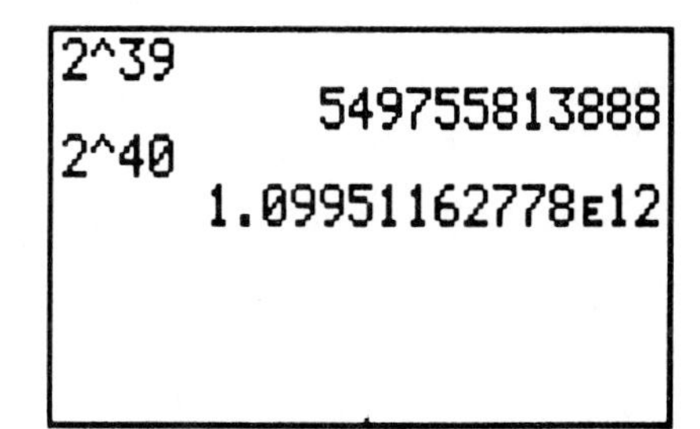

Figure 5.4

Small Numbers (Close to Zero) on the TI-85

Numbers that are close to zero are usually represented in scientific notation on the TI-85. Enter .000000000105 on the Home screen. See Fig. 5.5. The decimal is converted to scientific notation. This notation means 1.05×10^{-10}, which is the same as .000000000105. This number is very close to zero.

.000000000105
1.05E-10

Figure 5.5

Section 5.2 Polynomial Expressions

A New Method of Evaluating a Polynomial Expression on the Grapher

Section 1.3 described one method of evaluating a polynomial expression at a particular value of x. Now we will describe another method to evaluate a polynomial that uses the y(x)= menu. We will enter the polynomial in $y1$, store the value of x in the TI-85 variable 'x', and then print the value of $y1$ on the Home screen. We print the value of $y1$ by copying it to the Home screen.

The Method of Entering the Expression in the Function Defining Menu

Example 1: Evaluate $2x^3 - 4x^2 + 5x - 7$ at $x = -1$.

Enter $y1 = 2x \wedge 3 - 4x \wedge 2 + 5x - 7$ in the y(x)= menu, then exit to the Home screen and clear the Home screen. Store –1 in x by pressing

[(-)] **1** [STO▶] [X-VAR] [ALPHA].

Now copy the variable $y1$ to the Home screen by pressing

[2nd] [:] [2nd] [ALPHA] [Y] **1**.

and then print the value of $y1$ by pressing [ENTER]. See Fig. 5.6. The TI-85 evaluates $y1$ at the current value of 'x'.

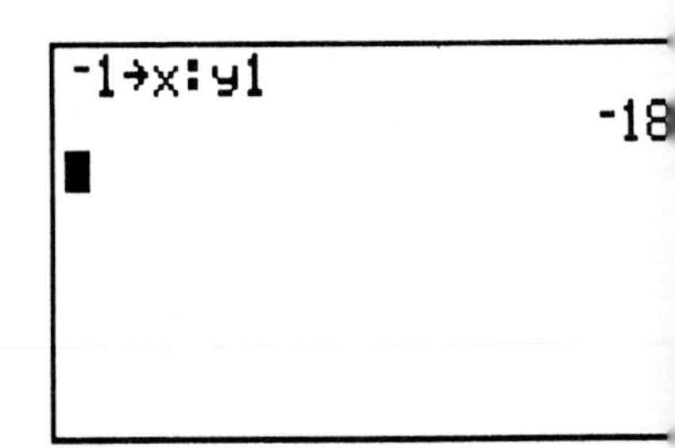

Figure 5.6

Using the Solver to Evaluate Expressions

Another method to evaluate polynomials is with the **Solver**. You can use the Solver feature to enter an equation with several variables and evaluate the equation at various values of those variables. Press

[2nd] [SOLVER] [CLEAR]

to see the Solver edit screen. See Fig. 5.7. Enter the equation

$2x \wedge 3 - 4x \wedge 2 + 5x - 7.$

Figure 5.7

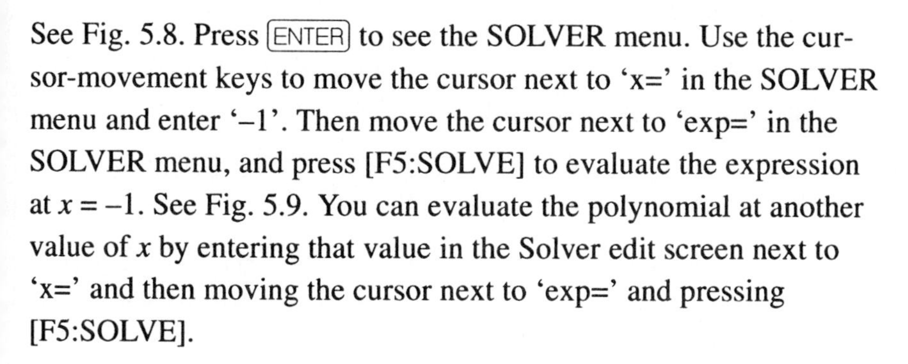

See Fig. 5.8. Press [ENTER] to see the SOLVER menu. Use the cursor-movement keys to move the cursor next to 'x=' in the SOLVER menu and enter '–1'. Then move the cursor next to 'exp=' in the SOLVER menu, and press [F5:SOLVE] to evaluate the expression at $x = -1$. See Fig. 5.9. You can evaluate the polynomial at another value of x by entering that value in the Solver edit screen next to 'x=' and then moving the cursor next to 'exp=' and pressing [F5:SOLVE].

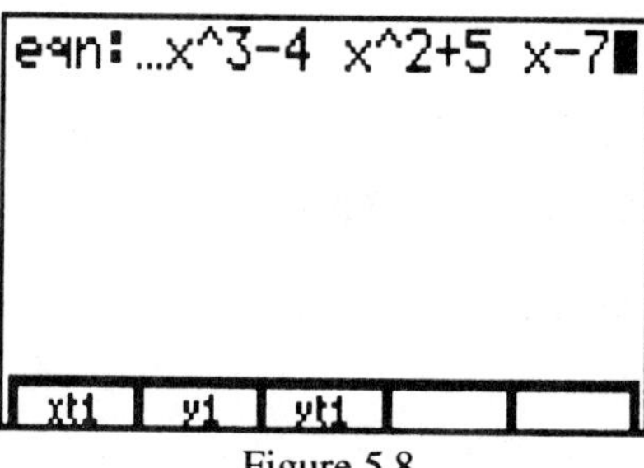

Figure 5.8

Table Building for Evaluating Polynomials

Tables are not possible on the TI-85.

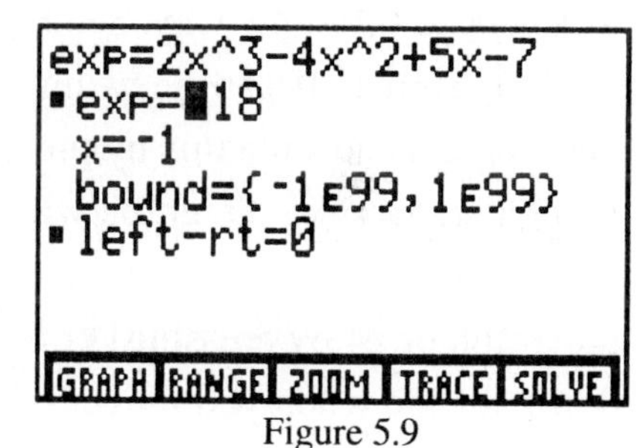

Figure 5.9

Graphical Support for the Addition of Polynomials

In Section 5.1 we used [GRAPH] and [F4:TRACE] to provide visual support for rules of exponents. In this section we will use a similar method to provide support for addition of polynomials. We will enter the two polynomials in $y1$ and $y2$. Then we will enter $y3 = y1 + y2$ and enter the sum we derived algebraically in $y4$. Next we will deselect $y1$ and $y2$ and then graph $y3$ and $y4$ in the same viewing window. If there appears to be only one graph in the window, we have support that $y3$ and $y4$ are the same. We can gather more support by Tracing both curves to see if they are the same.

Entering the Sum of Two Polynomials into a Grapher

Example 2: Find the sum of the polynomials $6x^4 + 2x^3 + 2x^2 + 10x + 1$ and $-5x^4 - 7x^3 + 3x^2 - 3x - 8$ and support your answer graphically.

The algebraic answer is $x^4 - 5x^3 + 5x^2 + 7x - 7$. Support this answer by clearing the y(x)= menu and entering

$y1 = 6x \wedge 4 + 2x \wedge 3 + 2x \wedge 2 + 10x + 1$

and

$y2 = -5x \wedge 4 - 7x \wedge 3 + 3x \wedge 2 - 3x - 8$.

Enter $y3 = y1 + y2$ by placing the cursor next to 'y3 =' and pressing

[F2:y] **1** [+] [F2:y] **2** [ENTER].

Then enter

$y4 = x \wedge 4 - 5x \wedge 3 + 5x \wedge 2 + 7x - 7$.

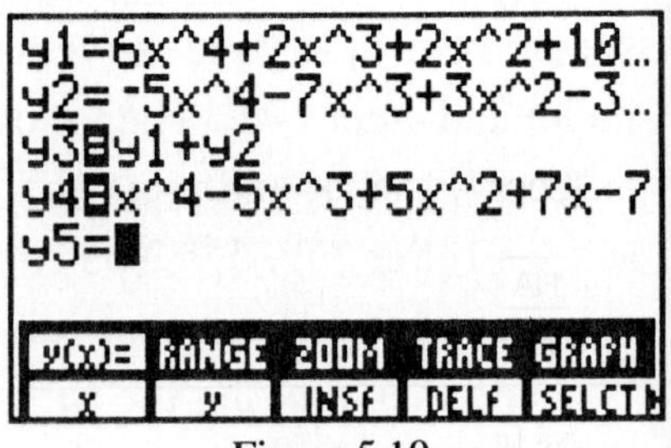

Figure 5.10

Deselect $y1$ by moving the cursor to the same line as $y1$ and pressing [F5:SELCT]. This should unhighlight the '=' symbol next to 'y1'. Deselect $y2$ the same way. See Fig. 5.10. Graph these polynomials in the [–5, 5] by [–10, 50] viewing window. There appears to be just one graph. Reinforce this by pressing [F4:TRACE] and moving the Trace cursor along the graph with the cursor-movement keys. As you move the Trace cursor along the graph, move from the graph of $y3$ to the graph of $y4$ by pressing [▼]. See Fig. 5.11. Since the coordinates don't change when you move from the graph of $y3$ to the graph of $y4$, we have support that the two polynomials are the same.

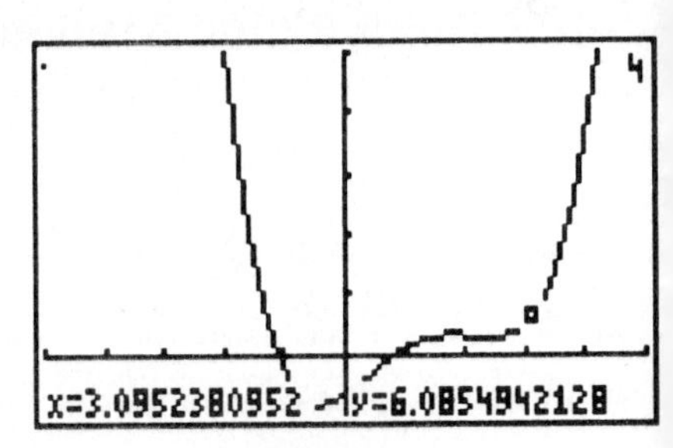

Figure 5.11

Evaluating Polynomials with Several Variables

We will use [STO ▶] to assign values to the variables in the polynomial and then evaluate the polynomial. This method was introduced in Section 1.3.

Example 3: The volume of a tennis ball container is given as the polynomial $\pi x^2 y + (2/3)\pi x^3$, where x and y are dimensions of the container. Find the volume of the container when $x = 3.75$ and $y(x) = 16.25$.

Clear the Home screen and press

3 [.] 75 [STO▶] [X-VAR] [ALPHA] [2nd] [:] 16 [.] 25 [STO▶] [Y] [ALPHA] [2nd] [:] [2nd] [π] [X-VAR] [x²] [ALPHA] [Y] [+] [(] 2 [÷] 3 [)] [2nd] [π] [X-VAR] [^] 3 [ENTER].

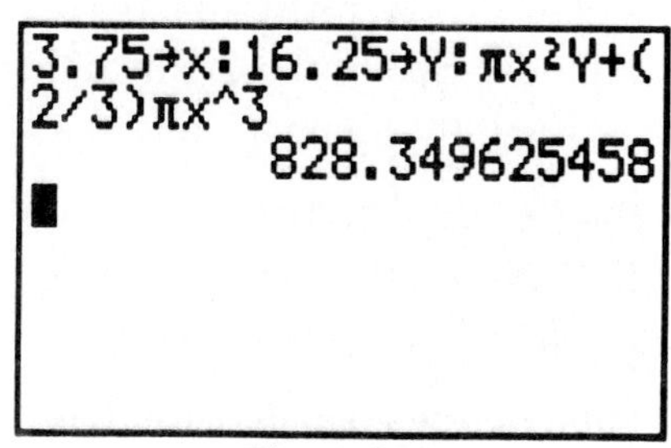

Figure 5.12

See Fig. 5.12. The volume is about 828.35.

The volume can also be found with the Solver. Press

[2nd] [SOLVER] [CLEAR]

to access the Solver edit screen. Then enter the equation by pressing

[ALPHA] [V] [ALPHA] [=] [2nd] [π] [X-VAR] [x²] [ALPHA] [Y] [+] [(] 2 [÷] 3 [)] [2nd] [π] [X-VAR] [^] 3.

Figure 5.13

Press [ENTER] to access the SOLVER menu. Enter 3.75 for ‘x’ and 16.25 for ‘Y’. Then move the cursor beside ‘V=’ and press [F5:SOLVE]. See Fig. 5.13. You could enter new values for x and Y and use [F5:SOLVE] to find the corresponding values for V.

Chapter 6 Rational Expressions and Equations

Section 6.1 Simplifying Rational Expressions

Using Parentheses in Rational Expressions

When evaluating and graphing rational expressions, you must use parentheses to enter the expression correctly.

Example 1: Graph $y = \frac{x-2}{x}$.

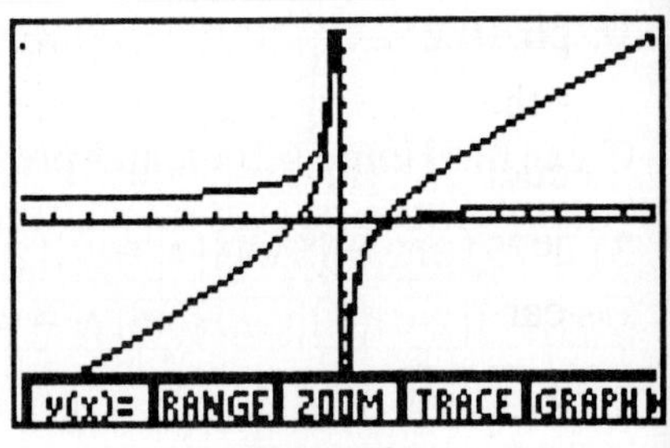

Figure 6.1

Enter $y1 = (x - 2)/x$ and $y2 = x - 2/x$. Graph in the Standard viewing window. See Fig. 6.1. Since there are two different graphs, we can see that $y1$ and $y2$ are not the same. The correct version is $y1$. This is because of the order of operations. Order of operations was discussed in Section 1.1. The parentheses in $y1$ force the subtraction to take place before the division. In $y2$ the division takes place before the subtraction since division has higher precedence than subtraction.

Math Errors and Rational Expressions

From the Home screen, press

[CLEAR] **0** [STO▶] [X-VAR] [ENTER] **1** [÷] [X-VAR] [ENTER].

You should see the same 'DIV BY 0' error that you saw in the Preliminary Chapter. This is because we have instructed the TI-85 to divide by zero. You may encounter this difficulty when evaluating other rational functions. For example, you cannot evaluate $1/(x - 2)$ at $x = 2$ because division by zero is undefined.

Section 6.3 Adding and Subtracting Rational Expressions

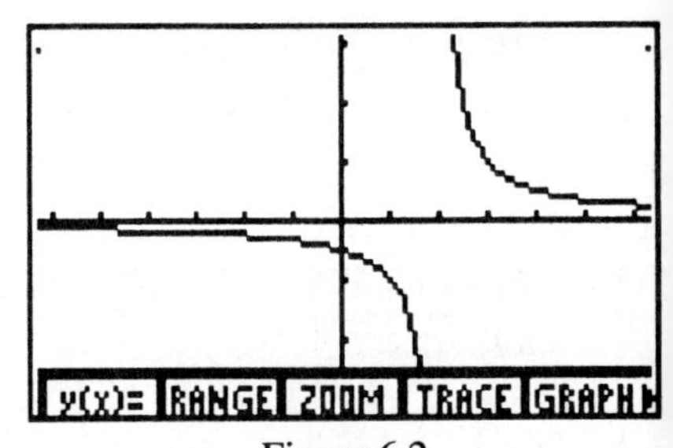

Figure 6.2

Resolution of the Screen and Graphing Rational Expressions

Example 1: Graph $y(x) = 1/(x - 2)$.

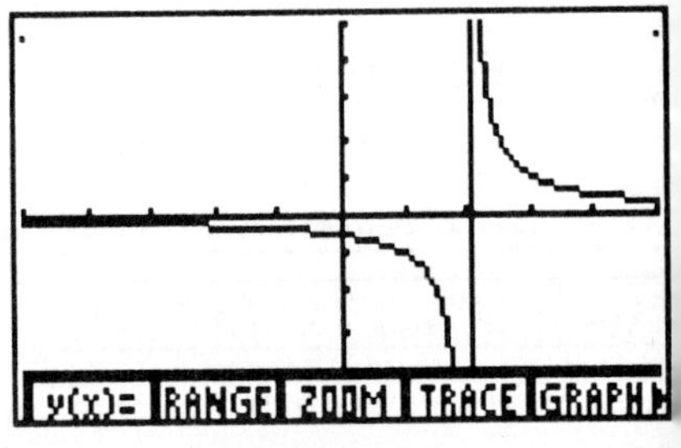

Figure 6.3

Enter $y1 = 1/(x - 2)$ and graph in the Decimal viewing window. See Fig. 6.2. Now regraph in the [–5, 5] by [–5, 5] viewing window. See Fig. 6.3. The vertical line that you see in Fig. 6.3 is not really a part of the graph of $1/(x - 2)$. In some viewing windows, the bottom branch of the graph of a rational expression may be connected to the top branch by a vertical line. This vertical line is not a part of the correct graph. The occasional connecting of the branches has to do with the way the TI-85 graphs functions. It plots some of the points in the function and then connects those points to produce the final graph. In some viewing windows, when the points are connected, a vertical line appears.

Section 6.4 Complex Fractions

Using Parentheses with Complex Fractions

Parentheses are required to enter expressions involving complex fractions. Make sure you have parentheses around the entire numerator and the entire denominator.

Example 1: Evaluate $\dfrac{\frac{1}{2}+\frac{2}{3}}{\frac{5}{6}-\frac{3}{4}}$.

Clear the Home screen, enter (1/2 + 2/3)/(5/6 – 3/4), and press [ENTER]. See Fig. 6.4. The answer is 14. The parentheses force the TI-85 to calculate the numerator and denominator first and then divide.

```
(1/2+2/3)/(5/6-3/4)
                 14
```

Figure 6.4

Chapter 7 Rational Exponents, Radicals, and Complex Numbers

Section 7.1 Roots and Radicals

Evaluating Square Roots of Negative Numbers

From the Home screen press

[CLEAR] [2nd] [√] [(-)] **3** [ENTER].

See Fig. 7.1. The answer looks like an ordered pair. This ordered pair is really a complex number. The TI-85 represents the complex number $0 + 1.73205080757i$ as (0, 1.73205080757). This complex number occurs because the square root of negative three is not a real number. $\sqrt{-3}$ is not real because no real number squared is –3.

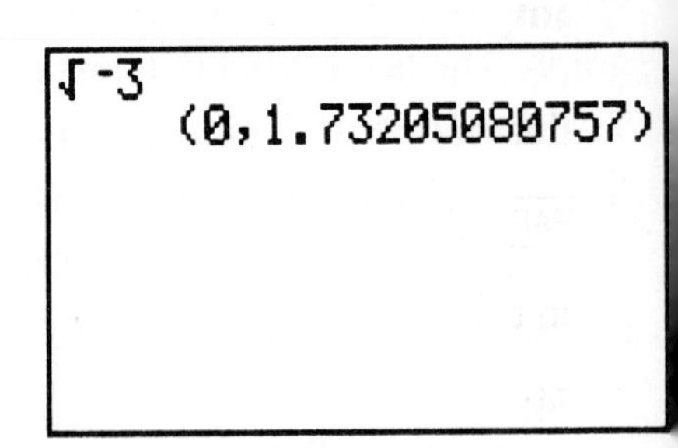

Figure 7.1

Evaluating Expressions with Fractional Exponents

Example 1: Evaluate $9^{1/2}$ and $125^{1/3}$.

Go to the Home screen and press

[CLEAR] **9** [^] [(] **1** [÷] **2** [)] [ENTER] **125** [^] [(] **1** [÷] **3** [)] [ENTER].

See Fig. 7.2. You must place parentheses around the exponent.

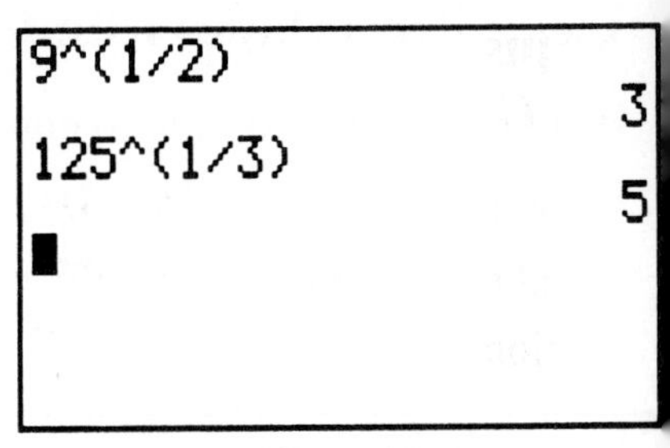

Figure 7.2

Example 2: Compare the graphs of $y = x \wedge (1/4) \wedge 2$ and $y = x \wedge 2 \wedge (1/4)$.

Enter $y1 = x \wedge (1/4) \wedge 2$ and graph in the Decimal viewing window. See Fig. 7.3. Now erase $y1$ and graph $y1 = x \wedge 2 \wedge (1/4)$ in the Decimal viewing window. See Fig. 7.4. Both graphs appear in the first quadrant, but only the second graph also appears in the second quadrant where x is negative. This is because the TI-85 evaluates the expression from left to right. When it evaluates the first expression, it takes the fourth root of x and then squares. The fourth root of a negative number is not real, so the first expression is not defined when x is negative. When the TI-85 evaluates the second expression, it squares and then takes the fourth root. The square of a negative number is positive, so the second expression is defined when x is negative.

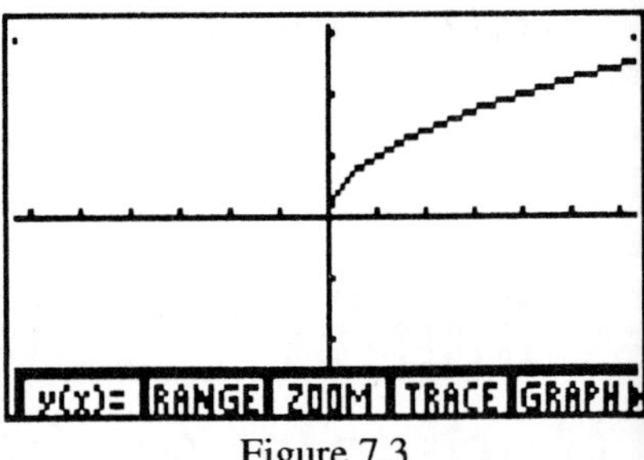

Figure 7.3

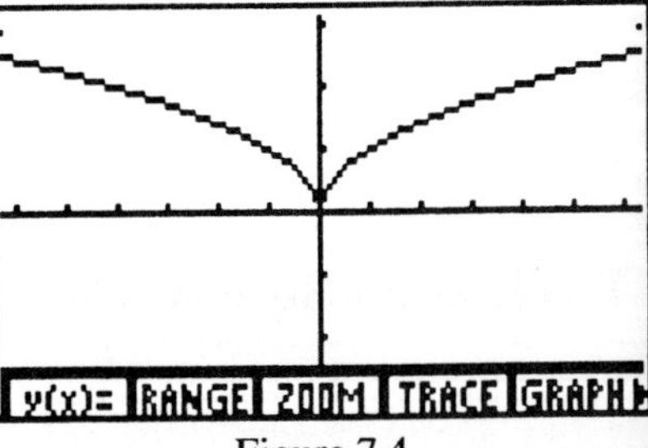

Figure 7.4

Finding Roots with the TI-85

Square roots can be evaluated with [√]. Other roots can be evaluated with the [ˣ√] option in the MATH [F5:MISC] menu.

Example 3: Evaluate $\sqrt[4]{8}$.

Evaluate the fourth root of eight by pressing

CLEAR **4** 2nd MATH [F5:MISC] MORE [F4: $\sqrt[x]{\ }$] **8** ENTER.

Then evaluate the cube root of –8.75 by pressing

3 [F4: $\sqrt[x]{\ }$] (-) **8** . **75** ENTER.

See Fig. 7.5.

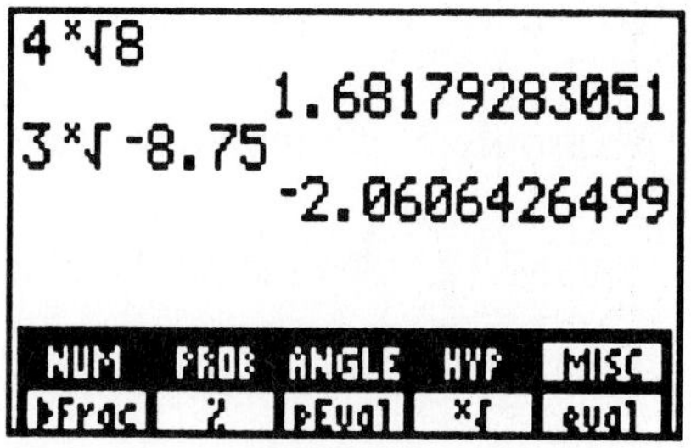

```
4ˣ√8
                1.68179283051
3ˣ√-8.75
               -2.0606426499

NUM  PROB  ANGLE  HYP  MISC
▸Frac  %  pEval  ˣ√  eval
```

Figure 7.5

Adjusting the Number of Decimal Places Displayed by the Grapher

We will adjust the number of decimal places displayed with the Fixed mode option in the Mode screen. This was discussed in Section 1.1.

Example 4: Evaluate $\sqrt[4]{8}$ using a Floating Decimal display setting and then a Fixed Decimal display setting of two digits.

Select Floating Decimal mode, and enter 8 ^ (1/4) on the Home screen. Next, select Fixed Decimal mode by highlighting the '2' on the second line of the Mode screen. Enter 8 ^ (1/4) on the Home screen. See Fig. 7.6. In Fixed Decimal (2) mode, only two digits to the right of the decimal are displayed.

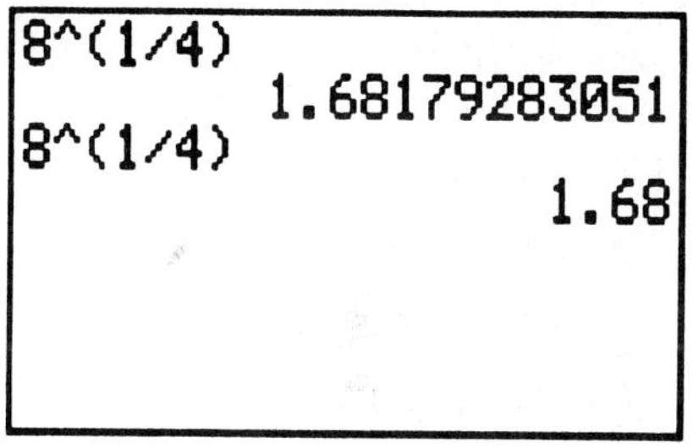

```
8^(1/4)
                1.68179283051
8^(1/4)
                         1.68
```

Figure 7.6

Section 7.6 Complex Numbers

Using the Grapher to Work with Complex Numbers.

We mentioned in Section 7.1 that complex numbers are represented by ordered pairs on the TI-85.

Example 1: Compute $(1 - i)(8 + 7i)$.

Select Floating Decimal mode and enter (1, –1)(8, 7) on the Home screen. See Fig. 7.7. The TI-85 result of (15, –1) means the answer is 15 – *i*.

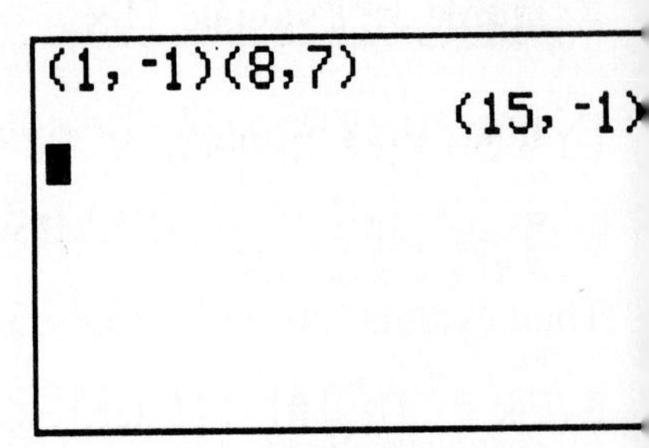

Figure 7.7

Example 2: Compute 3 ÷ (2 + 3*i*).

Enter 3/(2, 3)▶Frac on the Home screen by pressing

3 [÷] [(] **2** [,] **3** [)] [2nd] [MATH] [F5:MISC] [MORE] [F1:▶FRAC] [ENTER].

See Fig. 7.8. The answer is $^6/_{13}$ + $^9/_{13}i$.

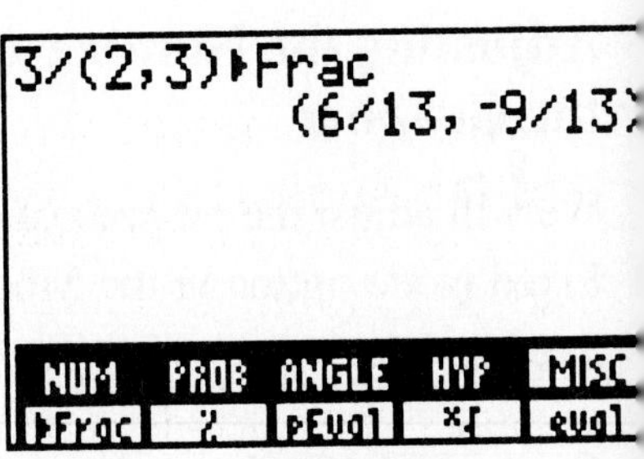

Figure 7.8

Chapter 8 Quadratic Functions

Section 8.4 Sketching Graphs of Quadratic Functions

See Section 2.1 for a discussion of the 0.2 window.

Section 8.6 Vertex and Symmetry of a Parabola

See Section 1.1 for a discussion of the FRAC command.

Chapter 9 Higher Order Systems of Equations and Matrices

Section 9.1 Solving Systems of Equations in Three Variables

Equations in Three Variables that Cannot Be Graphed and the Simultaneous Equation Solver

Equations in three variables cannot be graphed on the TI-85. The only support the TI-85 can provide for a solution to a system of equations in three variables is numerical. However the **Simultaneous equation solver** on the TI-85 can be used to find the numerical solution to a system of three equations in three variables.

Example 1: Solve the following system of equations:

$$\begin{cases} x + y + z = 13 \\ 3x + 2y + 2z = 28 \\ x - y + z = 6 \end{cases}.$$

Press

[2nd] [SIMULT].

Press **3** for the number of equations (see Fig. 9.1), then press [ENTER]. Enter the coefficients of the first equation by pressing

1 [ENTER] **1** [ENTER] **1** [ENTER] **13**.

See Fig. 9.2. Now press [ENTER] to bring up the screen to input the

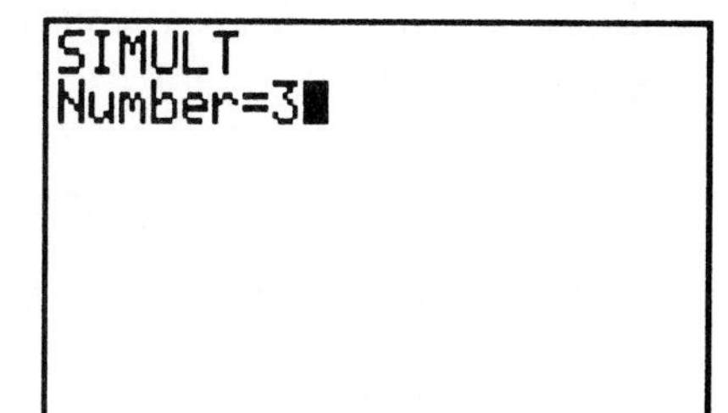

Figure 9.1

coefficients for the second equation. Press

3 [ENTER] **2** [ENTER] **2** [ENTER] **28** [ENTER].

This should bring up the screen for the last equation. Press

1 [ENTER] [(-)] **1** [ENTER] **1** [ENTER] **6**.

After you have entered the coefficients for all three equations, press [F5:SOLVE] to solve the system. See Fig. 9.3. The solution is $x = 2$, $y = 3.5$, $z = 7.5$.

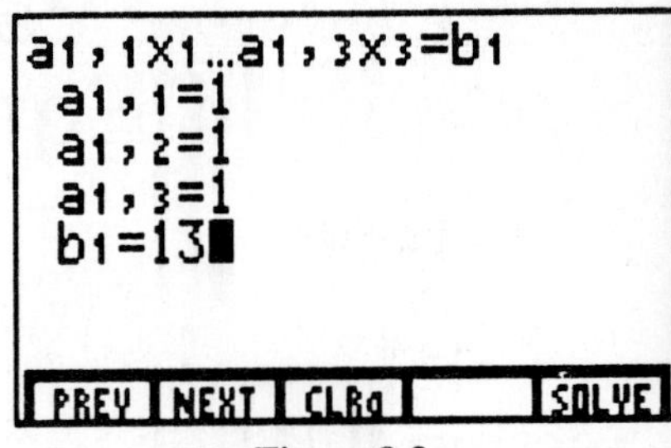

Figure 9.2

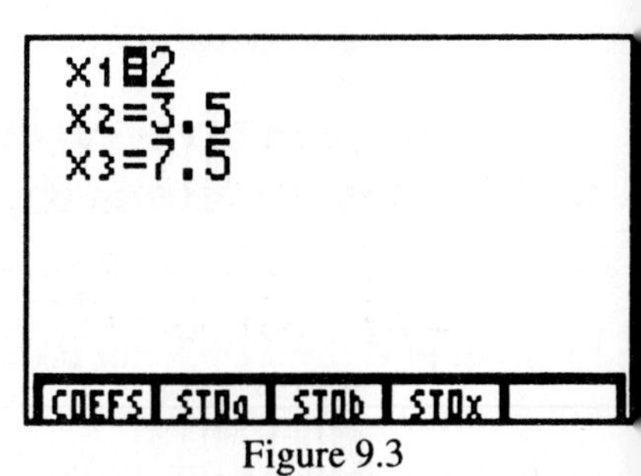

Figure 9.3

Section 9.2 Matrix Algebra

Entering a Matrix on the TI-85

The matrix key is the second function of the **7** key. Press

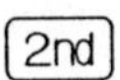

to see the **MATRIX menu**. See Fig. 9.4.

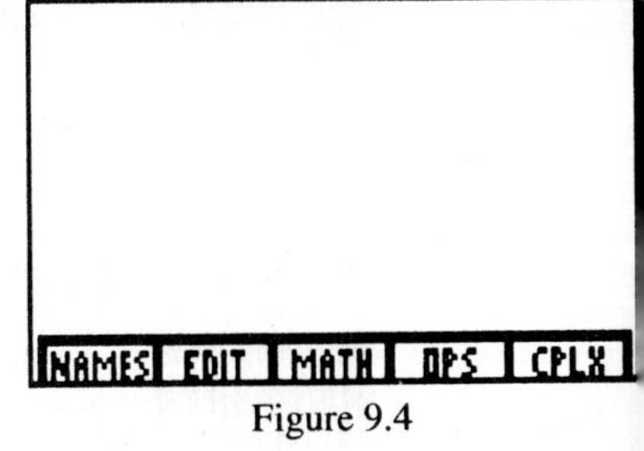

Figure 9.4

Example 1: Enter the 2×2 matrix $A = \begin{bmatrix} 4 & -3 \\ 2 & 6 \end{bmatrix}$ and the 2×2 matrix $B = \begin{bmatrix} 2 & 5 \\ 1 & 2 \end{bmatrix}$.

From the MATRIX menu, press [F2:EDIT]. The TI-85 asks for the name of the matrix. Notice the cursor indicates that the TI-85 is in the Alpha mode. Enter the name 'A' by pressing

[LOG] [ENTER].

Notice [LOG] has 'A' above it. Now we change the dimension of matrix $[A]$ to two rows and two columns by pressing

2 [ENTER] **2** [ENTER].

See Fig. 9.5. Enter the values in the matrix by pressing

4 [ENTER] [(-)] **3** [ENTER] **2** [ENTER] **6**.

Figure 9.5

See Fig. 9.6. Notice you enter the values of the matrix one row at a time even though the matrix edit screen does not show all the elements at the same time. Now create matrix [*B*] by pressing

[2nd] [F2:EDIT] [B] [ENTER]

([B] is the second function of [SIN]). Designate the dimension of matrix *B* by pressing

2 [ENTER] **2** [ENTER],

and then input the elements of the matrix by pressing

2 [ENTER] **5** [ENTER] **1** [ENTER] **2** [ENTER].

See Fig. 9.7. Press

[2nd] [QUIT]

to return to the Home screen. You can print the values in matrix *A* and matrix *B* by pressing

[CLEAR] [ALPHA] [A] [ENTER] [ALPHA] [B] [ENTER].

See Fig. 9.8.

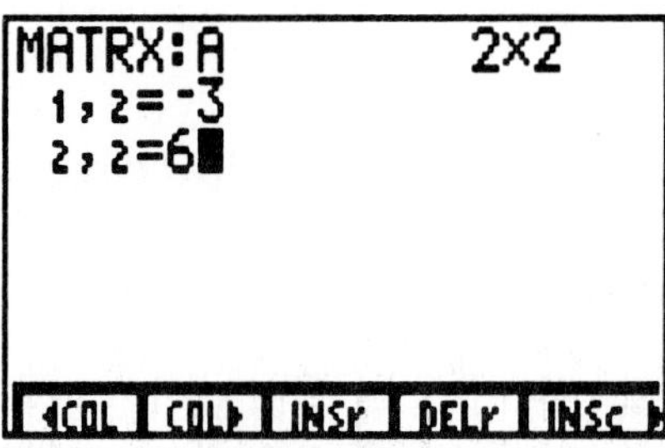

Figure 9.6

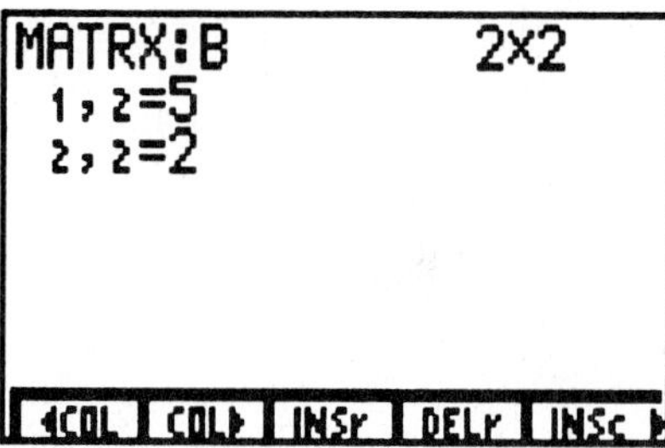

Figure 9.7

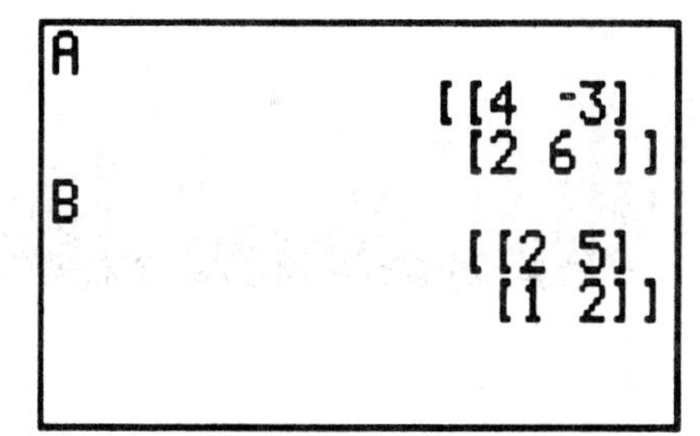

Figure 9.8

Adding and Subtracting Matrices

Example 2: Find $A + B$ and $A - B$, where A and B have the values assigned in Example 1.

From the Home screen press

[CLEAR] [ALPHA] [A] [+] [ALPHA] [B] [ENTER]

to see the matrix sum. Press

[ALPHA] [A] [−] [ALPHA] [B] [ENTER]

to see the difference. See Fig. 9.9.

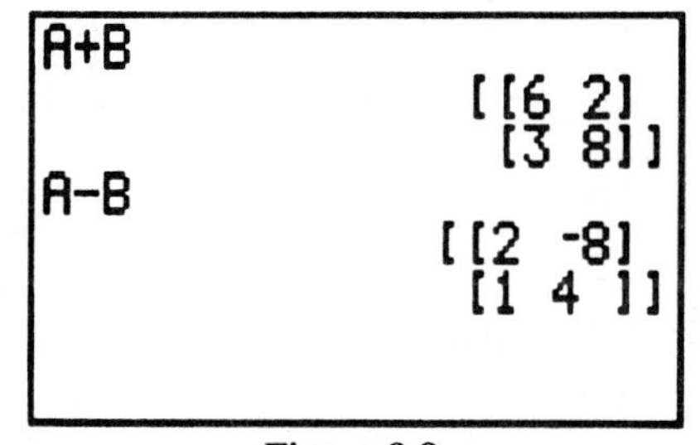

Figure 9.9

Scalar Multiplication of Matrices

Example 3: Find $3[A]$ where $[A] = \begin{bmatrix} 2 & 1 & -5 \\ 6 & -1 & 3 \end{bmatrix}$.

Enter matrix A by pressing

[2nd] [F2:EDIT] [A] [ENTER] **2** [ENTER] **3** [ENTER] **2** [ENTER] **1** [ENTER] **5** [ENTER] **6** [ENTER] [(-)] **1** [ENTER] **3** [ENTER].

See Fig. 9.10. Press

Figure 9.10

[2nd] [QUIT]

to return to the Home screen. Now perform the scalar multiplication by pressing

[CLEAR] **3** [ALPHA] [A] [ENTER].

See Fig. 9.11.

Figure 9.11

Section 9.3 Multiplication of Matrices

Once matrices have been entered on the TI-85, you can multiply them on the Home screen the same way you would any other variables.

Example 1: If $A = \begin{bmatrix} 3 & -4 \\ 2 & -8 \end{bmatrix}$ and $B = \begin{bmatrix} 4 & -2 \\ -1 & 9 \end{bmatrix}$, find $A \times B$.

Enter the matrices A and B with the Matrix edit screen. (Entering matrices was explained in Section 9.2.) Then from the Home screen press

[CLEAR] [ALPHA] [A] [×] [ALPHA] [B] [ENTER].

See Fig. 9.12.

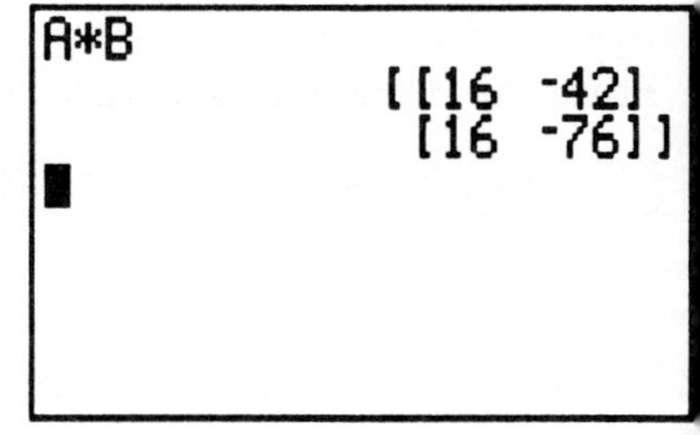

Figure 9.12

The Identity Matrix on the Grapher

Identity matrices of various dimensions can be easily created with the third option in the MATRX [F4:OPS] menu. To create a 3×3

identity matrix on the Home screen, press

[2nd] [F4:OPS] [F3:ident] **3** [ENTER].

See Fig. 9.13.

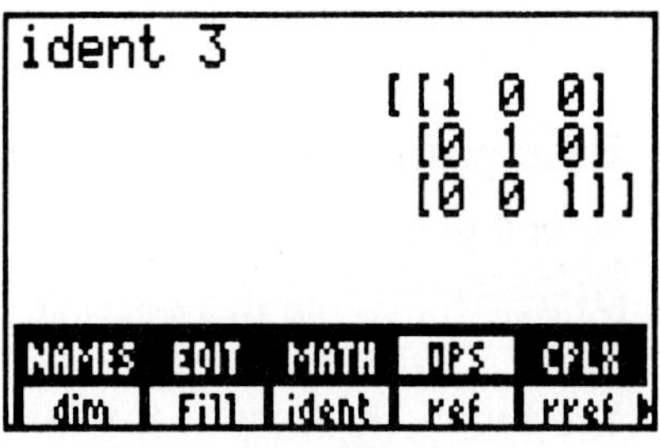

Figure 9.13

Inverse Matrices Using the [x^{-1}] Key

If a matrix can be inverted, you can usually find its inverse with [x^{-1}]. If the inverse doesn't exist, you will see an error message or an odd-looking matrix with entries that are very near zero.

Example 2: Find the inverse of $A = \begin{bmatrix} 5 & 1 \\ 4 & 1 \end{bmatrix}$.

Enter matrix *A* with the Matrix edit screen, and then from the Home screen press

[CLEAR] [ALPHA] [A] [2nd] [x^{-1}] [ENTER].

See Fig. 9.14.

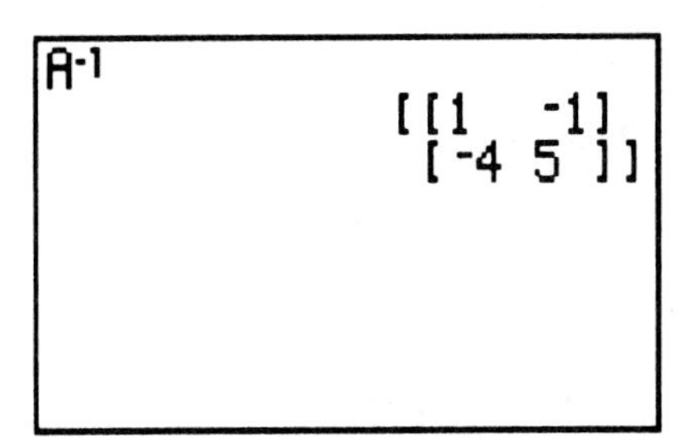

Figure 9.14

Example 3: Find the inverse of $A = \begin{bmatrix} 3 & 1 \\ 3 & 1 \end{bmatrix}$.

Enter matrix *A* with the Matrix edit screen, and from the Home screen press

[CLEAR] [ALPHA] [A] [2nd] [x^{-1}] [ENTER].

See Fig. 9.15. You can see the other entries by pressing [▶]. Notice all the entries are approximately zero. This matrix is incorrect since the inverse of *A* does not exist. If all the entries are very near zero when you attempt to find the inverse of a matrix, the inverse likely does not exist.

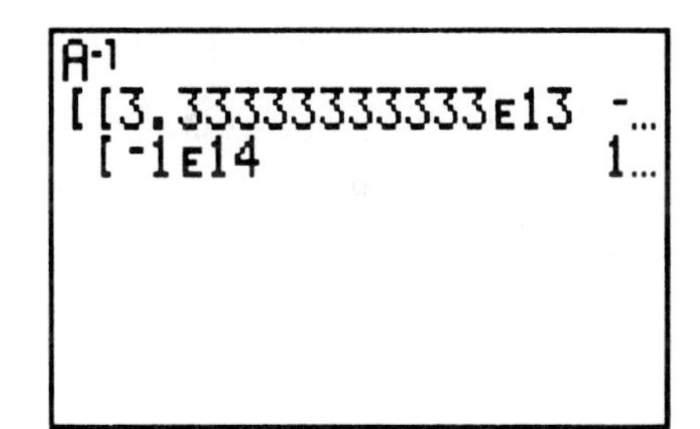

Figure 9.15

Determinants of a Square Matrix

The option [F1:det] (determinant) is the first option in the MATRX [F3:MATH] menu. This feature can be used to find determinants of square matrices.

Example 4: Find the determinant of $A = \begin{bmatrix} 3 & 1 \\ 3 & 1 \end{bmatrix}$.

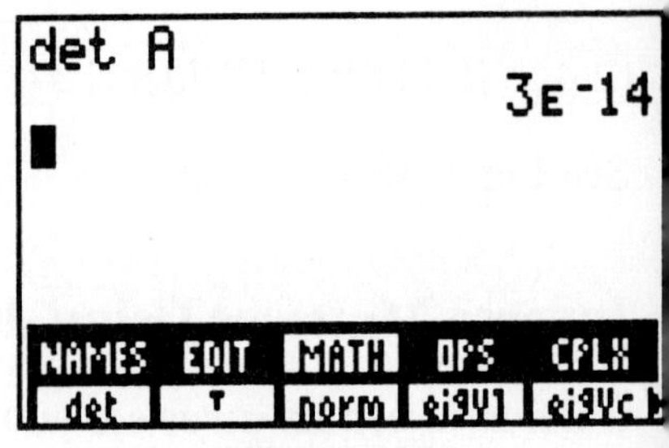

Figure 9.16

Enter matrix A with the Matrix edit screen. Then from the Home screen press

CLEAR 2nd [F3:MATH] [F1:det] ALPHA A ENTER.

See Fig. 9.16. A value this small means the determinant is probably zero. You can check with paper and pencil methods in this case to see that the exact determinant of A is zero.

Section 9.4 Solving a System of Equations Using Matrices

Solving a System of Equations with Matrices on the Grapher

Systems of equations can be expressed with matrices. When this is done, the solution of the system is a matrix. We can solve for this solution matrix with the inverse matrix feature of the TI-85.

Example 1: Solve the system $\begin{cases} 4x - 3y = -26 \\ 15x + 11y = 36 \end{cases}$.

This system can be expressed in matrix form as $AX = B$, where

$$A = \begin{bmatrix} 4 & -3 \\ 15 & 11 \end{bmatrix}, \quad X = \begin{bmatrix} x \\ y \end{bmatrix}, \quad \text{and} \quad B = \begin{bmatrix} -26 \\ 36 \end{bmatrix}.$$

We then solve the matrix equation $AX = B$ for X, obtaining $X = A^{-1}B$. Matrix X contains the solution to the original system of equations. Enter the values just described for the 2×2 matrix A and the 2×1 matrix B with the Matrix edit screen. Don't forget to change the dimension of B to two rows, one column. After you have entered A and B, return to the Home screen and find the product $A^{-1}B$ by pressing

CLEAR ALPHA A 2nd x^{-1} × ALPHA B ENTER.

See Fig. 9.17. The solution is $x = -2$, $y = 6$. Matrix multiplication is not commutative, so you must enter $A^{-1}B$, not BA^{-1}.

```
A-1*B
                [[-2]
                 [6 ]]
```

Figure 9.17

Chapter 10 Conic Sections

Conic Sections

In general, when we graph conic sections on the TI-85, we must rewrite the conic equation as $y = ...$ and then graph. While solving a conic equation for y, we usually obtain two new equations. A lengthy radical will probably be created in both equations. We will enter this radical into the $y1$ slot and enter the two equations in $y2$ and $y3$ with $y1$ used to represent the radical. Then we will deselect $y1$ and graph $y2$ and $y3$. See Section 3.1 to review how to select and deselect graphs. There may be gaps between the graphs of $y2$ and $y3$ because of screen resolution.

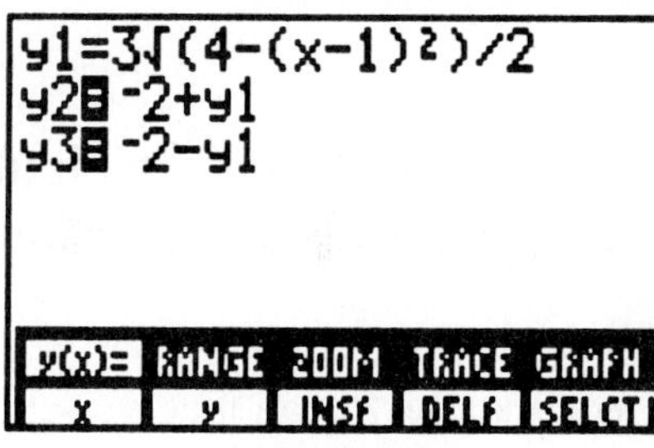

Figure 10.1

Example 1: Graph the ellipse $(x - 1)^2/4 + (y + 2)^2/9 = 1$.

We solve for y and obtain $y = -2 \pm 3\sqrt{(4-(x-1)^2)}/2$. Enter $y1 = 3\sqrt{(4-(x-1)^2)}/2$, $y2 = -2 + y1$, and $y3 = -2 - y1$. Deselect $y1$. See Fig. 10.1. Now graph in the Standard viewing window. See Fig. 10.2.

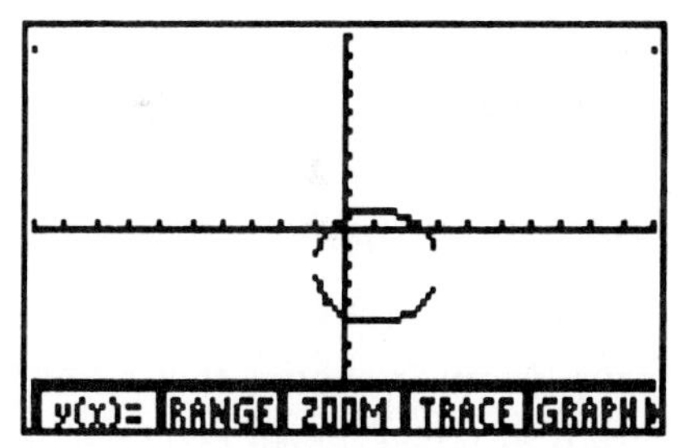

Figure 10.2

Section 10.1 Circles

Graph in the Square Window for a Circle

A circle must be graphed in a square window for the true shape of the circle to appear. In the Standard window, the circle will look distorted.

Example 1: Graph the circle $x^2 + y^2 = 25$.

Enter $y1 = \sqrt{(25 - x\wedge 2)}$ and $y2 = -y1$. Graph in the Standard window. See Fig. 10.3. The circle looks like an ellipse. Now press [F3:ZOOM] MORE [F2:ZSQR]. See Fig. 10.4. The proportions of the circle are correct in a square window.

Figure 10.3

Section 10.5 Nonlinear Systems of Equations

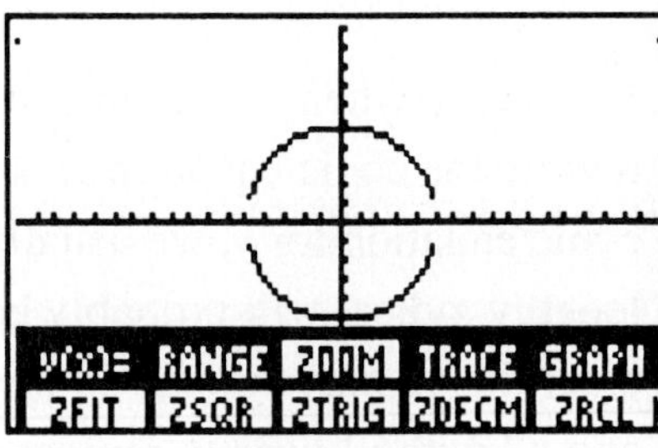

Figure 10.4

A Review of Solving Nonlinear Systems of Equations Graphically

See Chapter 3 for a review of how to solve systems of equations graphically. Recall that we set the scale marks to 0.01 to assist with a solution with at most 0.01 error.

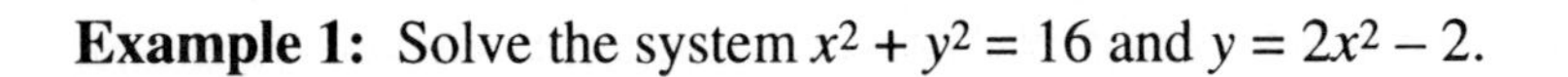

Example 1: Solve the system $x^2 + y^2 = 16$ and $y = 2x^2 - 2$.

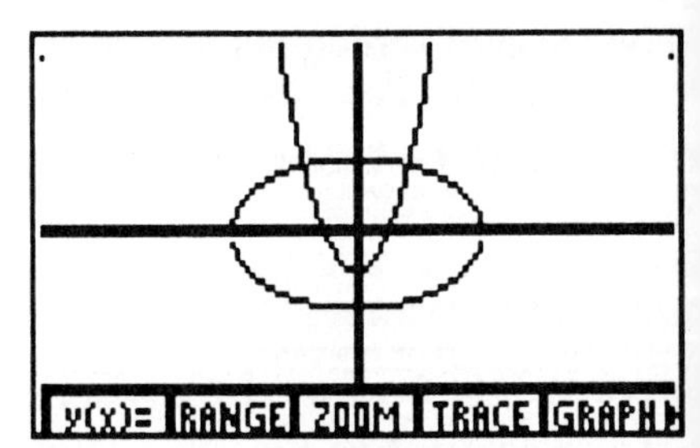

Figure 10.5

We must solve the first equation for y before we can solve graphically. This gives $y = \pm\sqrt{(16 - x^2)}$. Enter $y1 = \sqrt{(16 - x\wedge 2)}$ and $y2 = -y1$. Enter $y3 = 2x \wedge 2 - 2$, and graph all three in the [–10, 10] by [–10, 10] window with scale values of 0.01. See Fig. 10.5. Before we Zoom In, notice that $y2$ has no intersection points with $y3$, so we will deselect $y2$ for faster graphing. Now we Zoom In several times on the intersection in the first quadrant. See Fig. 10.6. Since the intersection is contained between the scale marks, we know that the error is at most 0.01. The other solution can be found by regraphing in the original window and then Zooming In on the intersection in the second quadrant.

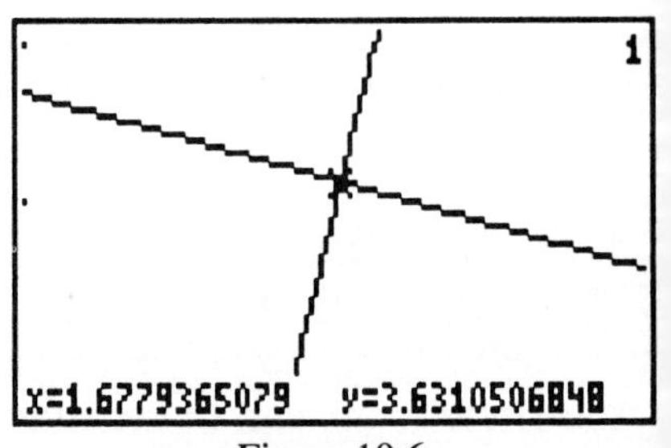

Figure 10.6

Chapter 11 Exponential and Logarithmic Functions

Section 11.3 Logarithmic Functions

The [LOG] and [LN] Keys

The base 10 log key [LOG] is on the fifth row, first column. The natural log key [LN] is just below the base 10 log key.

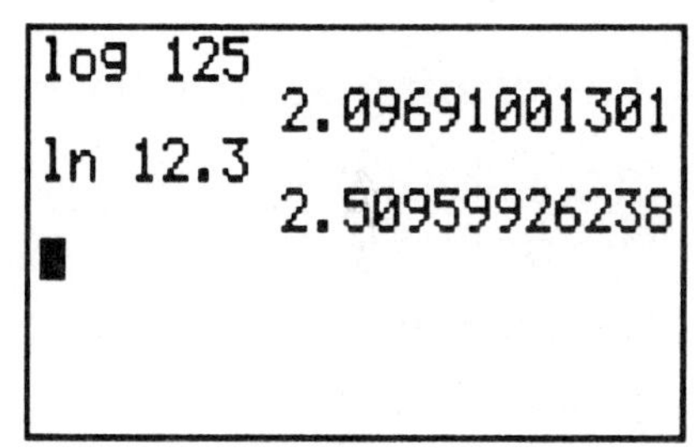

Figure 11.1

Example 1: Evaluate log 125 and ln 12.3.

From the Home screen press

[CLEAR] [LOG] **125** [ENTER] [LN] **12** [.] **3** [ENTER].

See Fig. 11.1.

Section 11.5 Solving Logarithmic Equations

Some Graphers Cannot Evaluate $y1$ at 3 by Using the Notation $y1(3)$

The TI-85 cannot evaluate $y1$ at 3 by using the notation $y1(3)$. You can, however, use the evalF function. To evaluate $y1$ at 3, enter evalF($y1$, x, 3). The function evalF is found in the CALC menu as well as the Catalog. See Section 5.2 to review how to evaluate an expression on the TI-85.

Revisiting the Grapher-Generated Table

Tables are not a built-in feature of the TI-85.

Chapter 12 Sequences and Series and the Binomial Theorem

Section 12.1 Arithmetic Sequences and Series

Generating Sequences with the seq(Command

It is possible to generate sequences with a **seq(** command on the TI-85. This command has five parameters. They are (1) the formula for the *n*th term of the sequence, (2) the index, (3) the initial and (4) final values of the index, and (5) the amount by which the index will increase.

Example 1: Generate the sequence 3,5,7,9,...,$(2n + 1)$,....

From the Home screen press

[CLEAR] [2nd] [L] [F5:OPS] [MORE] [F3:seq] **2** [ALPHA] [N] [+] **1** [,] [ALPHA] [N] [,] **1** [,] **10** [,] **1** [)] [ENTER].

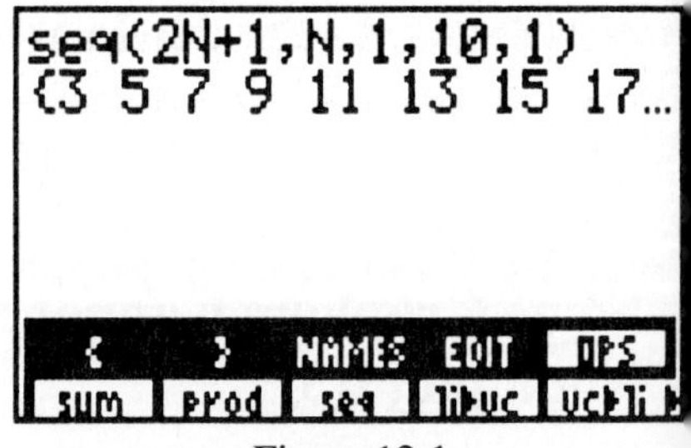

Figure 12.1

See Fig. 12.1. The first 10 terms of the sequence are generated. You can see the last terms in the sequence by pressing the right cursor-movement key.

Displaying a Graph of a Sequence

Sequences can be graphed like functions if we use the Dot mode and graph in the Integer window [0, 126] by [0, 62].

Example 2: Display the graph of the sequence 23,26,29,..., and determine the value of the sixth term.

This sequence is given by the formula 23 + (x – 1)3. Enter this formula in $y1$, change from DrawLine to DrawDot mode in the GRAPH [F3:FORMT] menu, and graph in the [0, 126] by [0, 62] window. Use x and y scales of 10. Now trace until $x = 6$. See Fig. 12.2. The sixth term is 38.

Figure 12.2

Section 12.3 Binomial Expansion

Factorials

The Factorial function is the first option in the MATH [Prob] menu.

Example 1: Evaluate 5! and 8!5!/(4!3!).

From the Home screen, press

[CLEAR] **5** [2nd] [MATH] [F2:PROB] [F1:!] [ENTER]

to evaluate the first expression, and then

8 [F1:!] **5** [F1:!] [÷] [(] **4** [F1:!] **3** [F1:!] [)] [ENTER]

to evaluate the second expression. See Fig. 12.3.

Figure 12.3

Evaluating a Binomial Coefficient on the Grapher

$\binom{n}{r}$ means the same thing as ${}_nC_r$. ${}_nC_r$ is notation for a combination of n things taken r at a time. This option is found in the MATH [F2:PROB] (probability) menu.

Example 2: Determine the value of the binomial coefficient $\binom{5}{3}$. $\binom{5}{3} = {}_5C_3$.

From the Home screen press

[CLEAR] **5** [2nd] [MATH] [F2:PROB] [F3:${}_nC_r$] **3** [ENTER].

See Fig. 12.4. The result is 10.

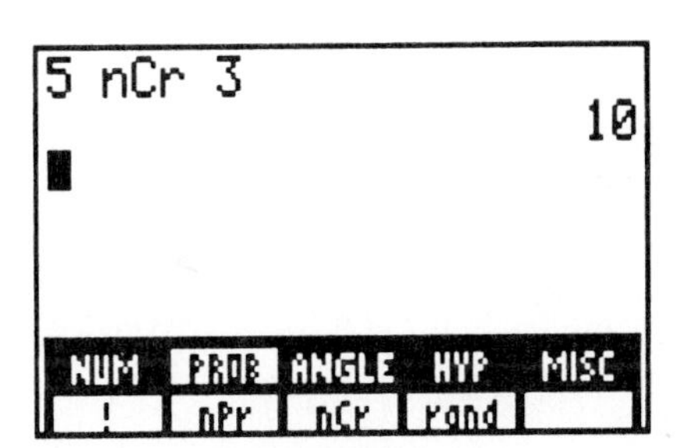

Figure 12.4

Section 12.4 Counting Principle, Permutations, and Combinations

Evaluating a Permutation $_nP_r$

This option is also found in the MATH [Prob] menu. The notation $_nP_r$ means a permutation of n things taken r at a time.

Example 1: Evaluate $_{26}P_6$.

From the Home screen, press

CLEAR **26** 2nd MATH [F2:PROB] [F2:$_n$P$_r$] **6** ENTER.

See Fig. 12.5

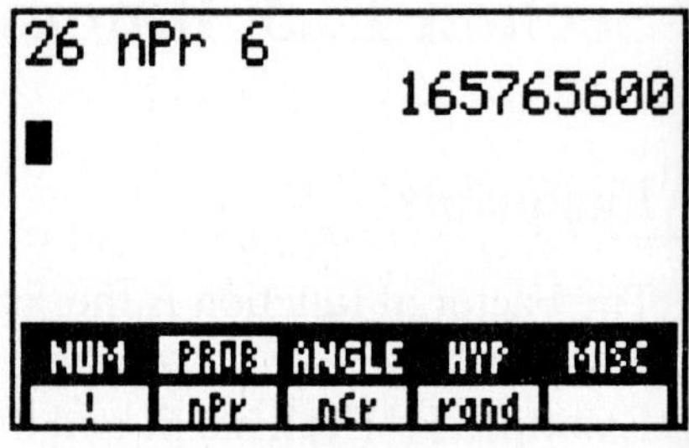

Figure 12.5

Evaluating a Combination $_nC_r$

This feature was described in Section 12.3.

Example 2: Evaluate $_4C_3$.

Press

4 2nd MATH [F2:PROB] [F3:$_n$C$_r$] **3** ENTER.

The answer is 4. See Fig. 12.6.

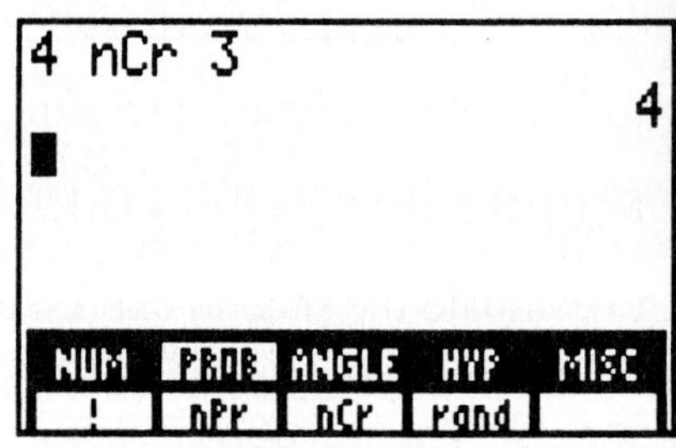

Figure 12.6

Preliminary Chapter Getting Started Using the Casio fx-7700G, fx-6300G, and fx-7000G

This material is written for the Casio 7700G family (7700G, 7700GB, 7800G), which all have the same keyboard, with notes about the 6300G and 7000G when needed. Both the 6300G and the 7000G (or 8000G) have all of their options displayed on the keyboard; there are no menu screens. The graphic illustrations are for all three calculators, and the text screens are for the Casio 7700. The screen on the Casio 6300 is smaller than that on the other calculators, but the graphic illustrations look the same. The text on the 6300 is written in a continuous line on the bottom of the screen. Use [◄] to see what has been written before. The text is the same on the 7000 without the menus and with the symbol differences described in the notes.

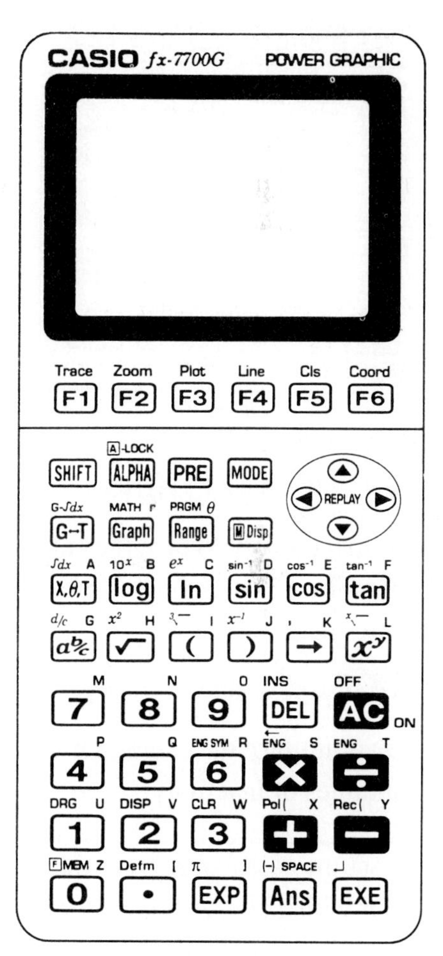

Figure 1

Section 1 The Keyboard

The keyboards for the different models are shown in Figs. 1, 2, and 3. The keyboard is color coded. The yellow [SHIFT] in the first row, first column accesses the operations written in yellow (or green) above the keys. The red [ALPHA] accesses the alphabetic symbols written in red above the keys. The keys in the top rows are black and are operation keys. The keys in the bottom rows are gray and are the arithmetic keys.

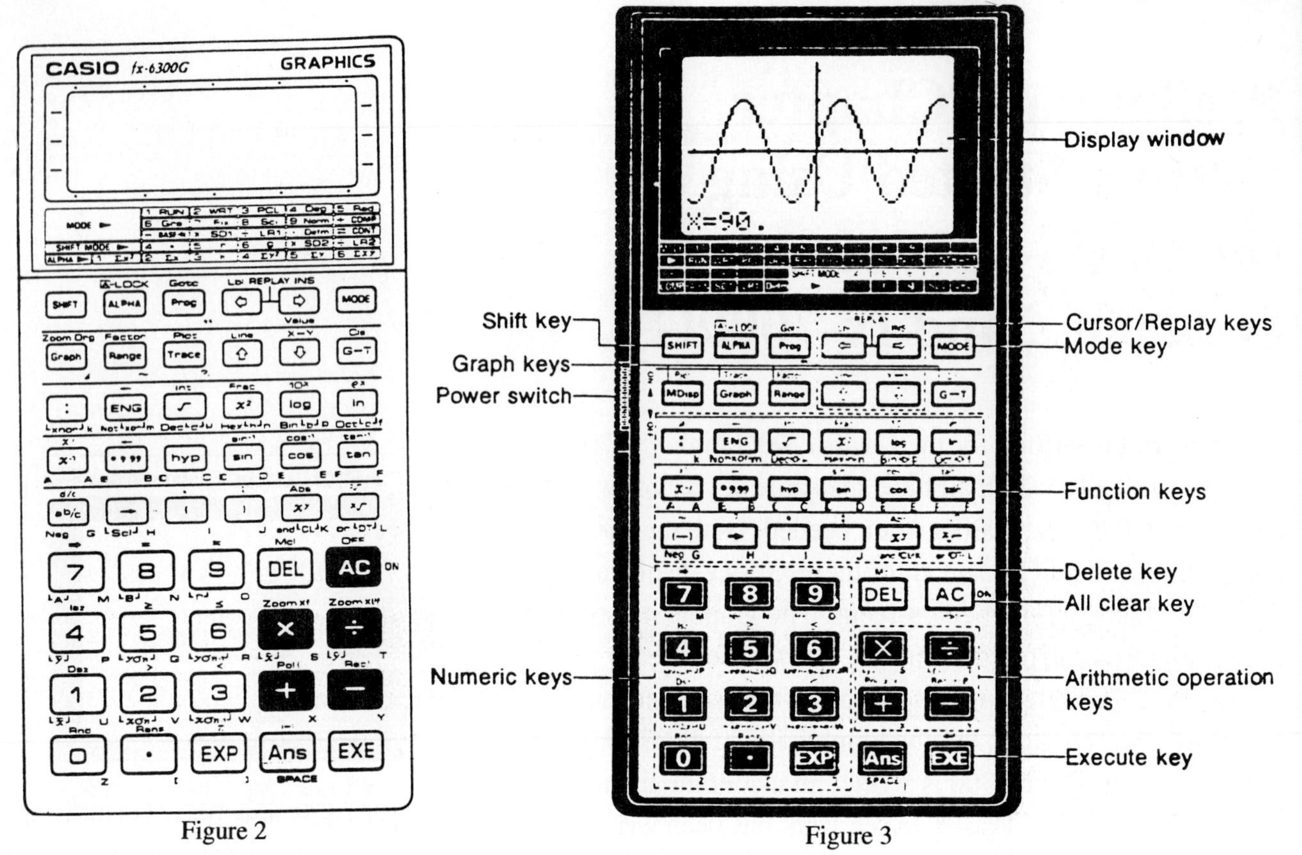

Figure 2

Figure 3

The [ON] and [OFF] Keys

Press the [AC] key to turn the calculator on. This key is in the right column, first row of the bottom gray set of keys. To turn the calculator off, press

[SHIFT] [AC].

If you don't turn the calculator off, it will turn off automatically after six minutes of inactivity.

To reset the calculator, press the reset button on the back of the calculator with a pen. All memories are erased and all of the settings are set at their default values.

NOTE for the 7000: This calculator has an **ON-OFF** switch on the left side of the calculator. It must be in the "on" position to use the calculator, and the "off" position is the only way to turn the calculator off. The automatic turnoff is activated after six minutes of inactivity. Press [AC] to reactivate after the automatic turnoff.

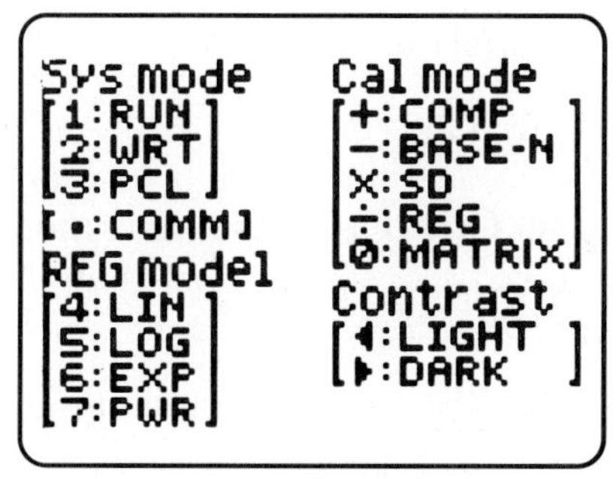

Figure 4

To access a menu, press [SHIFT] and the menu name, written in green or yellow with a green dot. A menu appears on the bottom of the screen. The menu option is accessed with the green keys directly under the viewing window. Press [PRE] to erase a menu.

Stat data
[1:STO]
[2:NON-]
Stat graph
[3:DRAW]
[4:NON-]
Draw type
[5:CONNECT]
[6:PLOT]
Graph type
[+:REC]
[−:POL]
[×:PARAM]
[÷:INEQ]

Figure 5

NOTE for the 6300 and 7000: These models are completely mode driven. There are no menus. The mode options are listed directly below the screen. The keys that are operational in the current mode are color coded with the mode options.

Adjusting the Contrast

To change the contrast press [MODE], then the left arrow key, [◄], to lighten or the right arrow key, [►], to darken the screen. Press [AC] to leave the Mode screen.

RUN / COMP
G-type : REC/CON
angle : Rad
display:Nrm2

Figure 6

MODES

A Casio is a mode-driven calculator. To get the results in this manual, the calculator must be set in the correct mode. There are two mode menus. Press the [MODE] key (Fig. 4) and the [SHIFT] [MODE] keys (Fig. 5). If the Mode screen is displayed (Fig. 4), press [SHIFT] to display Fig. 5.

Pressing [AC] deletes either mode menu. To do the examples in the beginning of this manual, turn the calculator [ON]. You should see the Mode display screen with the given settings (Fig. 6). If you do not, press the following keys to correct the screen.

To correct the first line press

[MODE] **1**

to get RUN. Press

[MODE] [+]

to set COMP (Computation mode). Press

[SHIFT] [MODE] [+]

to get REC (rectangular coordinates). Press

[SHIFT] [MODE] **5**

to set CON (connect plotted points). Press

[SHIFT] [DRG] [F2:Rad] [EXE]

to set Rad (radian measure of angle). Press

[SHIFT] [DISP] [F3:Nrm] [EXE]

to set Nrm 2 (Decimal mode).

You may check on the Mode menu at any time by pressing and holding down the [M Disp] key. Do not change these settings unless you are given directions in other sections. The new Mode display screen will be shown.

NOTE for the 7000: There is only one Mode screen. It should read as follows:

Sys Mode	:	RUN
Cal Mode	:	COMP
Angle	:	RAD
Display	:	Norm
Step	:	0

To correct Sys mode, press

[MODE] **1**.

To correct Cal mode, press

[MODE] [+].

To correct Angle, press

[MODE] **5** [EXE].

To correct Display, press

[MODE] **9** [EXE].

You may check on the Mode menu at any time by pressing and holding down the [M Disp] key. Do not change these settings unless you are given directions to do so in other sections.

NOTE for the 6300: There is no Mode screen. The current mode settings are shown in the upper right portion of the screen. Only the **R** on the second line should be highlighted. See your manual for setting modes. Do not change these settings (MODE 1, MODE +, MODE 6, MODE 9) unless you are given directions to do so in other sections.

Section 2 Screens

The System Mode Display

This screen is displayed when the calculator is first turned on. It shows the current mode settings. Pressing [M Disp] also displays this screen. Pressing any key erases this screen. Start using this manual with the mode settings described in Section 1 of this chapter.

The Text Screen

The Text screen is used to do calculations and to enter data. The [G↔T] key toggles between the Text screen and the Graph screen. Pressing [AC] returns to the Text screen. If a menu is displayed on the bottom of the screen, pressing [PRE] will erase it. All menus are displayed at the bottom of the Text screen.

NOTE for the 6300 and 7000: There are no menus at the bottom of the screen.

The Graph Screen

This screen displays graphs. It is displayed when the Graph option is executed or with the [G↔T] key. Pressing [G↔T] toggles between the Graph and Text screens.

The Range Parameter Screen

This screen displays the current settings for the range parameters of a graph. Press [RANGE], and the Rectangular Coordinate screen is displayed. Press [RANGE] again, and the Parametric-Polar screen is displayed. Press [RANGE] again to return to the Text screen.

NOTE for the 6300 and 7000: These calculators have only one Range Parameter screen as parametric-polar plotting is not a built-in option. When using these calculators, only press [RANGE] once, not twice.

The Display Memory

Press

[SHIFT] [Defm] [EXE].

The distribution of the 4,164 memory steps is shown. On a Casio, you may move memory to the mode where it is needed. Pressing [AC] returns to the Text screen.

NOTE for the 6300 and 7000: To access this screen press

[MODE] [.] [EXE].

There are only 422 memory steps.

Section 3 The Cursor

The **cursor** is a blinking line under the position where the next symbol entered will be placed on the screen. The arrow keys move the cursor on the display.

Press [SHIFT] and the cursor changes to a blinking **S**. Press [ALPHA] and the cursor changes to a blinking 'A.' When the cursor is positioned on a symbol, pressing

[SHIFT] [INS]

changes the cursor to an open blinking rectangle.

NOTE for the 6300: The cursor remains a blinking line. On the mode portion of the screen, an 'S' is highlighted when [SHIFT] is pressed, an 'M' is highlighted when [MODE] is pressed, and an 'A' is highlighted when [ALPHA] is pressed.

Section 4 The Basic Operation Keys

The bottom four rows contain the arithmetic keys. The operation keys on a Casio display exactly the same symbol on the screen that they show on the key. +, –, ×, ÷, x^y, and $\sqrt{\ }$ are the usual symbols.

To raise to powers use [x^y], where the x is the base number and y is the exponent. To enter 2^3, press

2 [x^y] **3**.

Evaluate 2 + 3 × 5 by pressing

2 [+] **3** [×] **5** [EXE].

The answer '17' is displayed on the right, below the expression (Fig. 7).

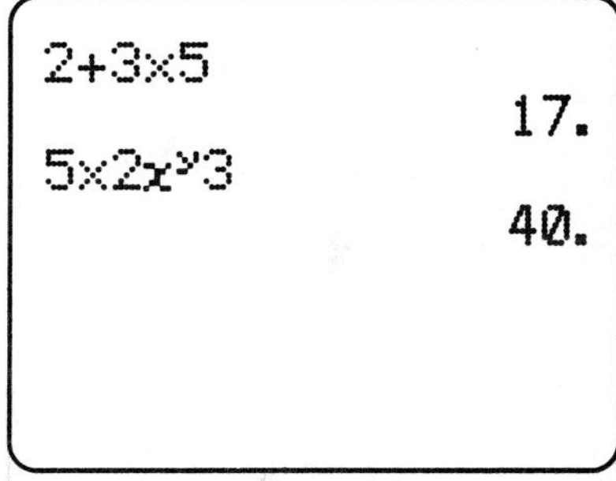

Figure 7

Evaluate 5 × 2^3 by pressing

5 [×] **2** [x^y] **3** [EXE].

The answer '40' is displayed (Fig. 7). Pressing [ANS] uses the previous answer in the next calculation. Evaluate 2 + 3, then subtract the answer from '9.' Press

2 [+] **3** [EXE] **9** [–] [ANS] [EXE].

The answer '4' is displayed (Fig. 8).

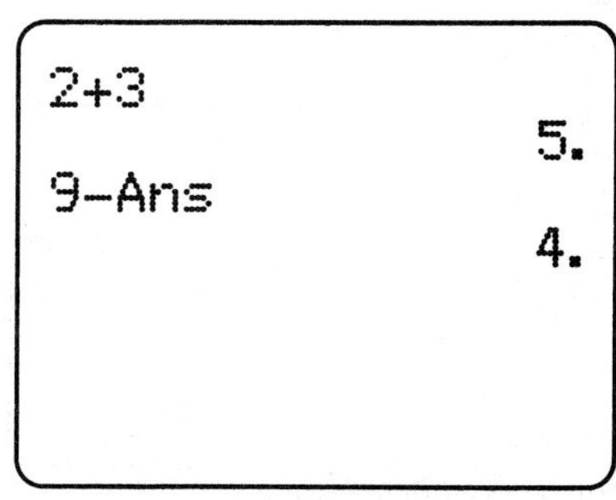

Figure 8

Fractions and mixed numbers can be displayed on the Text screen using the [a b/c] key. The symbol ⊥ is displayed on the screen. To display the mixed number

123 234/345

press

123 [a b/c] **234** [a b/c] **345**.

Press [EXE] and the fraction is reduced to 123 78/115 (Fig. 9).

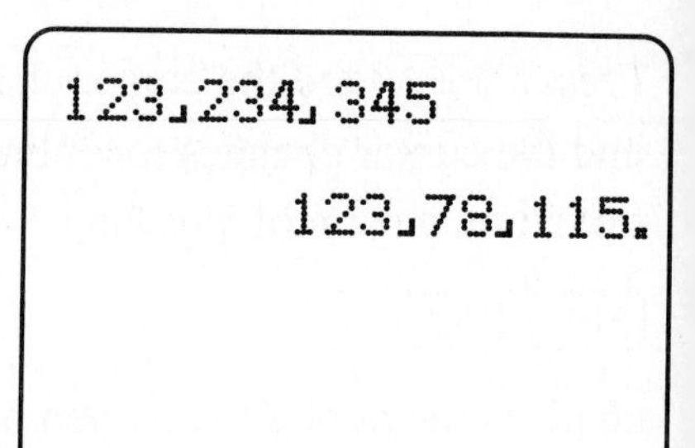

Figure 9

Press

[SHIFT] [a b/c]

to get the improper fraction 14223/115. Press [a b/c] to get a decimal representation 123.6782609. Press [a b/c] to return to the reduced mixed number (Fig. 9). Continue to change forms as often as you like. Fractions can be used in any operation, whether as exponents or in functions or matrices, etc.

Using more than three digits in mixed numbers, or four digits in fractions, will only return a decimal answer.

NOTE for the 7000: This calculator does not have any fraction capability.

The Opposite Key and Subtraction

The [–] key is used for both subtraction and negative numbers when doing arithmetic calculations. To do arithmetic with negative numbers, access the negative symbol (–) in the last row, fourth column by pressing [SHIFT] [(–)]. To evaluate 2 + –3, press

2 [+] [SHIFT] [(–)] **3** [EXE].

The answer '–1' is displayed (Fig. 10). Or press

2 [+] [–] **3** [EXE].

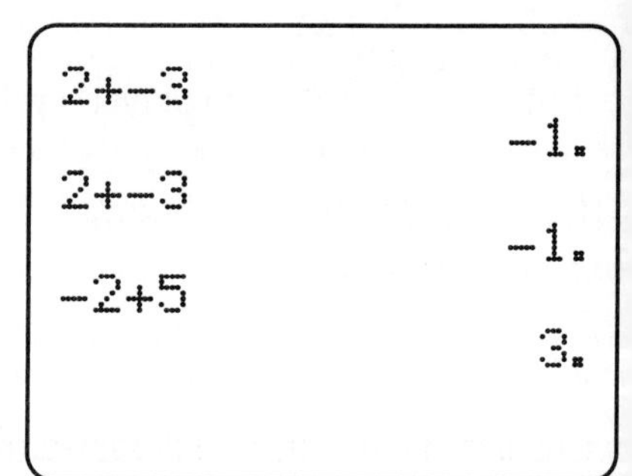

Figure 10

The answer '–1' is displayed (Fig. 10). If the negative number is the first symbol on a line and there are previous calculations on the Text screen, you must use the key [(–)]. The equation –2 + 5 is entered by pressing

[SHIFT] [(–)] **2** [+] **5** [EXE].

The answer '3' is displayed (Fig. 10).

NOTE for the 7000: The subtraction key [–] cannot be used for the opposite key, [(–)].

Example: To enter 2 + –3, press

2 [+] [(–)] **3** [EXE].

Using the [x^2] and [x^{-1}] Keys

When you press the [x^2] and [x^{-1}] keys, only the exponent is displayed. You must first enter a symbol for the base number so that the calculator knows which number to use.

If you enter

[SHIFT] [x^2] [EXE]

on a blank Text screen, you get an error message, 'Syn error.'

If you enter

2 [SHIFT] [x^2] [EXE],

the display reads **2**^{2}, and the answer '4.' is displayed.

If you enter

[X/θ/T] [SHIFT] [x^2] [EXE],

the display reads x^2 and the answer is the square of the number stored as x.

NOTE for the 6300 and 7000: The [x^2] and [x^{-1}] keys are not accessed with the Shift key.

How to Execute and Clear

When you want the calculator to do something, press [EXE], the blue key in the lower right corner. The Casio then will perform whatever operation is called for on the Text screen. To return to a blank Text screen, press [AC]. To erase a menu from the Text screen, press [PRE].

NOTE for the 7000: The [EXE] key is gray.

NOTE for the 6300 and 7000: There are no menus on the screens of these calculators.

The Editing Keys

The Casio has some of the same features as a word processor. A blinking cursor determines the current position on the screen. Moving the cursor onto text changes the cursor so that it blinks alternately with the symbol. The arrow keys are the cursor-movement keys.

Pressing a key overwrites a new symbol at the cursor. Pressing [DEL] deletes the symbol blinking at the cursor. Pressing [INS] inserts a symbol before the one blinking at the cursor.

The left and right arrow keys are "replay" keys. You can think of them as the "undo" or "whoops" keys. After you press the [EXE] key following input, press [◀] to move the cursor to the end of the input display or [▶] to move the cursor to the front of the input display. You can then execute the display again. This replay option is available until you execute again, but you must delete all text after the original calculation.

When an error message occurs, the replay keys move the cursor to the position of the error.

The [▲] and [▼] keys move the cursor up and down through the rows of text.

NOTE for the 7000: The replay option is not available if any key is pressed after pressing [EXE].

Error Messages

When an error message is displayed, the type of error and its position are displayed. Pressing a replay key will put the cursor at the position of the error on the Text screen. You can make the necessary corrections with the editing features described previously. A complete list of error types and things to look for is found in the Appendix of your Casio Owner's Manual. Examples for the common arithmetic errors follow.

'**Syn ERROR**' (Syntax error)

There is an error in the way the display is entered.

> **Example**: Entering [x^2] without entering x.
>
> **Example**: Entering a comma when entering large numbers (1,000 instead of 1000).

'**Ma ERROR**' (Math error)

The calculator is unable to perform the mathematics operation displayed on the Math screen.

> **Example**: Finding 25^{100} is out of the range of numbers possible for input.
>
> **Example**: Computing $2 \div 0$ is not a valid operation.

CHAPTER 1 NUMERICAL MATHEMATICS AND THE GRAPHING CALCULATOR

Section 1.1 Real Numbers and the Graphing Calculator

Refer to your Casio Owner's Manual to review methods of performing computations.

Changing Decimals to Fractions

If the decimal representation of a number has fewer than six digits and is not repeating, the calculator will change the decimal to a fraction and reduce the answer. You must enter the decimal as a fraction, then press [EXE].

Example: The decimal .44 equals the fraction $44/100$ (the number of digits in the decimal is the number of zeros in the denominator. Press **44** [a b/c] **100** [EXE].

The answer is $11/25$ (Fig. 1.1).

Example: The decimal 1.44 equals the fraction $144/100$. Press **144** [a b/c] **100** [EXE].

The answer is 1 $11/25$ (Fig. 1.1). By pressing

[SHIFT] [a b/c],

the answer changes to the improper fraction $36/25$.

```
44⌟100
                11⌟25
144⌟100
              1⌟11⌟25
```

Figure 1.1

Changing Fractions to Decimals

Divide the numerator of a fraction by its denominator to change the fraction to a decimal.

Example: Press

[AC] **11** [÷] **25** [EXE].

The answer is 0.44 (Fig. 1.2). Or use the fraction key

11 [a b/c] **25** [EXE] [a b/c].

The answer is 0.44 (Fig. 1.2).

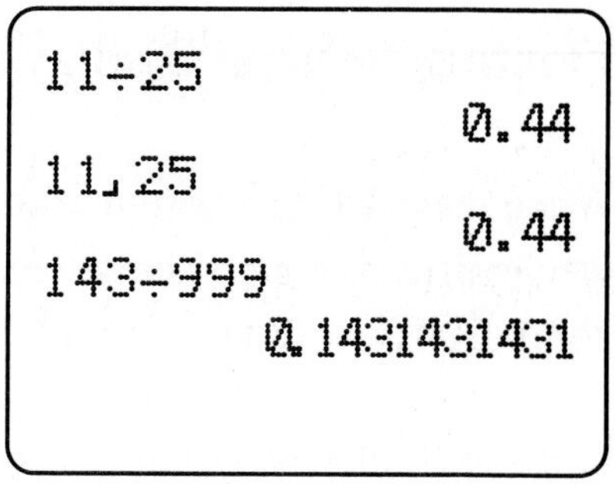

```
11÷25
                 0.44
11⌟25
                 0.44
143÷999
         0.1431431431
```

Figure 1.2

Example: Press

143 [÷] **999** [EXE].

The answer is 0.1431431431 (Fig. 1.2).

Repeating and Terminating Decimals

Fractions written in decimal form are either terminating or repeating decimals. The decimal representation of 11/25 is exact because it is a terminating decimal.

Decimal Representation

The representation of 143/999 is an approximation because it is impossible to display all of the digits of a repeating decimal.

Example: Press

[AC] **1** [÷] **17** [EXE].

The displayed answer is 0.05882352941. The calculator does not display enough digits to show whether this is a terminating or repeating decimal. (It is a repeating decimal. Compute

1 [÷] **17, 2** [÷] **17, 3** [÷] **17, 4** [÷] **17.**

If you look for the pattern, you can see the repeating digits are 0.0588235294117647.)

NOTE for the 7000: The 7000 does not have fraction capabilities. Use [÷] instead of [a b/c].

Floating Decimal Display versus Fixed Point Display

When the calculator is set in floating point display, it displays only the significant digits. The calculator can display 12 digits. In floating point display, it shows only the digits needed.

Example: First, clear the screen using [AC]. Then enter

2 ÷ 4 = 0.5, 11 ÷ 25 = 0.44, and **143 ÷ 999 = 0.1431431431.**

To change the number of digits displayed to some fixed number, use the display menu, by pressing

[SHIFT] [DISP].

NOTE for the 6300 and 7000: Use

[MODE] **72** [EXE]

to fix to two places. Press

[MODE] **9** [EXE]

to return to floating point.

Example: Press

143 [÷] **999** [EXE].

The answer is 0.1431431431 (Fig. 1.3). Now press

[SHIFT] [DISP] [F1:Fix] **2** [EXE].

The answer is 0.14 (Fig. 1.3). Press

[F1] **4** [EXE].

The answer is 0.1431. Pressing

[F3:Nrm] [EXE]

returns to the floating decimal point mode. The answer is 0.1431431431 (Fig. 1.3). Pressing [PRE] erases the menu from the screen.

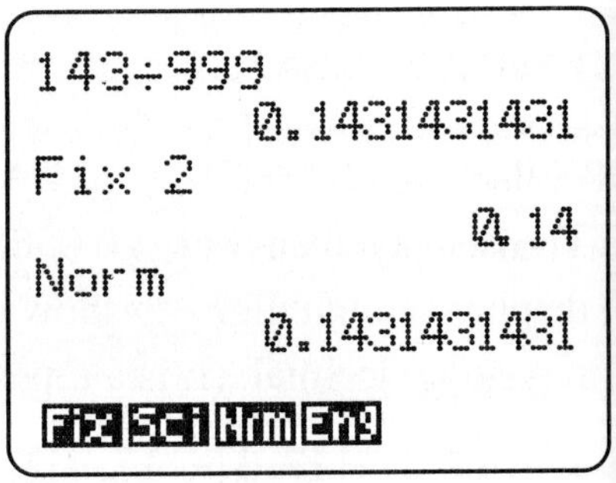

```
143÷999
            0.1431431431
Fix 2
                    0.14
Norm
            0.1431431431
Fix Sci Nrm Eng
```

Figure 1.3

NOTE for the 6300 and 7000: Press

143 [÷] **999** [EXE] [MODE] **72** [EXE] [MODE] **74** [EXE] [MODE] **9** [EXE]

(Fig. 1.3). This first fixes the decimal point at two places, then four places, and then returns to the floating point display.

Finding Square Roots

Example: To find $\sqrt{25}$ press

[AC] [√] **25** [EXE].

The answer is 5 (Fig. 1.4). Notice that the square root sign is not over the number. To evaluate the square root of a sum, you must use parentheses. $\sqrt{2+3}$ is entered by pressing

[√] [(] **2** [+] **3** [)] [EXE].

The answer is 2.2361.

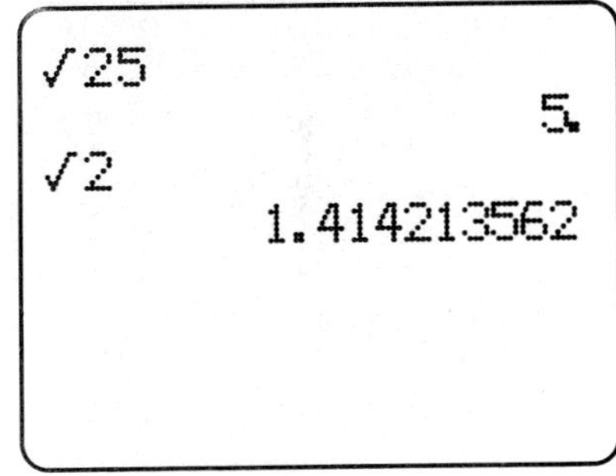

Figure 1.4

Example: To find $\sqrt{2}$ press

[√] **2** [EXE].

The answer is 1.414213562 (Fig. 1.4).

When the decimal form of a square root is not terminating, it is called a **non-terminating decimal**. The digits continue to have a random pattern forever.

Parentheses and Exponentiation

The [x^y] key uses the number before it as the base, x, and the number after it as the exponent, y.

Example: To find 3^5 press

[AC] **3** [x^y] **5** [EXE]

(Fig. 1.5).

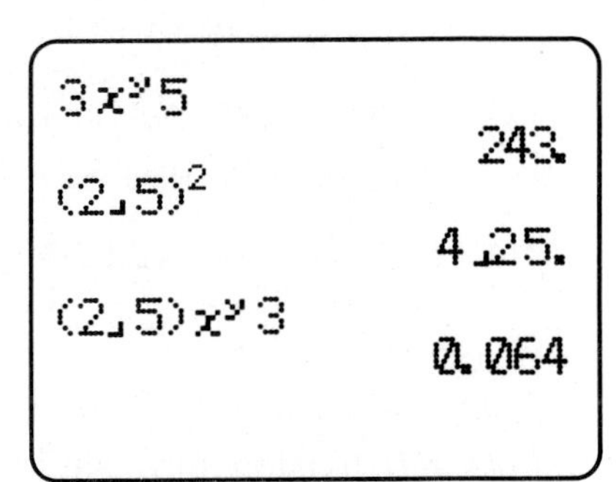

Figure 1.5

Example: To find $(2/5)^2$ you must use parentheses so that the calculator will square the entire fraction.

Enter

[(] 2 [a b/c] 5 [)] [SHIFT] [x^2] [EXE]

(Fig. 1.5). Note that to find $(2/5)^3$, you must enter

[(] 2 [a b/c] 5 [)] [x^y] 3 [EXE]

(Fig. 1.5). This will return a decimal answer.

Example: To find -3^2 press

[AC] [SHIFT] [(-)] 3 [SHIFT] [x^2] [EXE]

(Fig. 1.6).

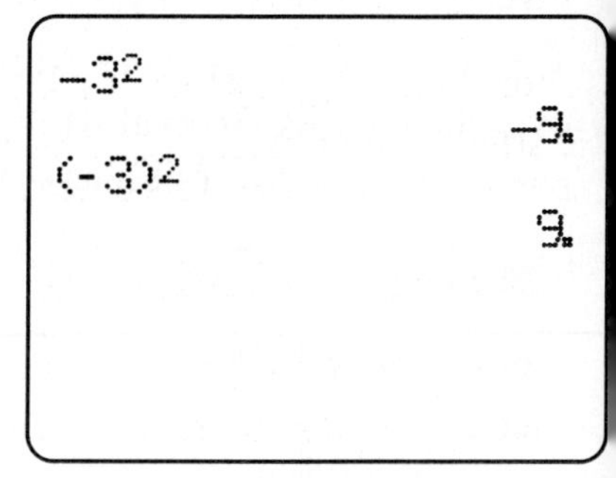

Figure 1.6

Example: To find $(-3)^2$ press

[(] [-] 3 [)] [SHIFT] [x^2] [EXE]

(Fig. 1.6).

NOTE: Pressing

[SHIFT] [x^2]

prints only the exponent 2.

You need the negative sign [(-)] at the beginning of a line; any other time the negative can be written as a subtraction [-].

NOTE for the 7000: Use [÷] instead of [a b/c].

The Order of Operations in Arithmetic

The calculator will do its calculations in algebraic order. Expressions in parentheses are evaluated first, powers and roots second, multiplication and division third, addition and subtraction fourth. If more than one operation of the same priority is in the expression, calculation is done from left to right.

Example: $2 + 3 \times 4^2 - 6 \div 2$ is entered by pressing

[AC] **2** [+] **3** [×] **4** [SHIFT] [x^2] [−] **6** [÷] **2** [EXE].

The answer is 47 (Fig. 1.7).

2+3×4²−6÷2
47.

Figure 1.7

Section 1.2 Properties of Real Numbers and the Basic Rules of Operation

The Properties of Opposites

There is no Opposite key on the 7700. See Section 4 of the "Getting Started" chapter.

The Absolute Value Key

Geometrically, the absolute value of a number represents the distance from the number to the origin, so the absolute value is always a positive number. Press [ABS] followed by the number to calculate an absolute value. If the number is written as a sum, you will need parentheses around it.

NOTE for the 6300 and 7000: The [SHIFT] [x^y] keys are used instead of the [ABS] key.

Example: Evaluate $-|-7|$. Press

[AC] [−] [SHIFT] [MATH] [F3:Num] [F1:Abs] [−] **7** [EXE].

The answer is −7 (Fig. 1.8).

Example: Evaluate $|5.2 - 17.8|$. Press

[F1:Abs] [(] **5** [.] **2** [−] **17** [.] **8** [)] [EXE].

The answer is 12.6 (Fig. 1.8). To erase the menu from the screen press

[PRE] [PRE].

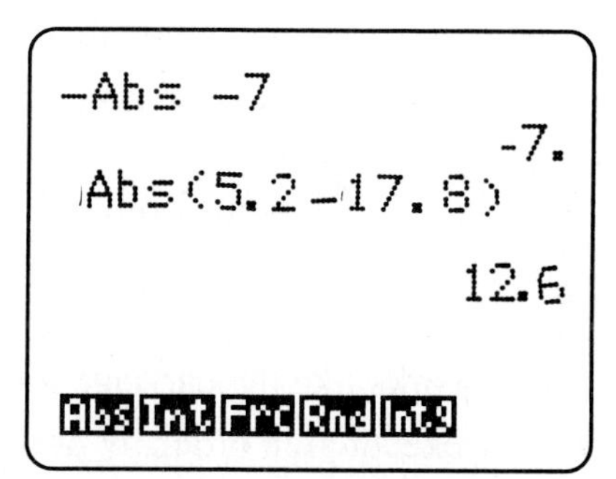

Figure 1.8

Section 1.3 Algebraic Expressions and Problem Situations

Storing Values and Evaluating Expressions

To store a value, use [→], which is on the last line of black keys.

NOTE for the 6300 and 7000: The [→] key is the second key in the fifth row.

Example: Store 25 in the variable x. Press

[AC] **25** [→] [X/θ/T] [EXE].

The number 25 is displayed (Fig. 1.9). Now when you enter an expression containing x, the expression will be evaluated with $x = 25$.

Enter

[X/θ/T] [x^y] **3** [+] **17** [EXE].

The answer '15642' is displayed (Fig. 1.9).

NOTE for the 6300 and 7000: Use the [ALPHA] [X] keys for x.

```
25→X
                25.
X x^y 3+17
             15642.
```

Figure 1.9

Evaluating Expressions for More than One Choice for the Variable

Concatenation

If an expression is to be evaluated for more than one choice for the variable, there are two ways to do it.

Method 1. Connect the commands with

[SHIFT] [↵],

which works like the carriage return on a typewriter. All of the steps will be executed in order by pressing [EXE]. The replay keys can be used to change a variable.

NOTE for the 6300: Use [:] to connect expressions. There is no carrage return.

Example: Evaluate $x^3 + 17$ at $x = 25, 39$, and 52. Press

25 [→] [X/θ/T] [SHIFT] [↵] [X/θ/T] [x^y] **3** [+] **17** [EXE].

The answer '15642' is displayed (Fig. 1.9).

Press [▶] to return to the beginning of the text. Press

39

to replace 25 with 39 and then press [EXE]. The answer '59336' is displayed (Fig. 1.10). Press [▶] to return to the beginning of the text. Press

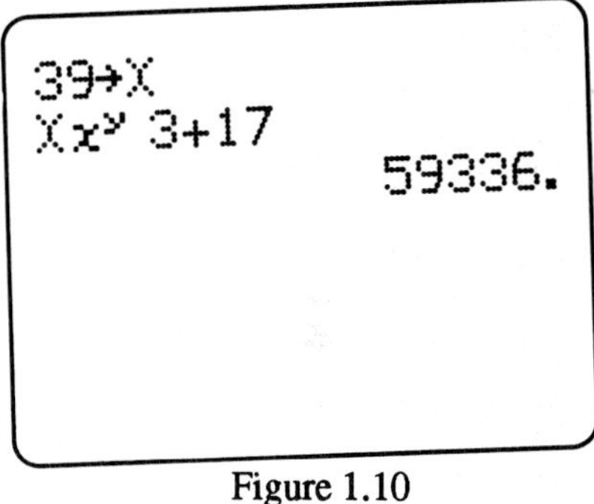

Figure 1.10

52

to overwrite 39 and then press [EXE]. The answer '140625' is displayed.

Method 2. Use [ⒻMEM] to store the expression. The [ⒻMEM] function is a storage area for six expressions. Press

[SHIFT] [ⒻMEM] [F4:LIST]

to see what is being stored.

Pressing

[F1:STO] **1**

will store what is on in the Text screen into f_1. Pressing

[F2:RCL] **1**

will place what is in f_1 onto the screen.

Example: Evaluate $x^3 + 17$ at $x = 25, 39$, and 52.

You must start with a blank screen. Press

[AC] [X/θ/T] [x^y] **3** [+] **17** [F1:STO] **1**

to store $x^3 + 17$ in f_1. (If there was another function in f_1, $x^3 + 17$ replaces it.) Pressing [AC] returns the Text screen. Press

25 [→] [X/θ/T] [SHIFT] [↵] [F2:RCL] **1** [EXE].

The answer is 15642 (Fig. 1.9). Now evaluate at $x = 39$. Press

39 [→] [X/θ/T] [SHIFT] [↵] [F2:RCL] **1** [EXE].

The answer is 59336 (Fig. 1.10). Now evaluate at $x = 52$.

52 [→] [X/θ/T] [SHIFT] [↵] [F2:RCL] **1** [EXE].

The answer is 140625.

Evaluating Multivariable Expressions

First you store the expression in the [ⒻMEM] function. Then store all the variables in the proper letter storage.

Next, recall the expression and evaluate.

Example: Evaluate $4T - 3S^3/5R$ for $R = 13.5$, $S = -23.7$, $T = 3.28$. Remember, the fraction line is a grouping symbol so the numerator must be in parentheses.

Press

[AC] [SHIFT] [ⒻMEM]

if necessary. Entering

[AC] [(] **4** [ALPHA] [T] [−] **3** [ALPHA] [S] [x^y] **3** [)] [÷] **5** [ALPHA] [R] [SHIFT] [↵] [F1:STO] **1** [AC]

stores the expression in f_1. Press

[AC] **13** [.] **5** [→] [ALPHA] [R] [SHIFT] [↵] [−] **23** [.] **7** [→] [ALPHA] [S] [SHIFT] [↵] **3** [.] **28** [→] [ALPHA] [T] [SHIFT] [↵] [F2:RCL] **1** [EXE].

The answer is 591.8411704 (Fig. 1.11).

You may now use the replay keys to return to the text and change the value of the variables.

NOTE for the 6300 and 7000: There is no function memory in these models. Evaluate expressions using method 1. On the 6300, instead of [SHIFT] [↵], use the [:] key.

```
13.5→R
-23.7→S
3.28→T
(4T-3Sx^y 3)÷R
        591.8411/04
STO RCL fn LIST
```

Figure 1.11

Implied Multiplication

When multiplying by a fraction, using the [a b/c] key will return the correct answer. If you enter the fraction using [÷], the fraction must be in parentheses.

Example: Evaluate $5/9 \cdot (41 - 22)$. Enter

[AC] **5** [a b/c] **9** [(] **41** [−] **32** [)] [EXE].

The correct answer is 5. An alternative is to use

[(] **5** [÷] **9** [)] [(] **41** [−] **32** [)] [EXE].

Example: Evaluate $4T - 3S^3/5R$. If the numerator is $4T - 3S^3$, it must be put in parentheses or $3S^3/5R$ will be evaluated first (Fig. 1.11).

Section 1.5 Numerical Representations of Problem Situations

Making Tables

There is no table command on the Casio, but by using the [F MEM] storage on the Casio 7700G (this is discussed in Section 1.3), the values to be written in the table are easily calculated.

NOTE for the 6300 and 7000: Only one expression can be evaluated at a time. See method 1 in Section 1.2.

Example: Given the Width x, find the Length $24 - 2x$ and the Area $x(24 - 2x)$ for different values of x. Enter

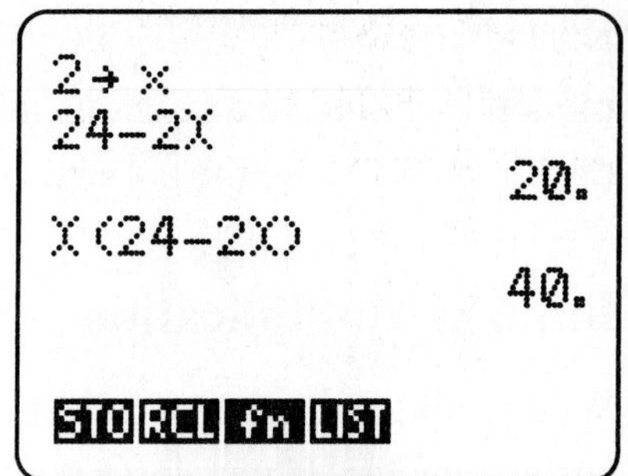

Figure 1.12

[AC] **24** [−] **2** [X/θ/T] [SHIFT] [ⒻMEM] [F1:STO] **1** [AC] [X/θ/T] [(] **24** [−] **2** [X/θ/T] [)] [F1:STO] **2.**

The two functions are stored in f_1 and f_2. Press

[AC] **2** [→] [X/θ/T] [SHIFT] [↵] [F2:RCL] **1** [EXE] [F2:RCL] **2** [EXE].

The answers are Length = 20 and Area = 40 (Fig. 1.12).

Repeat these calculations with $x = 3$ by pressing

3 [→] [X/θ/T] [SHIFT] [↵] [F2:RCL] **1** [EXE] [F2:RCL] **2** [EXE].

The answers are Length = 18, and Area = 54. Repeat for as many values of x as you want.

Finding Percents

Example: Markups at 20%. Given the wholesale price, x, in dollars, find the markup, $.2x$, in dollars and the retail price, $x + .2x$, in dollars. To get dollars and cents, set the Casio at the two-digit display setting using the display menu by pressing

[AC] [SHIFT] [DISP] [F1:Fix] 2 [EXE] [SHIFT] [ⒻMEM]

[AC] [.] **2** [X/θ/T] [F1:STO] **1**

[AC] [X/θ/T] [+] [.] **2** [X/θ/T] [F1:STO] **2** [AC]

10 [→] [X/θ/T] [SHIFT] [↵] [F2:RCL] **1** [EXE] [F2:RCL] **2** [EXE].

The answers are Markup = $2.00 and Retail Price = $12.00 (Fig. 1.13). Do not press [AC]. Now press

11 [→] [X/θ/T] [SHIFT] [↵] [F2:RCL] **1** [EXE] [F2:RCL] **2** [EXE].

The answers are Markup = $2.20 and Retail Price = $13.20. Pressing

[SHIFT] [DISP] [F3] [EXE]

returns the display to floating point. Press [PRE] to erase the menu from the screen.

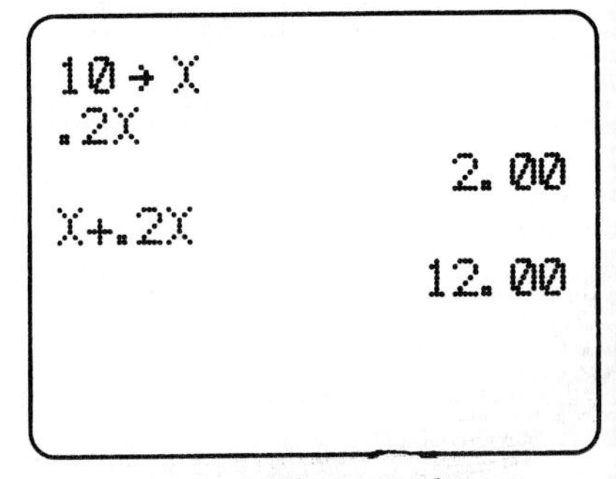

Figure 1.13

NOTE for the 6300 and 7000: Pressing

[MODE] **72** [EXE]

changes the display to two decimals. Press

10 [→] [ALPHA] [X] [:] [.] **2** [ALPHA] [X] [EXE] [ALPHA] [X] [+] [.] **2** [ALPHA] [X] [EXE].

The answers are Markup = $2.00 and Retail price = $12.00. By pressing

[MODE] **9** [EXE]

this changes back to floating point mode.

Section 1.6 Algebraic Representations of Problem Situations

Parametric Simulation

Change to Parametric Graphing mode by pressing

[MODE] [SHIFT] [X].

Hold down the [M Disp] key (Fig. 1.14).

Figure 1.14

Set the size of the window by pressing

[AC] [RANGE] **0** [EXE] **150** [EXE] **10** [EXE] **0** [EXE] **8** [EXE] **1** [EXE] **0** [EXE] **5** [.] **5** [EXE] **0** [.] **01** [EXE].

The *T* range stays the same.

Enter

[GRAPH] **30** [X/θ/T] [SHIFT] [,] **5** [SHIFT] [↵] [GRAPH] **28** [X/θ/T] [+] **10** [SHIFT] [,] **2** [EXE]

(Fig. 1.15). The comma is above the [→] key. Casio does not plot graphs simultaneously.

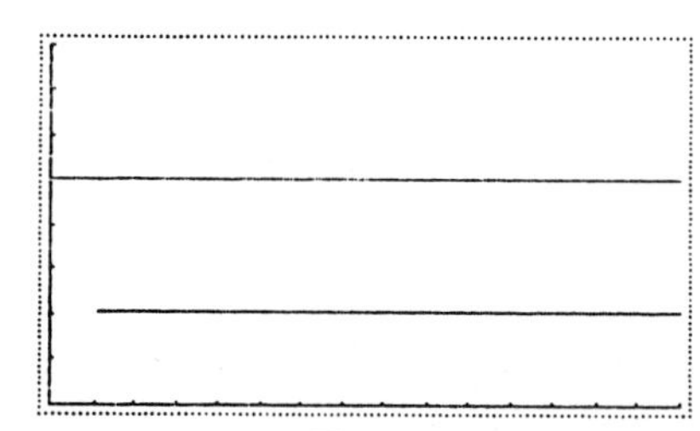

Figure 1.15

Press [F1] to activate the Trace cursor (a star). Use [▶] to move the cursor along the bottom line. Whenever T is a whole number, press [▲] to read the value of T that makes the x values the same. Press [▼] to return to the bottom line. You may hold [▶] down as the cursor moves very slowly. At $T = 5$, $X = 150$ on both lines and $Y = 2$.

To change back to the Rectangular Coordinate mode, press

[MODE] [SHIFT] [MODE] [+].

NOTE for the 6300 and 7000: A program must be written to plot in the Parametric mode.

Chapter 2 GRAPHING EQUATIONS AND FUNCTIONS

Section 2.1 The Rectangular Coordinate Plane

The Window

The screen at the top of the calculator is called a **window,** and it displays part of the coordinate plane. To see a graph of a function on this screen, called a **Graphing window**, either execute the command [GRAPH] or press [G↔T] to change, or toggle, between the Graphing window and Text screen.

The Range

The size of the Graphing window is set using the [RANGE] key. ***X*min** and ***X*max** are the smallest and largest numbers that can be plotted on the horizontal axis. ***Y*min** and ***Y*max** are the smallest and largest numbers that can be plotted on the vertical axis. ***X*scl** and ***Y*scl** tell where the reference tick marks are located on the axis. The size of a window is given by [*X*min, *X*max], *X*scl by [*Y*min, *Y*max], *Y*scl. If no values are given for scale, use 1.

Pixels

The Graphing window is made up of little dots, called **pixels**. A graph is drawn by connecting these pixels. Using different settings for the range will change the value associated with a pixel.

Example: Set the Graphing window using [–20, 10], *X*scl = 5 by [–15, 30], *Y*scl = 10. (If you have not changed from the Parametric Graphing mode back to the Rectangular Coordinate mode you need to now. Press

[MODE] [SHIFT] [+].)

Press

[RANGE] [−] **20** [EXE] **10** [EXE] **5** [EXE] [−] **15** [EXE] **30** [EXE] **10** [EXE].

Press [RANGE] [RANGE] to return to the Text screen.

Press

[SHIFT] [F3:PLT] [EXE]

to activate the **free-moving cursor**. This cursor is a small cross blinking in the middle of the Graph screen. Its coordinates are shown at the bottom of the screen. Using the cursor keys, move it as near the point (0, 0) as possible. Write down the coordinates. They are small, but not (0, 0). Change the Range screen to [–10, 10] by [–10, 10]. Press

[SHIFT] [F3:PLT] [EXE]

to engage the free-moving cursor. Note the coordinates are smaller than before, but they are still not equal to zero.

NOTE for the 6300 and 7000: Press

[SHIFT] [PLOT] [EXE]

to activate the free-moving cursor. It is a small dot.

The Integer Window

If the range is set at [–47, 47] by [–31, 31], the Graph window is called the **Integer** window. Activate and move the free-moving cursor to see that all the pixels have coordinates that are integers. If both *X*scl and *Y*scl are 1, the axes look like double lines. This happens because there is a tick mark on every pixel. To get rid of the extra lines, change both *X*scl and *Y*scl to 5. Any window where *X*max – *X*min = 94, or a multiple of 94, and *Y*max – *Y*min = 62, or a multiple of 62, is an Integer window. On an Integer window the origin is (0, 0) and it lies on a pixel.

NOTE for the 6300: The Integer window is [–38, 38] by [–22, 22].

The 0.1 Decimal Window

On the **Decimal** window each pixel is 0.1 unit. Pictures are their true shape, and the origin is (0, 0). This window is sometimes called the 0.1 window. On a Casio, this window is the **Default** window. When the calculator is reset these are the initial values. This is the only window that can be set with an option. When the Range screen is displayed, press [F1:INIT]. Any multiple of these settings will result in a Decimal window.

NOTE for the 6300 and 7000: To set the default range press

[SHIFT] [Mcl]

on the 6300 and the 7000. The default range on the 7000 is the 0.1 window and on the 6300 the default is the 0.2 window.

The 0.2 Window

To get the 0.2 window, set the range at [–9.4, 9.4] by [–6.2, 6.2]. These values are twice the values of the 0.1 window. Each pixel is 0.2 units. This window will be used later in Chapter 6.

NOTE for the 6300: When the range is set at default, it is an 0.2 window with a range of [–3.8, 3.8] by [–2.2, 2.2].

Section 2.2 Equations in Two Variables and Their Graphs

Plotting Points

There are two ways to plot points on the Graphing screen.

Method 1. You used the free-moving cursor in Section 2.1. If the cursor is located at the point you want to plot, press [EXE], which marks the point on the screen. The free-moving cursor has returned to the middle of the screen.

Method 2. Use the plot option, [F3:PLT], with the coordinates of the point.

Example: Plot the following solution pairs to

$14x + 28y = 168$.

Set the range to [–20, 20], *X*scl = 5 by [–15, 15], *Y*scl = 5. The *T* Range stays the same, with the points (–15, 13.5), (–10, 11), (0, 6), and (15, –1.5). Press

[RANGE] [–] **20** [EXE] **20** [EXE] **5** [EXE] [–] **15** [EXE] **15** [EXE] **5** [EXE]
[RANGE] [RANGE] [SHIFT] [F3:PLT] [–] **15** [SHIFT] [,] **13** [.] **5** [EXE].

This plots the first point. Pressing

[F3:PLT] [–] **10** [SHIFT] [,] **11** [EXE]

plots the second point. Pressing

[F3:PLT] **0** [SHIFT] [,] **6** [EXE]

plots the third point. Pressing

[F3:PLT] **15** [SHIFT] [,] [–] **1** [.] **5** [EXE]

plots the last point (Fig. 2.1). The third point does not show as it is on the *y*-axis.

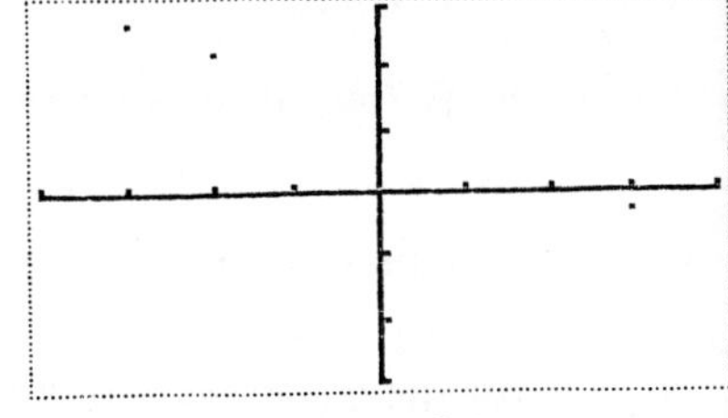

Figure 2.1

Graphing Functions

Enter the function, adjust the range, and execute. Use [GRAPH] or store the function in [F MEM].

Example: Graph $y = 2x - 8$ in the Integer window.

Refer to Section 2.1 for a discussion of the Integer window.

Method 1. Press

[GRAPH] **2** [X/θ/T] [–] **8**.

Set the range to [–47, 47], Xscl = 10 by [–31, 31] Yscl = 10. Return to the Text screen by pressing [RANGE] and press [EXE] (Fig. 2.2).

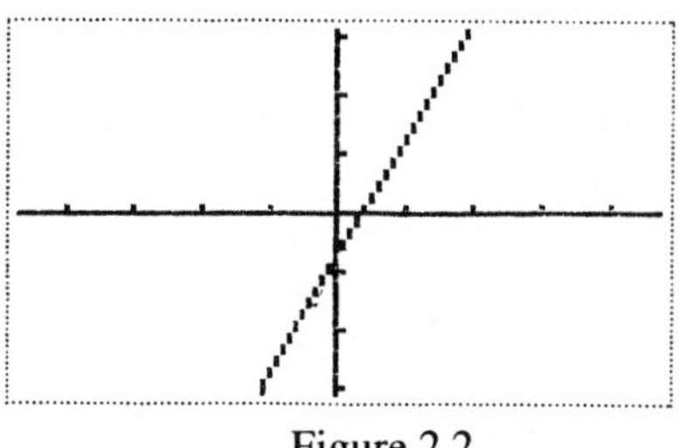
Figure 2.2

Method 2. Press

[AC] **2** [X/θ/T] [–] **8** [SHIFT] [ⒻMEM] [F1:STO] **1** [AC] [GRAPH] [F2:RCL] **1** [EXE]
(Fig. 2.2).

NOTE for the 6300 and 7000: Use method 1. Use [ALPHA] [X] for x.

Multiple Graphs

Graphing more than one function is possible on the same screen. By using [ⒻMEM] six functions can be graphed on the screen. Using the option [GRAPH] allows as many functions to be graphed as fit in the screen buffer.

NOTE for the 6300 and 7000: Use [ALPHA] [X] for x.

TRACE

The **Trace** cursor stays on the last function graphed. The right and left cursor keys move the cursor along the graph. It is not free moving. The up and down cursor keys move the cursor between the graphs of different functions. The x-coordinate remains the same, but the y-coordinate changes. Activate Trace with [F1] from the Graph screen or with

[SHIFT] [F1:TRC]

from the Text screen.

NOTE for the 6300 and 7000: Use [SHIFT] [TRACE] to activate Trace on the 7000. On the 6300 use the [TRACE] key.

Example: Graph y = 3/4, and x – 6 and use Trace to find the y values at x = –9, –12, –11, and 3. If necessary, set range values to the Integer window.

Enter

[GRAPH] **3** [a b/c] **4** [X/θ/T] [−] **6** [EXE] [F1] (Fig. 2.3).

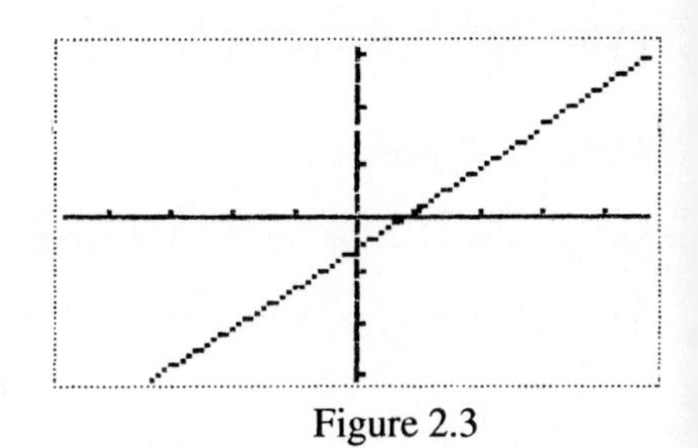
Figure 2.3

The values $X = -33$, $Y = -30.75$ appear on the screen.

Use [▶] to trace to $X = -9$, read $Y = -12.75$.

Use [▶] to trace to $X = -12$, read $Y = -15$.

Use [▶] to trace to $X = -11$, read $Y = -14.25$.

Use [▶] to trace to $X = 3$, read $Y = -3.75$.

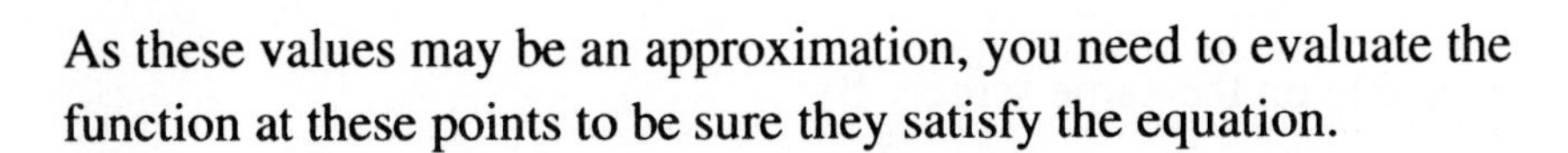

As these values may be an approximation, you need to evaluate the function at these points to be sure they satisfy the equation.

Press

[AC] [−] **9** [→] [X/θ/T] [SHIFT] [↵] **3** [a b/c] **4** [X/θ/T] [−] **6** [EXE] [a b/c].

Change −9 to −12. Press

[▶] [▶] **1** [SHIFT] [INS] **2** [EXE].

Check all the other points.

Remember: When you press **3** [a b/c] **4** [X/θ/T] this is $(3/4)x$ or $(3x/4)$. Pressing

3 [÷] **4** [X/θ/T]

is $3/(4x)$.

NOTE for the 6300 and 7000: Only the x value is displayed on the screen using Trace. Use

[SHIFT] [X↔Y]

to get the y value.

Section 2.3 Linear Equations and the Slope of a Line

Graphing Vertical Lines

There are two ways to graph a vertical line on a Casio.

Method 1. Plot the two end points and join them with a line. Use [F3] or the free-moving cursor and [F4].

Example: Graph the line $x = 3$. First press

[SHIFT] [F5:CLS] [EXE]

to clear the Graph screen. Then pressing

[RANGE] [F1:INIT] [RANGE] [RANGE]

initializes the Graph screen. Pressing

[SHIFT] [F3:PLT] [EXE]

activates the free-moving cursor.

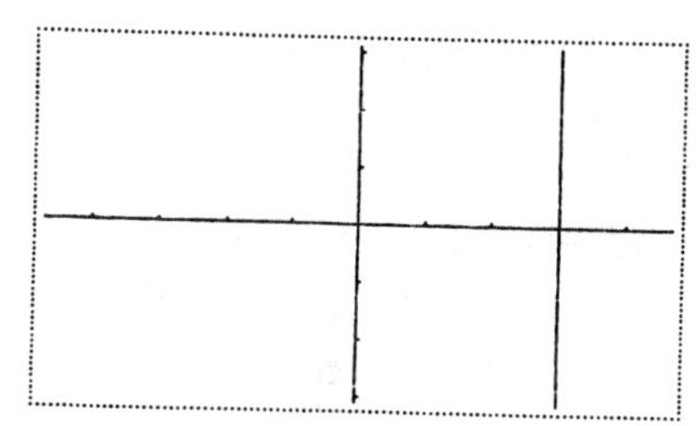

Figure 2.4

Move to (3, 3.1) and press [EXE]. Then move to (3, –3.1) and press

[F4] [EXE]

(Fig. 2.4).

Method 2. Change to parametric coordinates by pressing

[RANGE] [F1:INIT] [RANGE] [–] **3** [.] **1** [EXE] **3** [.] **1** [RANGE] [MODE] [SHIFT] [X].

Then press

[GRAPH] **3** [SHIFT] [,] [X/θ/T] [EXE]

(Fig. 2.4). Press

[MODE] [SHIFT] [+]

to return to the rectangular coordinates.

NOTE for the 6300 and 7000: Pressing

[SHIFT] [PLOT] [EXE]

activates the free-moving cursor. Use method 1 on these calculators.

Section 2.4 The Slope-Intercept Form and Point-Slope Form of Linear Equations

Friendly Windows

Unless the **tick** marks for a unit on the x-axis and the y-axis are the same size, the graph will be distorted. For example, perpendicular lines will not appear to cross at right angles (Figs. 2.5 and 2.6).

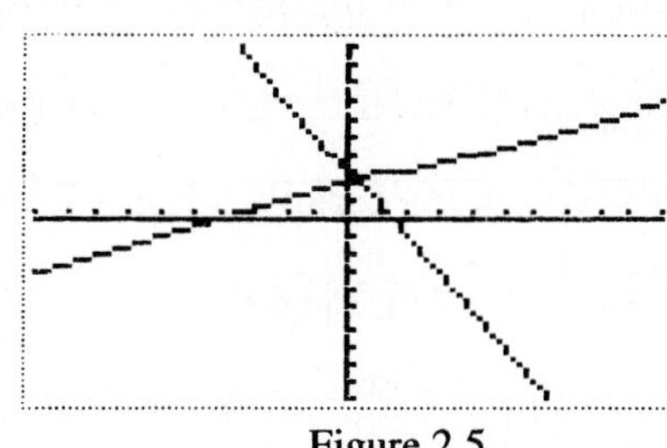

Figure 2.5

A friendly window is one that does *not* distort the graph. The default and integer ranges are friendly windows. The range [–15, 15] by [–10, 10] also makes graphs appear their correct shape.

Example: Graph $y = 0.5x + 2$ and $y = -2x + 3$ in the window [–10, 10] by [–10, 10] (Fig. 2.5) and the window [–15, 15] by [–10, 10] (Fig. 2.6).

Press

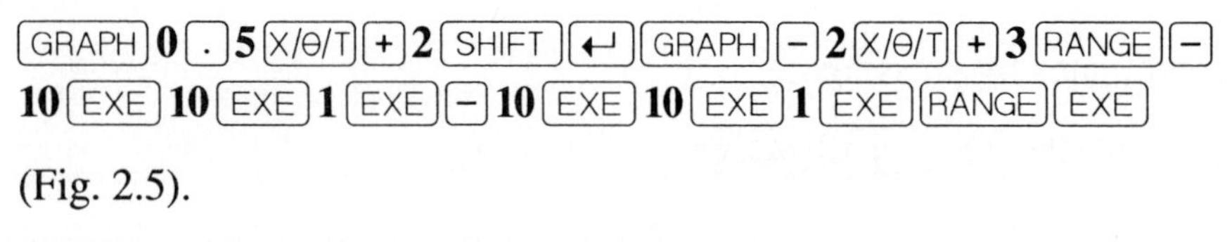

[GRAPH] **0** [.] **5** [X/θ/T] [+] **2** [SHIFT] [↵] [GRAPH] [–] **2** [X/θ/T] [+] **3** [RANGE] [–] **10** [EXE] **10** [EXE] **1** [EXE] [–] **10** [EXE] **10** [EXE] **1** [EXE] [RANGE] [EXE]

(Fig. 2.5).

[RANGE] [–] **15** [EXE] **15** [EXE] **1** [EXE] [RANGE] [RANGE] [EXE]

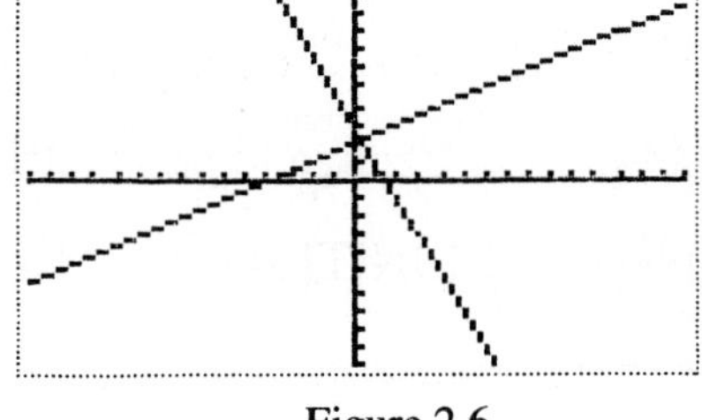

Figure 2.6

(Fig. 2.6). Are the lines perpendicular?

NOTE for the 6300 and 7000: Use [ALPHA] [X] for [X/θ/T] (Fig. 2.6).

Sec. 2.5 Functions and Graphs of Functions

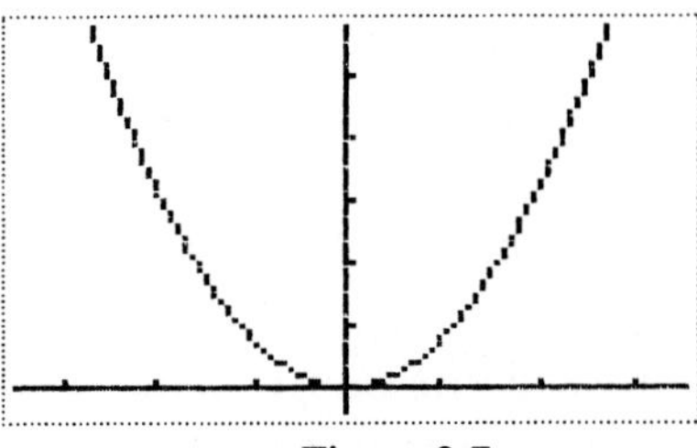
Figure 2.7

Graphing Functions and Built-in Graphics

The functions x^2, x^{-1}, and $\sqrt{x}$ can be graphed with a range programmed into the calculator. These ranges are picked to show a complete graph. These are called built-in graphs.

Press

[GRAPH] [SHIFT] [x^2] [EXE]

(Fig. 2.7). Press

[GRAPH] [SHIFT] [x^{-1}] [EXE]

(Fig. 2.8). Press

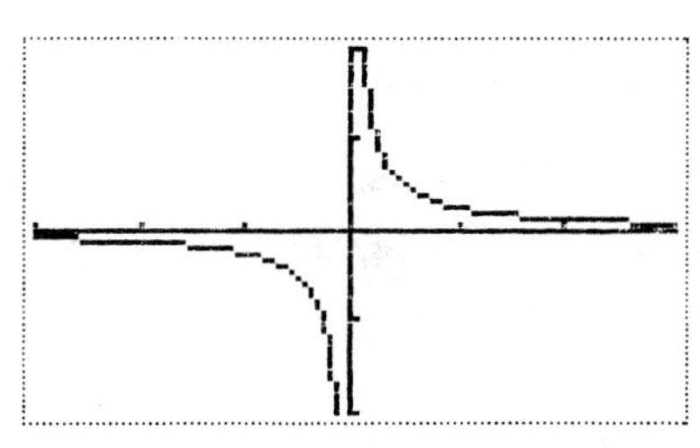
Figure 2.8

[GRAPH] [$\sqrt{\ }$] [EXE]

(Fig. 2.9).

Note that the Graph screen is cleared and the range changed before each graph is drawn. No variable is needed with the option.

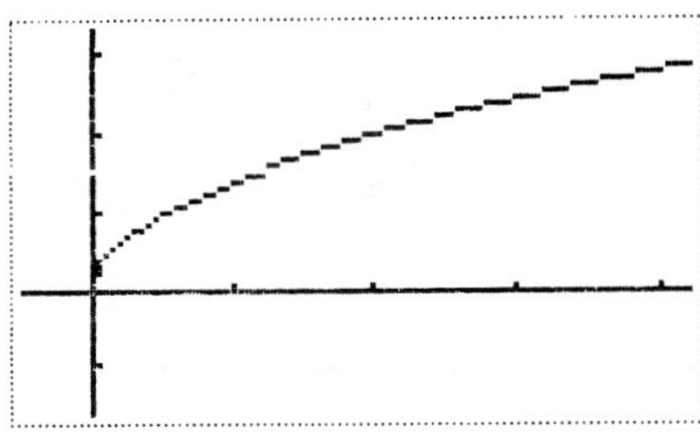
Figure 2.9

NOTE for the 6300 and 7000: The keys [x^2] and [x^{-1}] are not Shift keys.

Clearing the Graph screen

Press

[SHIFT] [F5:CLS] [EXE]
or make any change in the Range menu.

Example: Graph $y = -x^2 - 3x$.

Use the default range. Graphs use the notation $Y = -X^2 - 3X$ to represent the function $f(x) = -x^2 - 3x$. Press

[GRAPH] [(−)] [X/θ/T] [SHIFT] [x^2] [−] **1** [X/θ/T] [RANGE] [F1] [RANGE] [RANGE] [EXE]

to display the graph (Fig. 2.10).

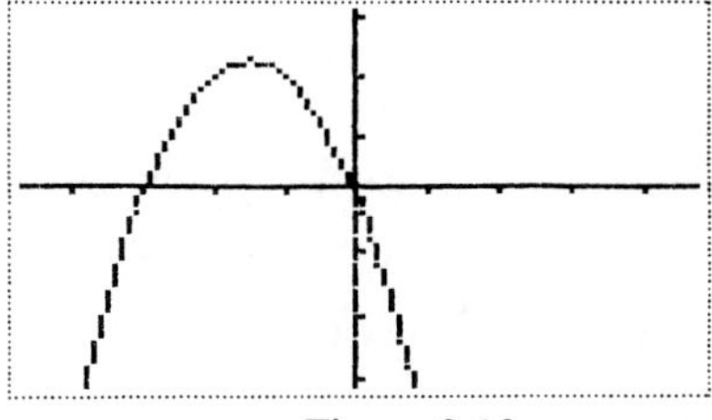
Figure 2.10

Tracing

Trace was discussed in Section 2.2. You can use the Trace option to help determine the domain and range of a function.

Example: Find the domain and range of $f(x) = \sqrt{x-3}$.

Start with the integer range. Set the range to [–47, 47], *X*scl = 10 by [–31, 31], *Y*scl = 5. Then press

[GRAPH] [√] [(] [X/θ/T] [–] **3** [)] [EXE]

(Fig. 2.11). Pressing [F1] activates the Trace cursor. The cursor appears at the far left point of the function, (3, 0), and it implies the domain is $x > 3$. The graph lies entirely above the x-axis, which implies the range is $y > 0$. Press [▶] to move the Trace cursor along the curve. When the Trace cursor approaches the right side of the window, the display will scroll so that you can see the graph for larger values of x. Scroll several times to confirm that the graph is continuing to rise, with the x-and y-coordinates increasing.

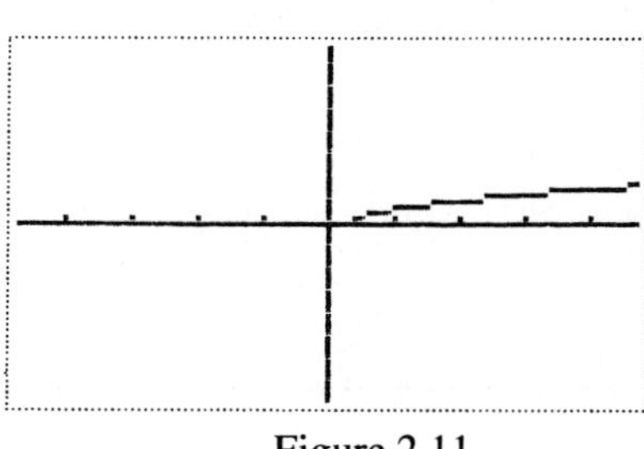

Figure 2.11

Example: Determine the range of $f(x) = x^4 - 3x^2 + 5x - 6$, starting with the window [–5, 5] by [–10, 10].

Set the Range menu by pressing

[RANGE] [–] **5** [EXE] **5** [EXE] **1** [EXE] [–] **10** [EXE] **10** [EXE] **1** [EXE] [RANGE].

Then press

[GRAPH] [X/θ/T] [x^y] **4** [–] **3** [X/θ/T] [SHIFT] [x^2] [+] **5** [X/θ/T] [–] **6** [EXE]

(Fig. 2.12).

Figure 2.12

Press [▼] to move the graph up onto the screen. Using this feature is called **scrolling.** Scrolling will redraw a translated graph. It is activated by pressing an arrow key immediately after the graph is drawn. Press [▼] again (Fig. 2.13).

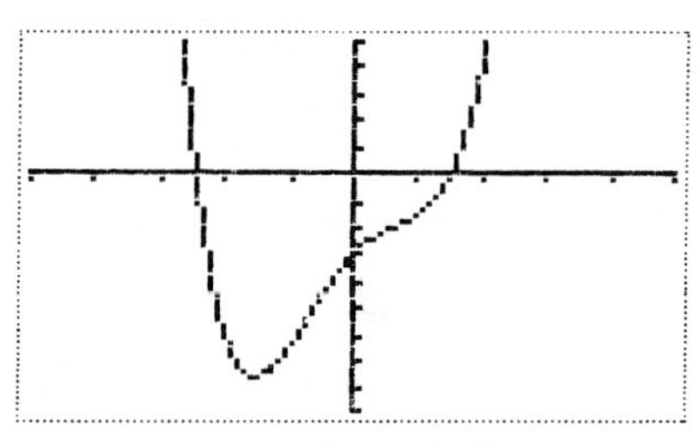

Figure 2.13

NOTE for the 7000: This calculator does not scroll.

How to Graph a Relation that Is Not in Function Form

A relation must be solved for y before it can be graphed on the Casio.

Chapter 3 SOLVING EQUATIONS AND SYSTEMS OF EQUATIONS

Section 3.1 Finding Graphical Solutions

Features for Viewing Multiple Graphs

A Casio will only draw graphs sequentially in order of entry. To view a graph, its function must appear on the Text screen in the form Graph $Y = f(x)$ using [GRAPH]. The function $f(x)$ may be recalled from the [F MEM] or entered directly onto the Text screen. If you enter onto the screen, there is a 127-step area available for execution of computations. By dividing the calculations into parts, you may graph as many functions as you want on one screen. Refer to Section 2.1 of this manual for discussions of the basic windows.

Example: Graph $y = 2x + 5$ and $y = 21$ on the same screen.

Use the Integer window to show the graph of $y = 21$. Set the range to [–47, 47], Xscl = 10 by [–31, 31], Yscl = 10 by pressing

[RANGE] [(-)] **47** [EXE] **47** [EXE] **10** [EXE] [(-)] **31** [EXE] **31** [EXE] **10** [EXE] [RANGE].

Press

[GRAPH] **2** [X/θ/T] [+] **5** [SHIFT] [↵] [GRAPH] **21** [EXE]

(Fig. 3.1).

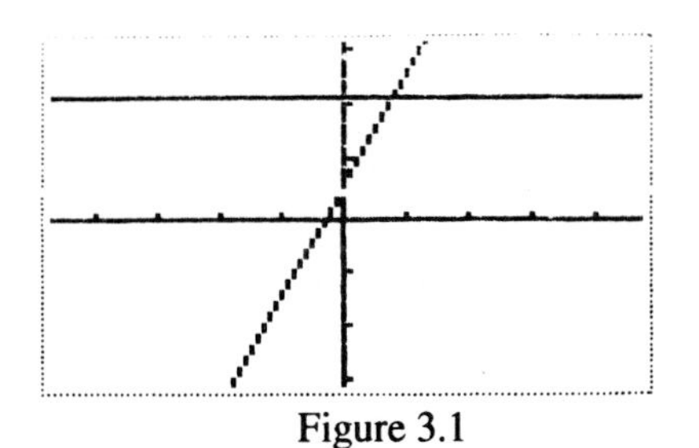

Figure 3.1

NOTE for the 6300: Use [–38, 38] by [–22, 22] for the Integer screen. Use the [:] key instead of pressing [SHIFT] [EXE].

The Multigraph Method for Solving Linear Equations

Graph each side of the equation separately, then use [TRACE] to find the intersection. Start with the Integer window for the range.

Example: Solve $17 - 1.5x = -10$.

Clear the graph screen by pressing

[SHIFT] [F5:CLS] [EXE].

Enter the two functions to graph by pressing

[GRAPH] **17** [−] **1** [.] **5** [X/θ/T] [SHIFT] [↵] [GRAPH] [−] **10** [EXE]

(Fig. 3.2). Press [F1] to activate the Trace cursor.

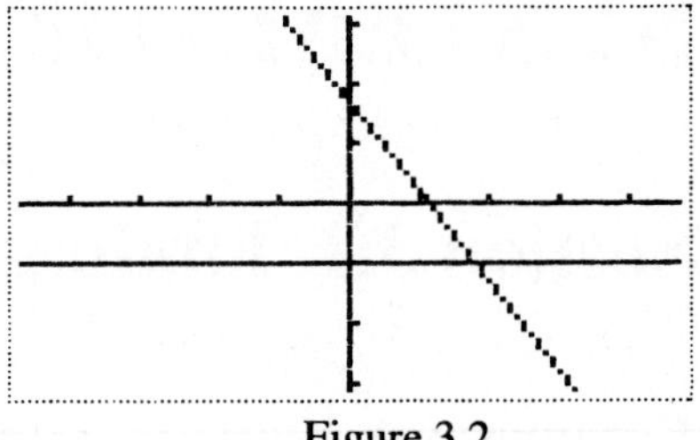

Figure 3.2

Use [▶] to move the cursor to the intersection of the lines. Use [▲] and [▼] to toggle between the two lines. If the y value does not change, you have found the intersection. The y-coordinate of the intersection is the solution of the equation. It is stored as x and can be used in calculations. When the y values are the same on both graphs, $x = 18$. Return to the Text screen by pressing [G↔T] and enter

17 [−] **1** [.] **5** [X/θ/T] [EXE].

The result is –10 as desired.

NOTE for the 6300 and 7000: You cannot toggle between the two lines.

Section 3.2 Solving Equations Using Zoom

Solving Equations Using Zoom and the 0.1 Window

The Default window range settings on a Casio give the 0.1 settings discussed in Section 2.1. To set the 0.1 window press

[RANGE] [F1:INIT] [RANGE] [RANGE].

Zooming is the option that allows you to **zoom in** to enlarge a portion of your graph for detailed analysis or **zoom out** for a broader view.

Using the Zoom Menu

To activate the Zoom menu, press [F2] from the Graph screen or press

[SHIFT] [F2:ZM]

from the Text screen. A submenu appears at the bottom of the screen.

(a) Pressing [F2:FCT] displays the factor input screen. The factors are the magnification factor when zooming. The default is 2.

(b) Pressing [F3:xf] zooms in on the graph by the amount of the zoom factors.

(c) Pressing [F4:$x^{1/f}$] zooms out on the graph by the amount of the zoom factors.

(d) Pressing [F5:ORG] returns the screen to the settings when the graph was originally drawn.

NOTE for the 6300 and 7000: To set the 0.1 window press

[SHIFT] [Mcl].

To zoom in by a factor of 2, press

[SHIFT] [X];

to zoom out by a factor of 2, press

[SHIFT] [÷].

NOTE for the 7000: To set a factor of 5 insert

[SHIFT] [FACTOR] **5**

before the graph command.

NOTE for the 6300: Press

[SHIFT] [FACTOR]

to activate the Factor screen. To change to a factor of 5 in both directions, press

5 [EXE] **5** [EXE].

Press

[SHIFT] [Zoom x|f]

to zoom in.

Using Zoom to Solve Equations

Set the range, graph the function, and trace to the intercept; then zoom to find the answer to the required accuracy.

Example: Find a solution to $2.7x + 3.28 = 0$, with an error of at most 0.01.

Use the Default range with *X*scl and *Y*scl set to 0.01 (changing the range will clear the Graph screen). Press

[RANGE] [F1:INIT] [EXE] [EXE] [.] **01** [EXE] [EXE] [EXE] [.] **01** [EXE] [RANGE].

NOTE for the 6300 and 7000: To clear the Graph screen use

[SHIFT] [Mcl]

instead of [F1:INIT]. Do *not* press [RANGE] at the end.

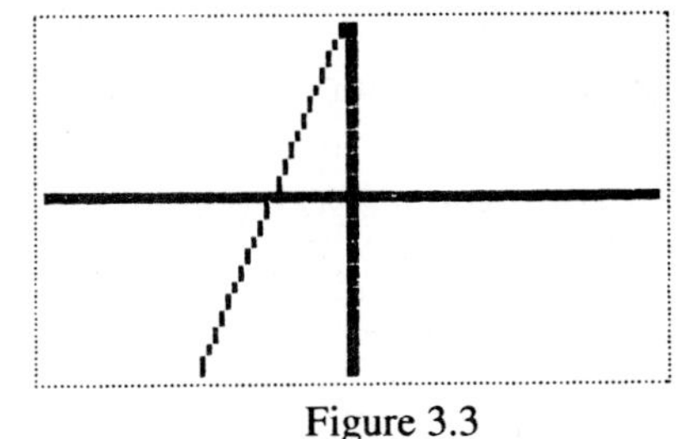

Figure 3.3

Enter

[GRAPH] **2** [.] **7** [X/θ/T] [+] **3** [.] **28** [EXE].

The axis appears as double marks because the tick marks are very close together (Fig. 3.3).

On the Casio 7700 change both Zoom factors to 10 from the Graph screen by pressing

[F2] [F2:FCT] **10** [EXE] **10** [EXE].

Trace to the intersection and zoom in. Press

[SHIFT] [F1:TRC]

and use the [▶] key to move to the intersection. Press

[F2] [F3:xf],

and repeat. Press [F1] and use the [▶] key to move to the intersection. Press

[F2] [F3:xf].

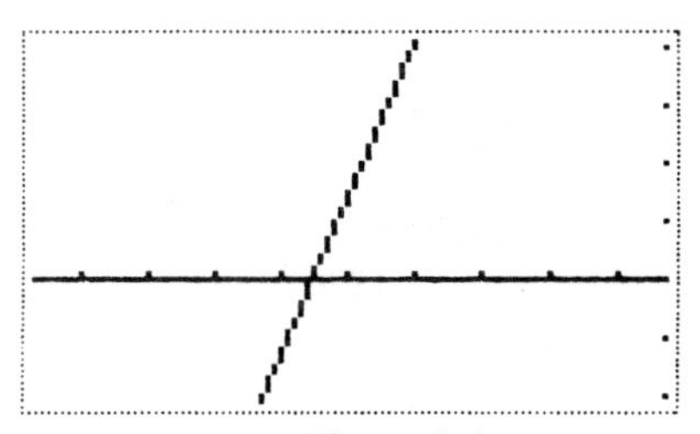

Figure 3.4

You can see the tick marks on the x-axis (Fig. 3.4).

Press the [F1] key and use the [▶] key to move to the intersection; it reads $x = -1.215$. Finding the intersection between two tick marks set 0.01 pixels apart satisfies the error requirement of the problem.

Note that the tick marks for the y-axis appear on the right side of the screen, even though the axis does not.

NOTE for the 7000: Press

[GRAPH] **2** [.] **7** [ALPHA] [X] [+] **3** [.] **28** [EXE] [SHIFT] [TRACE].

Trace to the intersection. To change both zoom factors to 10, press

[G↔T] [▶] [SHIFT] [INS] [SHIFT] [FACTOR] **10** [:] [EXE] [G↔T] [▶] [EXE] [SHIFT] [TRACE].

Trace to the intersection. It reads $x = -1.215$.

NOTE for the 6300: Change the zoom factors both to 5 by pressing

[SHIFT] [FACTOR] **5** [EXE] **5** [EXE].

Then Press

[GRAPH] **2** [.] **7** [ALPHA] [X] [+] **3** [.] **28** [EXE] [TRACE].

Trace to the intersection ($x = -1.2$). Then press

[SHIFT] [Zoom x|f] [SHIFT] [Zoom x|f] [SHIFT] [Zoom x|f].

Trace to the intersection ($x = 1.2144$). Then press

[SHIFT] [Zoom x|f].

Trace to the intersection ($x = 1.215$) which is correct to three places. The factors are set to 5 rather than 10 to get a more accurate answer.

Problems with the Accuracy of a Solution on the Calculator Using Zoom

Graphs are used to approximate answers, not find them exactly.

Example: Solve $3x - 2 = 0$. The answer is obviously $2/3$.

Solve the equation graphically. Start with the Default range, with *X*scl and *Y*scl both reset at 0.01. Enter

[GRAPH] **3** [X/θ/T] [−] **2** [EXE].

Press [F1], and use the [▶] key to trace to the intersection, and then press

[F2] [F3].

Continue to press

until you can see the tick marks on the axis.

The intersection appears between $x = 0.666$ and $x = 0.667$. No matter how many times you zoom, you can only approximate the true answer $x = 2/3$.

NOTE for the 7000: Trace to the intersection ($x = 0.6$) by pressing

[G↔T] [▶] [INS] [SHIFT] [FACTOR] **10** [:] [EXE].

Trace to the intersection by pressing

[G↔T] [▶] [EXE].

Repeat until you see the tick marks on the axis.

NOTE for the 6300: Set the factors to 5 by pressing

[SHIFT] [FACTOR] **5** [EXE] **5** [GRAPH] **3** [ALPHA] [X] [−] **2** [EXE].

Trace to the intersection by pressing

[SHIFT] [Zoom x||f].

Repeat until you see the tick marks on the axis.

Finding an Appropriate Viewing Window

Start with the Default or Integer window and adjust the range values.

Example: Find a window to solve $y = 32000 + 960x$.

Set the range for the Integer window. Note that $y = 32000$ when $x = 0$. Reset the y range values $Y\text{min} = -1000$, $Y\text{max} = 40000$, $Y\text{scl} = 1000$. Press

[GRAPH] **32000** [+] **960** [X/θ/T] [EXE].

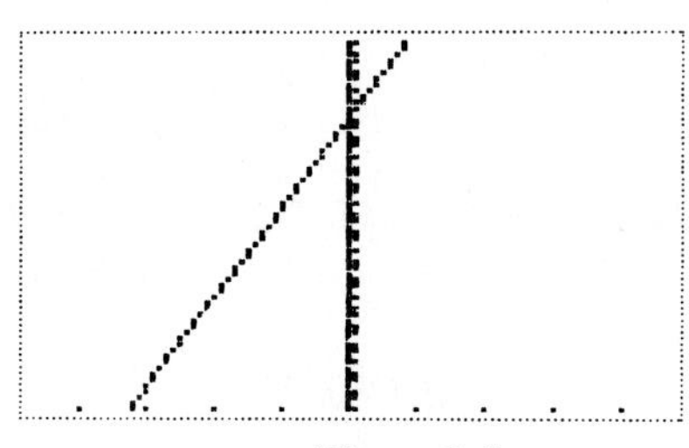

Figure 3.5

You now have a complete graph (Fig. 3.5).

Section 3.3 Some Nonlinear Equations

Using Zoom with the Multigraph Method

You must be careful when using the multigraph method of solving equations; sometimes a graph can be deceiving.

Example: Solve $x^3 = x - 0.5$ using the multigraph method.

Start in the Default window. Press

[RANGE] [F1:INIT] [RANGE] [RANGE].

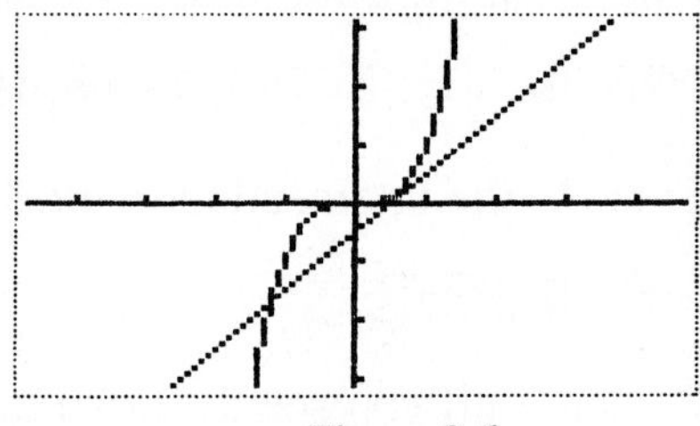
Figure 3.6

Enter

[GRAPH] [X/θ/T] [x^y] **3** [SHIFT] [↵] [GRAPH] [X/θ/T] [−] **0** [.] **5** [EXE].

There appear to be two intersections (Fig. 3.6).

Trace by pressing

[SHIFT] [F1]

to the intersection between $x = 0$ and $x = 1$ using the [▲] and [▼] keys to toggle between curves. You see that you do not have an exact value for the intersection. Zoom in on that point by pressing

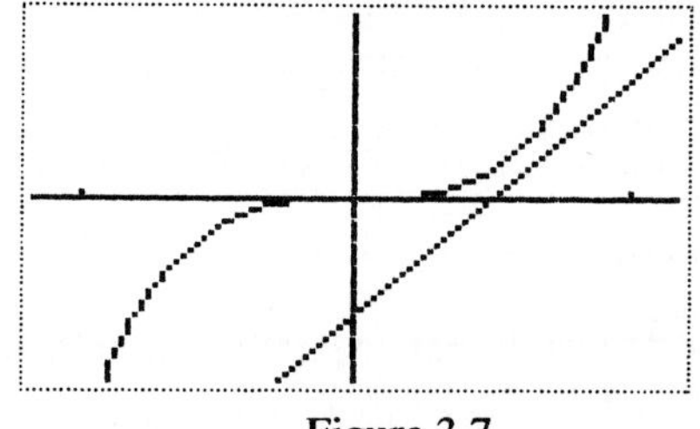
Figure 3.7

[F2] [F3].

You can now see that the graphs do not intersect on that interval (Fig. 3.7). Press [PRE] to delete the menu.

Section 3.4 Graphical Representations of Problem Situations

Motion Simulation

In order to Set the parametric solution to the baseball problem, you must set the Casio to Parametric mode by pressing

[MODE] [SHIFT] [X].

Press [M Disp] (Fig. 3.8).

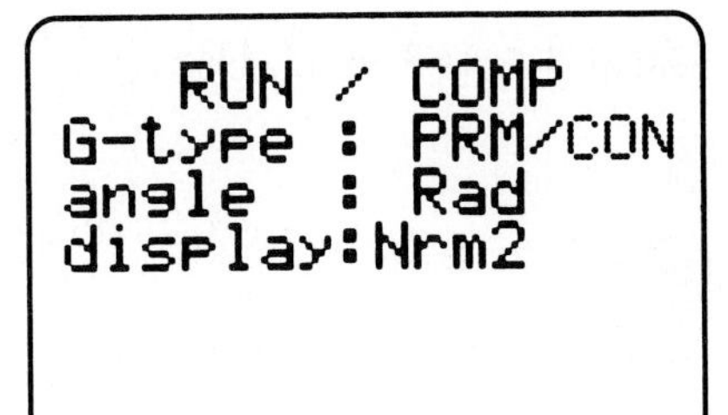

Figure 3.8

Set the viewing window by pressing [RANGE]. Then enter *X*min = 0, *X*max = 5.5, *X*scl = 1, *Y*min = 0, *Y*max = 150, *Y*scl = 10, *T*min = 0, and *T*max = 5.5. Press [RANGE] to return to the Text screen.

Press

Use the [▶] cursor and watch the *y* value carefully to find the maximum value. The approximate maximum value of $Y = 121$ occurs at about $T = 2.76$ seconds.

Change back to Rectangular mode by pressing

[MODE] [SHIFT] [+].

NOTE for the 6300 and 7000: These calculators need a program written to graph in Parametric mode.

Point Plotting with Function Drawing

Basic plotting was discussed in Section 2.2 of this manual.

Example: $y = 500/x$. Plot points for $x = 10, 15, 20, 30, 40,$ and 50. Then graph $y = 500/x$ in the same window. Set the window to $[-10, 60]$, *X*scl = 10 by $[-10, 60]$, *Y*scl = 10. Plot the points by pressing

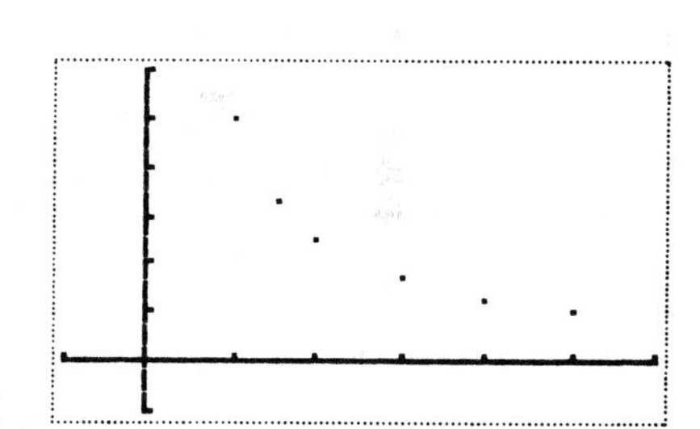

Figure 3.9

[SHIFT] [F3:PLT] **10** [SHIFT] [,] **500** [÷] **10** [EXE] [F3] **15** [SHIFT] [,] **500** [÷] **15** [EXE].

Continue with this option and plot all the points (Fig. 3.9). Press

[GRAPH] **500** [÷] [X/θ/T] [EXE]

and watch the graph go through the points that you plotted (Fig. 3.10).

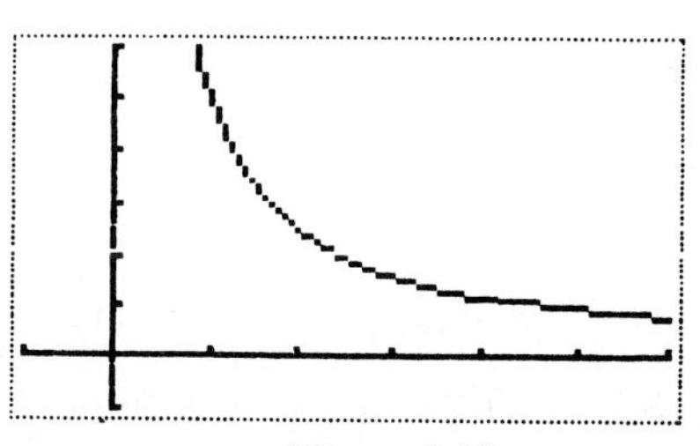

Figure 3.10

NOTE for the 6300 and 7000: Use

[SHIFT] [PLOT]

for SHIFT [F3:PLT]. Use

ALPHA X

for X/θ/T.

Section 3.5 Solving Systems of Equations by Graphing

Using Trace on Two Functions

Use the multigraph method of solution with Trace.

Example: Solve $y = 3x - 2$ and $y = -0.5x + 5$.

Use the Default window by pressing

RANGE [F1:INIT] RANGE RANGE.

Press

GRAPH **3** X/θ/T − **2** SHIFT ↵ GRAPH − **0** . **5** X/θ/T + **5** EXE.

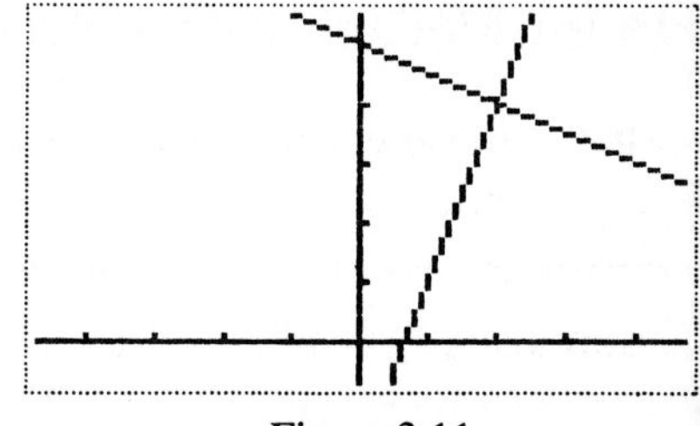

Figure 3.11

The intersection is off the screen. Press ▲ to scroll up. Press ▲ to scroll up again (Fig. 3.11). The intersection is now on the screen. Use F1 and ▶ to trace to the intersection. Use ▼ and ▲ to see if the value of y changes, at $x = 2$, $y = 4$ on both lines.

Solving Systems with Zoom

Example: Solve $y = -2x + 175$ and $y = 3x - 320$.

Set the Integer window [–47, 47] $X\text{scl} = 5$ by [–31, 31] $Y\text{scl} = 5$. Press

GRAPH − **2** X/θ/T + **175** SHIFT ↵ GRAPH **3** X/θ/T − **320** EXE.

The graphs do not appear on the screen. You need to zoom out to find the intersection. When zooming within the Integer range it is convenient to set the zoom factors at 10 so the window remains friendly. Press

[F2] [F2:FCT] **10** [EXE] **10** [EXE].

Now zoom out to see the intersection by pressing [F4:x1/f] (Fig. 3.12). Use [F1] and [▶] to trace to the intersection. Now zoom in to get a better accuracy by pressing [F2] [F3:xf] (Fig. 3.13). Use [F1] and [▶] to trace to the intersection. For $x = 99$ use [▲] and [▼] to toggle between the lines. The y values are the same. The solution is $x = 99$, $y = -23$. Setting the zoom factors at 5 or 10 from the Integer window keeps clean values for the variables.

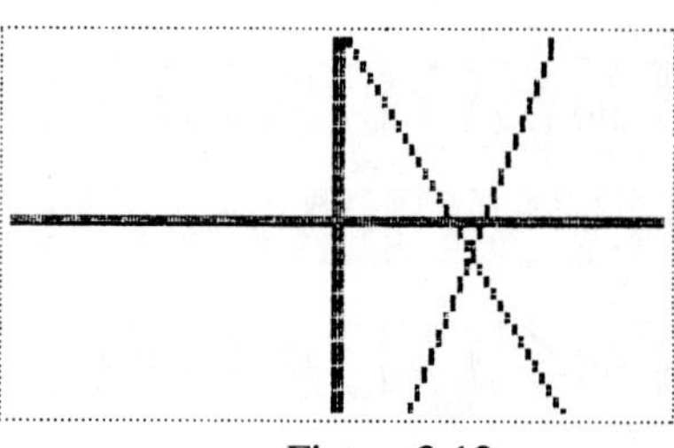

Figure 3.12

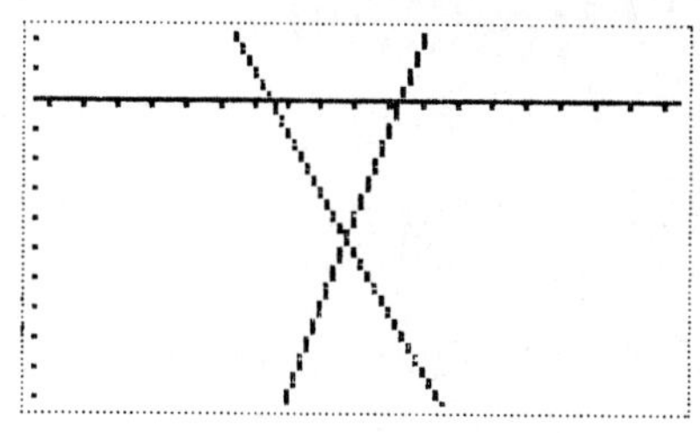

Figure 3.13

NOTE for the 7000: Pressing

[▶] [SHIFT] [INS] [SHIFT] [FACTOR] **10**

before [GRAPH] will zoom out by a factor of 10 on both the x- and y-axes. Use a factor of 1/10 to zoom in by a factor of 10.

NOTE for the 6300: The Integer window is [–38, 38], *X*scl = 10 by [–22, 22], *Y*scl = 10. Change the factors by pressing

10 [SHIFT] [FACTOR] **10** [EXE] **10** [EXE] [SHIFT] [Zoom xI|f].

This keystroke sequence zooms in. Pressing

[SHIFT] [Zoom xI|f]

zooms out.

CHAPTER 4 LINEAR INEQUALITIES AND SYSTEMS OF LINEAR INEQUALITIES

Section 4.1 Solving Linear Inequalities Algebraically

Refer to Section 1.3 of this manual for a discussion of evaluating an expression with the replay keys and edit features.

Section 4.4 Solving Absolute Value Equations and Inequalities

Refer to Section 1.5 of this manual for a discussion of generating tables.

Section 4.5 Linear Inequalities in Two Variables

The Casio calculator has an Inequality mode with shading to solve inequalities. To access this option press

[MODE] [SHIFT] [÷].

Press [M Disp]. Your screen should show Figure 4.1.

```
   RUN / COMP
G-type : INQ/CON
angle  : Rad
display:Nrm2
```

Figure 4.1

Shading

The Shade command is automatic on this calculator.

Example: Graph $y > 4x - 5$.

Press [GRAPH]. A submenu appears on the screen. Press

[F1:Y>] **4** [X/θ/T] [−] **5** [EXE].

All points in the shaded region satisfy the inequality. To find the point where the shaded area crosses the *x*-axis, trace and zoom by a factor of 2 by pressing

[F1] [F2] [F2:FCT] **2** [EXE] **2** [EXE].

Press

[SHIFT] [F1:TR] [EXE] [F1]

and use [▶] to trace the intersection. Press

[F2:FCT] [F3:xf] [F1]

and trace to the intersection. This value is $x = 1.25$ (Fig. 4.2).

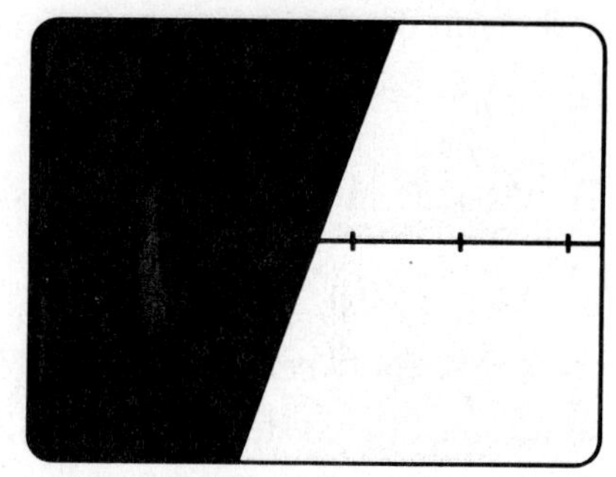

Figure 4.2

NOTE for the 6300 and 7000: There is no inequality option on these models. Graph the boundary and use numeric methods.

Section 4.6 Solving a System of Inequalities

Graphing Systems of Inequalities Using Shading

The automatic shading in the Casio Inequality mode makes solving systems of inequalities very easy.

Example: Sketch the graph of the solution of the system $y < -x + 9$ and $y > 3x - 15$.

Start by setting the Default window. Press

[RANGE] [F1:INIT] [RANGE] [RANGE].

Then press

GRAPH [F2:Y<] (−) X/θ/T + **9** SHIFT ↵ GRAPH [F1:Y>] **3** X/θ/T − **5** EXE.

The intersection does not show on the screen. Set the zoom factors to 5 by pressing

F2 [F2:FCT] **5** EXE **5** EXE

and zoom out by pressing

[F4:x^1/f].

Trace to the intersection by pressing

and toggle between the graphs. The intersection is at (6, 3) and the solution of the system of inequalities is the shaded area (Fig. 4.3). To return to rectangular mode, press

MODE SHIFT +.

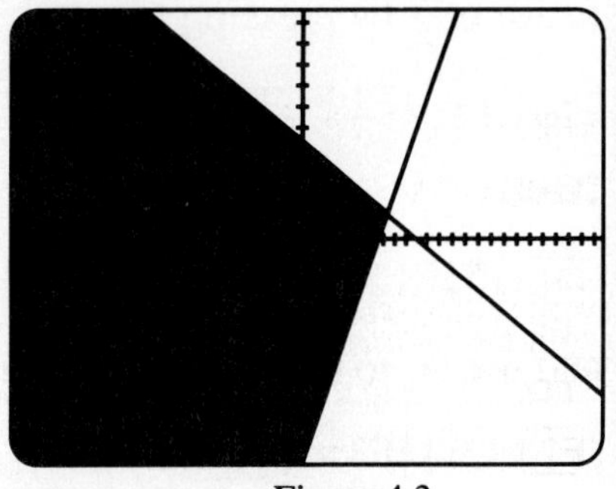

Figure 4.3

NOTE for the 6300 and 7000: There is no inequality option on these models. Inequalities must be solved algebraically.

CHAPTER 5 POLYNOMIALS

Section 5.1 Exponents and Their Properties

Using Trace to Show that Two Graphs Appear to Be Identical

You can find out if two graphs "appear to be identical" using the **Trace** option. After the graphs are drawn activate Trace by pressing [F1]. A blinking star appears on the second graph drawn. It can be moved along the curve with [▶] and [◀]. It can be moved between the two graphs with [▲] and [▼]. As the cursor moves, the coordinates of the lighted pixel are shown on the bottom of the screen. If, for a given value of x, the y values of the two graphs are the same, both graphs go through the same point. Refer to Section 3.1 of this manual for a discussion of moving the Trace cursor between the two graphs.

To show that $(2x^3)(3x^2) = 6x^5$, use the Default range. Press

[RANGE] [F1:INIT] [RANGE] [RANGE].

Then press

[AC] [GRAPH] **2** [X/θ/T] [x^y] **3** [×] **3** [X/θ/T] [SHIFT] [x^2] [SHIFT] [↵] [GRAPH] **6** [X/θ/T] [x^y] **5** [EXE].

Then press [F1] and with the [▶] cursor trace to $x = -0.5$, $y = -0.1875$. Use the [▲] key and note that y does not change. The value of y on both curves is the same for each cursor location (Fig. 5.1).

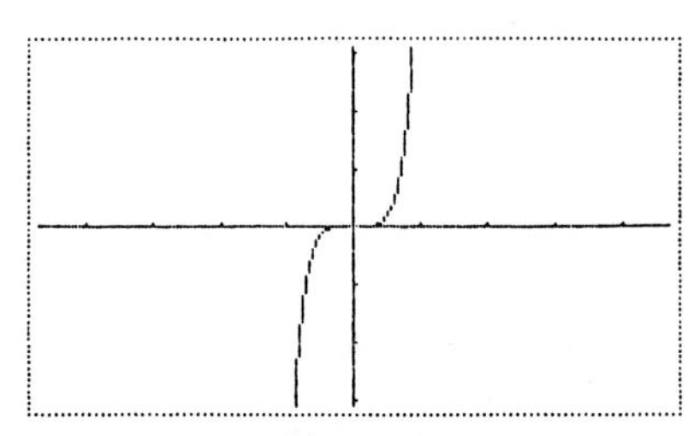

Figure 5.1

NOTE for the 6300 and 7000: On the 6300 and the 7000, only the x value is displayed with [TRACE]. The y value is displayed by pressing [X↔Y]. The cursor will not move between graphs.

Scientific Notation

To enter a number in scientific notation from the keyboard, use the [EXP] key to enter 1.23×10^3. Press

1 [.] **23** [EXP] **3** [EXE].

The result is 1230.

To set the calculator to give all the answers in scientific notation, press

[SHIFT] [DISP] [F2:Sci]

and choose the number of digits to display.

Example: Write 1230 in scientific notation with a three-digit display and a four-digit display by pressing

1230 [EXE] [F2:Sci] **3** [EXE] [SHIFT] [DISP] [F2:Sci] **4** [EXE]

(Fig. 5.2). Pressing

[F3:Nrm] [EXE]

return you to the floating decimal point mode.

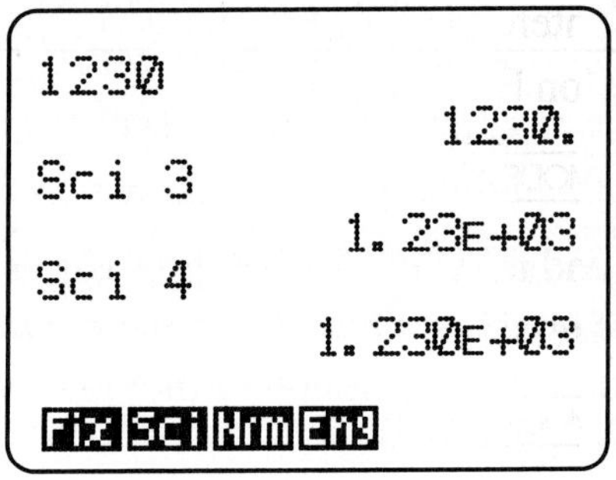

1230
1230.
Sci 3
1.23E+03
Sci 4
1.230E+03
Fix Sci Nrm Eng

Figure 5.2

To toggle between scientific and decimal notation, use the [F3:Nrm] key. [F3:Nrm] **2** is decimal, [F3:Nrm] **1** is scientific. Press and hold [M Disp]. If the display does not read 'Nrm **2**', press

[F3:Nrm] [EXE].

Enter

[.] **0052** [EXE] [F3:Nrm] [EXE] [F3:Nrm] [EXE].

(Fig. 5.3). To return to the floating point display, press

[SHIFT] [DISP] [F3:Nrm] [EXE].

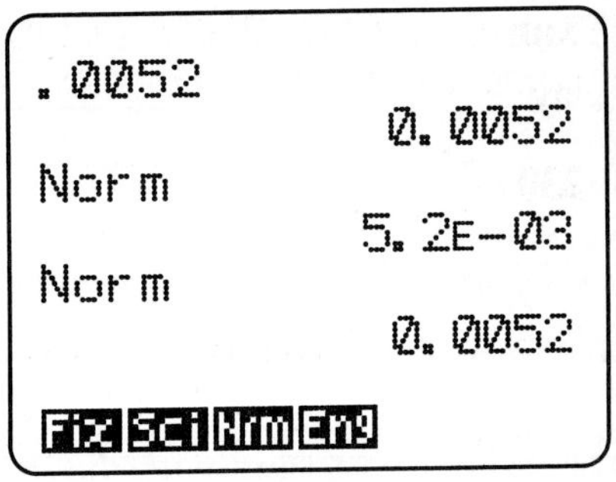

.0052
0.0052
Norm
5.2E-03
Norm
0.0052
Fix Sci Nrm Eng

Figure 5.3

Large Numbers on the Calculator

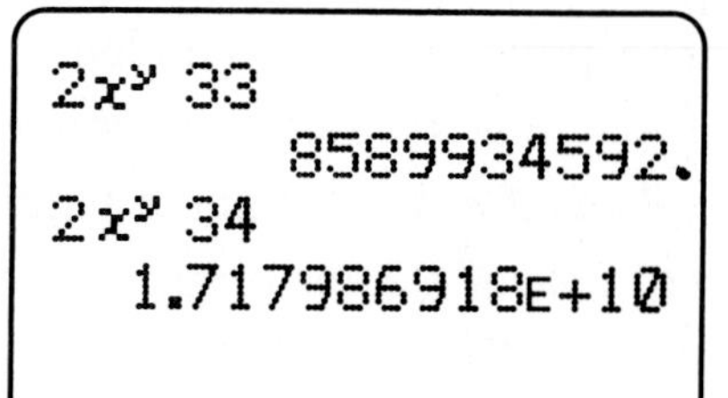

Figure 5.4

When a number is larger than ten digits, it will be represented in scientific notation. For example, 2^{33} is an eleven-digit number, and 2^{34} is represented in scientific notation. Press

2 [x^y] **33** [EXE] **2** [x^y] **34** [EXE]

(Fig. 5.4). To change 2^{33} from decimal to scientific notation, press

2 [x^y] **33** [EXE] [SHIFT] [DISP] [F2:Sci] **9** [EXE].

The number in scientific notation is 8.58993459E+09. Press

[F3:Nrm] [EXE]

to return to the Decimal mode.

NOTE for the 6300 and 7000: A number in scientific notation is entered by pressing the [EXP] key. The display is set to scientific notation by pressing

[MODE] **8**

and the number of digits by pressing the number and then [EXE]. Pressing

[MODE] **9** [EXE]

returns to decimal notation.

Example: Write 1230 in scientific notation with a three-digit display. Press

1230 [EXE] [MODE] **83** [EXE].

Numbers Close to Zero on the Calculator

When you pick the size of a Graphing window, zero may not be on a pixel. For example, graph $y = x$ on [–10, 10] by [–10, 10] and trace to the origin. The closest values you can get to zero are $x = -4\text{E} - 13$ and $y = -4\ \text{E} - 13$.

This problem can be avoided on a Casio by using the Default range or any multiple (either zooming in or zooming out) of the Default values. If Xmax – Xmin is a multiple of 9.4 and Ymax – Ymin is a multiple of 6.2, the origin is at $x = 0$, $y = 0$. The 0.1 window, 0.2 window, and Integer window are examples of windows where a pixel is on the origin.

Section 5.2 Polynomial Expressions
Evaluating a Polynomial Expression and Table Building Using Function Memory

Evaluating a polynomial expression was introduced in Section 1.3 in this manual. Here we will review the method using the function memory. Evaluate

$2x^3 - 4x^2 + 5x - 7$

for $x = -1$. Press

[AC] [SHIFT] [F MEM] **2** [X/θ/T] [x^y] **3** [−] **4** [X/θ/T] [SHIFT] [x^2] [+] **5** [X/θ/T] [−] **7**.

Pressing

[F1:STO] **1** [AC]

stores the function in f_1. Now store –1 in 'X' by pressing

[−] **1** [→] [X/θ/T] [SHIFT] [↵] [F3:fn] **1** [EXE].

The answer is –18 (Fig. 5.5).

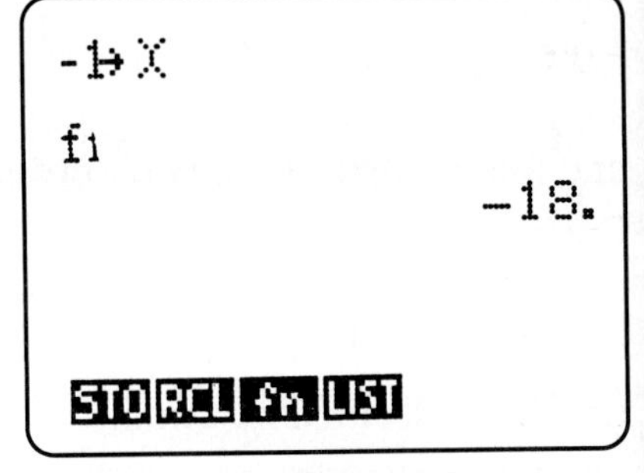

Figure 5.5

To evaluate the polynomial for a different value of x, change the value and execute. Pressing

[▶] **2** [DEL] [EXE]

evaluates the polynomial at $x = 2$, and the answer is 3 (Fig. 5.6).

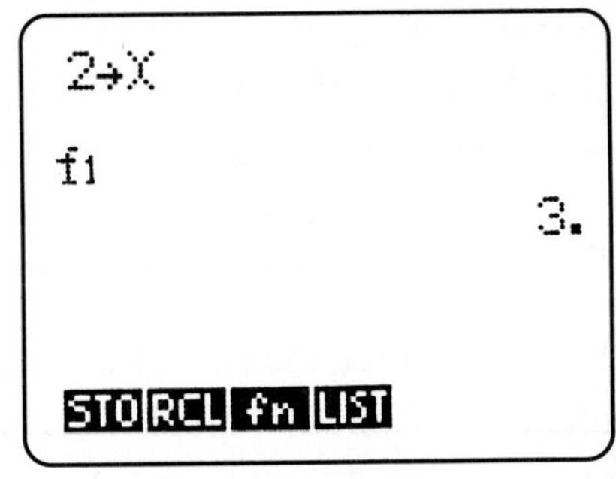

Figure 5.6

NOTE: Make the changes, then insert or delete since you do not want to remain in Edit mode after the change. Table building is quite efficient with this method.

NOTE for the 6300 and 7000:

Example: Evaluate

$2x^3 - 4x^2 + 5x - 7$

for $x = -1$ and $x = 2$.
On the 7000 press

[(-)] **1** [→] [ALPHA] [X] [SHIFT] [EXE] **2** [ALPHA] [X] [x^y] **3** [−] **4** [ALPHA] [X] [x^y] **2** [+] **5** [ALPHA] [X] [−] **7** [EXE].

The answer is –18. Press

[▶] **2** [DEL] [EXE].

The answer is 3.

Note for the 6300: Press

[−] **1** [ALPHA] [X] [:] **2** [ALPHA] [X] [x^y] **3** [−] **4** [ALPHA] [X] [x^2] [+] **5** [ALPHA] [X] [−] **7** [EXE].

The answer is –18. Press

[▶] **2** [DEL] [EXE].

The answer is 3.

When a function is stored in [F MEM] it may be recalled by pressing [F2:RCL] and its position or by pressing [F3:fn], which recalls the symbolic representation of the function. Algebraic operations can be used on the symbolic forms.

Example: Find the sum of the polynomials

$6x^4 + 2x^3 + 2x^2 + 10x + 1$

and

$-5x^4 - 7x^3 + 3x^2 - 3x - 8$,

and use a graph to show that the resulting quartics appear identical. Use [–4.7, 4.7] by [–10, 10]. Refer to Section 2.1 for a discussion of the 0.1 window with the vertical dimension changed. Press

[SHIFT] [F MEM] [AC] **6** [X/θ/T] [x^y] **4** [+] **2** [X/θ/T] [x^y] **3** [+] **2** [X/θ/T] [SHIFT] [x^2] [+] **10** [X/θ/T] [+] **1** [F1:STO] **1** [AC] [–] **5** [X/θ/T] [x^y] **4** [–] **7** [X/θ/T] [x^y] **3** [+] **3** [X/θ/T] [SHIFT] [x^2] [–] **3** [X/θ/T] [–] **8** [F1:STO] **2** [AC] [F3:fn] **1** [+] [F3:fn] **2** [F1:STO] **3** [AC] [X/θ/T] [x^y] **4** [–] **5** [X/θ/T] [x^y] **3** [+] **5** [X/θ/T] [SHIFT] [x^2] [+] **7** [X/θ/T] [–] **7** [F1:STO] **4** [RANGE] [F1:INIT] [▼] [▼] [▼] [–] **10** [EXE] **10** [EXE] [RANGE] [RANGE] [AC] [GRAPH] [F2:RCL] **3** [SHIFT] [↵] [GRAPH] [F2:RCL] **4**

Figure 5.7

(Fig. 5.7). Press [EXE] (Fig. 5.8). Use [F1], [▲], and [▼] to show that for each x, the y values on the two graphs are the same.

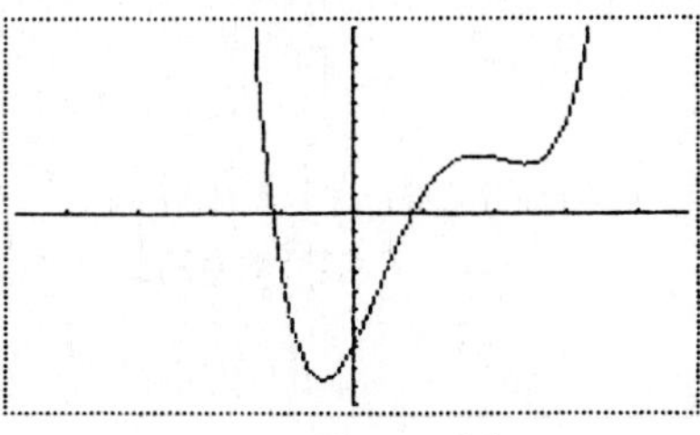

Figure 5.8

Notice that you call up the functions that you need. There is no turning on and off.

NOTE for the 6300 and 7000: Press

[GRAPH] **6** [ALPHA] [X] [x^y] **4** [+] **2** [ALPHA] [X] [x^y] **3** [+] **2** [ALPHA] [X] [x^2] [+] **10** [ALPHA] [X] [+] **1** [+] [(] [SHIFT] [(-)] **5** [ALPHA] [X] [x^y] **4** [–] **7** [ALPHA] [X] [x^y] **3** [+] **3** [ALPHA] [X] [x^2] [–] **3** [ALPHA] [X] [–] **8** [)] [:] [GRAPH] [ALPHA] [X] [x^y] **4** [–] **5** [ALPHA] [X] [x^y] **3** [+] **5** [ALPHA] [X] [x^2] [+] **7** [ALPHA] [X] [–] **7** [EXE].

Use [-4.7, 4.7] [-10, 10] for the range values. If the graphs are the same, the resulting quartics will appear identical. On these calculators you cannot move from one graph to the other. Remember there is no function memory on these calculators.

Evaluating Expressions that Have Two or More Variables

Example: The volume of a tennis ball container is given by the formula

$\pi x^2 y + (2\pi/3)x^3$.

Evaluate at $x = 3.75$, $y = 16.25$. Press

[AC] [PRE] 3 [.] 75 [→] [X/θ/T] [SHIFT] [↵] 16 [.] 25 [→] [ALPHA] [Y] [SHIFT] [↵] [SHIFT] [π] [X/θ/T] [SHIFT] [x^2] [ALPHA] [Y] [+] 2 [SHIFT] [π] [a b/c] 3 [X/θ/T] [x^y] 3 [EXE]

(Fig. 5.9).

```
3.75→X
16.25→Y
πX²Y+2π⌟3X x^y 3
          828.3496255
```

Figure 5.9

NOTE for the 6300 and 7000: Use the [:] key instead of [SHIFT] [↵]. The [x^2] key is not a Shift key. Use [ALPHA] [X] for [X/θ/T] and use [÷] for the [a b/c].

Section 5.4 Polynomial Equations and Factoring

Refer to Section 2.1 of this manual for a discussion of the 0.1 window.

CHAPTER 6 RATIONAL EXPRESSIONS AND EQUATIONS

Section 6.1 Simplifying Rational Expressions

Using Parentheses in Rational Expressions

Parentheses are necessary in rational expressions.

Example: Graph $y = (x - 2)/x$ and $y = x - (2/x)$ on the Default range. What is the difference?

[AC] [RANGE] [F1:INIT] [RANGE] [RANGE] [AC] [GRAPH] [(] [X/θ/T] [−] 2 [)] [÷] [X/θ/T] [SHIFT] [↵] [GRAPH] [X/θ/T] [−] 2 [÷] [X/θ/T] [EXE]

(Fig. 6.1).

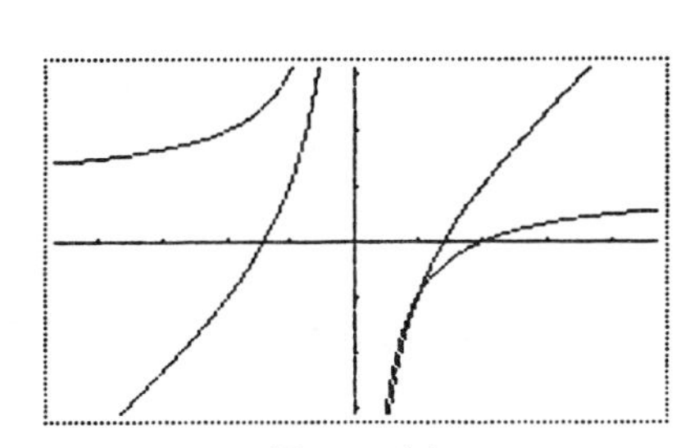

Figure 6.1

The graphs of the two functions are not identical. Remember the order of operations. In

$y = (x - 2)/x$,

the subtraction is done first, then the division. The x^{-1} key can be used in rational functions. Then the function

$1 \div (x - 2)$

can be entered by pressing

GRAPH (X/θ/T − 2) SHIFT x^{-1} EXE

(Fig. 6.2).

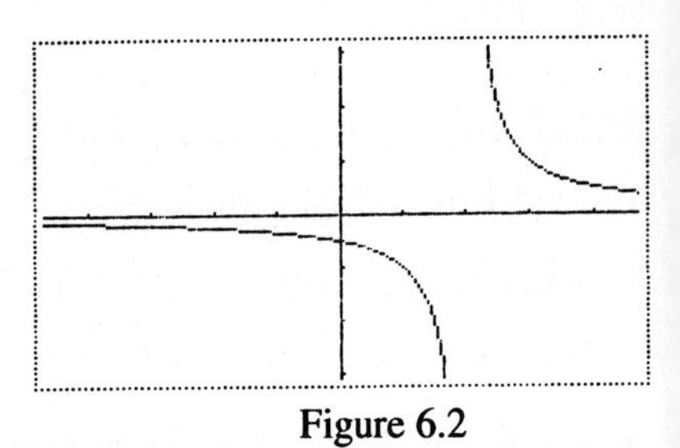

Figure 6.2

Error Messages with Rational Expressions

Division by zero is not defined in arithmetic or algebra. If you try to divide by zero, a math error occurs.

Example: Evaluate $1/x$ for $x = 0$. Press

AC **0** → X/θ/T SHIFT ↵ **1** ÷ X/θ/T EXE

(Fig. 6.3).

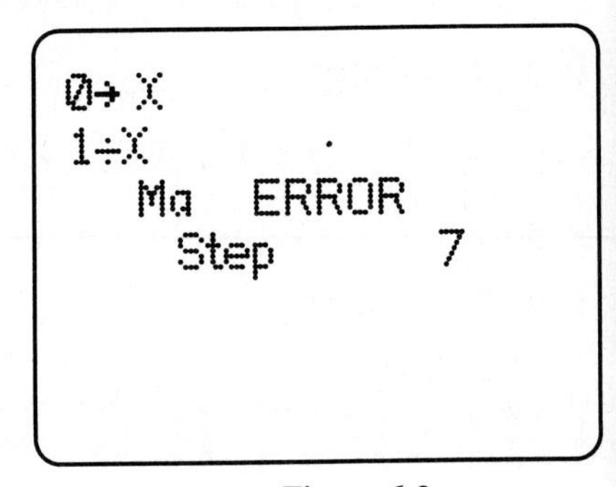

Figure 6.3

Section 6.2 Multiplying and Dividing Rational Expressions

See Section 3.1 of this manual for a discussion of selecting and deselecting graphs.

Section 6.3 Adding and Subtracting Rational Expressions

Extra Lines when Graphing Rational Expressions

With some settings of the range parameters, a vertical line that does not belong on the graph connects points on a rational function. Using the default range (decimal –0.1 range) or any multiple of it avoids this problem. Using the [–10, 10] by [–10, 10] setting introduces the extra line. Start with the equation

$y = 1 \div (x - 2)$

(Fig. 6.2). Press

[GRAPH] **1** [÷] [(] [X/θ/T] [–] **2** [)] [RANGE] [F1:INIT] [RANGE] [RANGE] [EXE]

(Fig. 6.2). Press

[RANGE] [–] **10** [EXE] **10** [EXE] [EXE] [–] **10** [EXE] **10** [EXE] [RANGE] [RANGE] [EXE]

(Fig. 6.4).

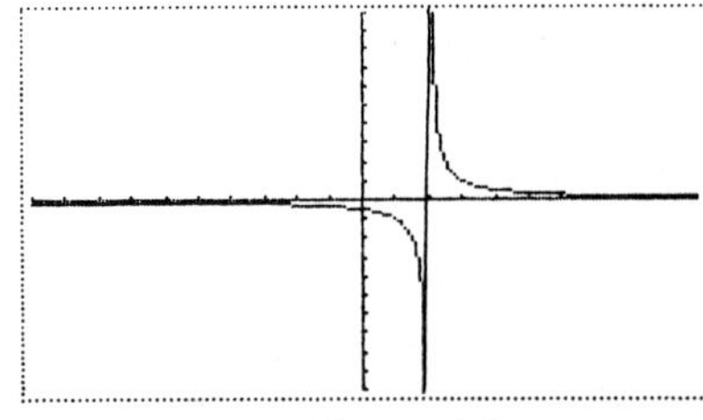

Figure 6.4

Section 6.4 Complex Fractions

Using Parentheses with Complex Fractions

Parentheses are needed in complex fractions. The simplification of a complex fraction requires that the fractions in the numerator be combined, then the fractions in the denominator are combined, and then the division can be done. As this is not the usual order (division before addition), parentheses must be used. Put parentheses around the entire numerator and around the entire denominator.

Example. Evaluate

$$\frac{\frac{1}{2}+\frac{2}{3}}{\frac{5}{6}-\frac{3}{4}}.$$

Press

[AC] [(] **1** [a b/c] **2** [+] **2** [a b/c] **3** [)] [÷] [(] **5** [a b/c] **6** [−] **3** [a b/c] **4** [)] [EXE]

(Fig. 6.5).

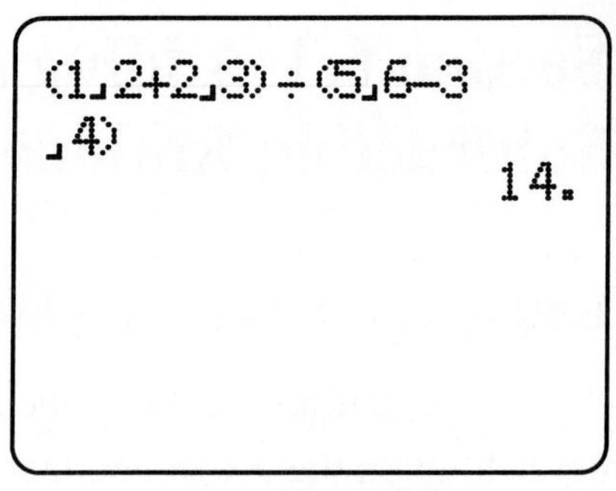

Figure 6.5

NOTE for the 6300 and 7000: Use [÷] in place of [a b/c].

Section 6.6 Solving Equations Containing Rational Expressions

Refer to Section 2.1 of this manual for a discussion of the 0.2 window and the Integer window.

CHAPTER 7 RATIONAL EXPONENTS, RADICALS, AND COMPLEX NUMBERS

Section 7.1 Roots and Radicals

Even Roots of Negative Numbers

The Casio family of graphic calculators does operations only with real numbers. If the answer to a calculation is not a real number, an error message occurs.

Example: Evaluate $\sqrt{-3}$, by pressing

[AC] [√] [(] [SHIFT] [(−)] **3** [)] [EXE]

(Fig. 7.1).

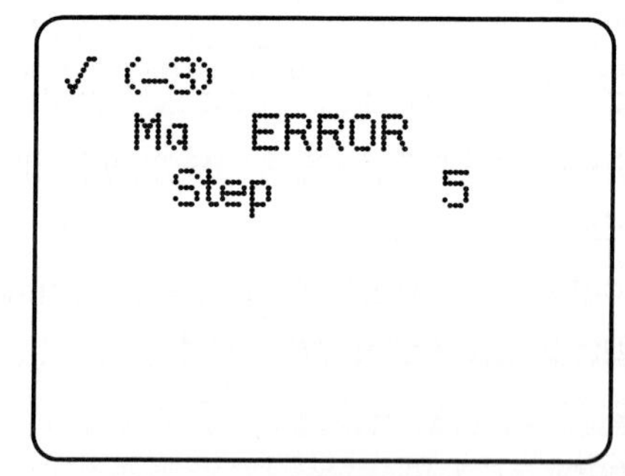

Figure 7.1

The square root (or other even roots) of negative numbers are not real numbers, since multiplying positive numbers by themselves or negative numbers by themselves an even number of times results in a positive number. Unfortunately, $\sqrt{-3}$ is not a real number, but $\sqrt[3]{-8}$ is a real number. Press

[AC] [SHIFT] [$\sqrt[x]{\ }$] [(] [SHIFT] [(-)] **8** [)] [EXE].

The answer is –2.

Evaluating Expressions with Fractional Exponents

Radicals can be evaluated using fractional exponents, an alternate notation for a radical.

Example: Evaluate $\sqrt{36} = 36^{1/2}$ and $\sqrt[3]{125} = 125^{1/3}$. Press

[AC] **36** [x^y] [(] **1** [a b/c] **2** [)] [EXE] **125** [x^y] [(] **1** [a b/c] **3** [)] [EXE]

(Fig. 7.2).

```
36x^y(1┘2)
                 6.
125x^y(1┘3)
                 5.
```

Figure 7.2

Finding Roots

Roots can be evaluated using either fractional exponents or the [$\sqrt[x]{\ }$] key.

Example: Evaluate $\sqrt[4]{8}$ and $\sqrt[3]{-8.75}$, by pressing

[AC] **4** [SHIFT] [$\sqrt[x]{\ }$] **8** [EXE] **3** [SHIFT] [$\sqrt[x]{\ }$] [–] **8** [.] **75** [EXE]

(Fig. 7.3).

```
4ˣ√ 8
        1.681792831
³√ -8.75
        -2.06064265
```

Figure 7.3

Setting the Number of Decimal Places with FIX

The number of decimal places displayed can be set using the display menu. Press

[AC] **4** [SHIFT] [$\sqrt[x]{\ }$] **8** [EXE] [SHIFT] [DISP] [F1:Fix] **2** [EXE] [F3:Nrm] [EXE]

(Fig. 7.4).

```
4ˣ√ 8
        1.681792831
Fix 2
               1.68
Norm
        1.681792831
Fix Sci Nrm Eng
```

Figure 7.4

Section 7.2 Rational Exponents and Radicals

Refer to Section 2.1 of this manual for a discussion of evaluating an expression with rational exponents on a Casio.

Section 7.6 Complex Numbers

Using the Graphs to Work with Complex Numbers

The Casio fx-7000G, fx-7700G, and fx-8700G do not have complex number capabilities. However, the scientific models fx-1500 and fx-300 have complex arithmetic capabilities and can be used. Graphing capabilities are not necessary for complex arithmetic.

Chapter 8 Quadratic Functions

Section 8.4 Sketching Graphs of Quadratic Functions

Refer to Section 2.1 of this manual for a discussion of the 0.2 window.

Section 8.6 Vertex and Symmetry of a Parabola

Refer to Section 1.1 for a discussion of fractions on a Casio.

CHAPTER 9 HIGHER-ORDER SYSTEMS OF EQUATIONS AND MATRICES

Section 9.1 Solving Systems of Equations in Three Variables

Systems Involving Two and Three Variables

Equations in two variables can be solved in several ways. Equations in three variables cannot be graphed in two dimensions. You can only find the solution of a system of equations in three variables numerically. A graphical solution is not possible.

Section 9.2 Matrix Algebra

Entering a Matrix into a Casio

To do problems containing matrices, the calculator must be put into Matrix mode. Press

MODE **0**.

The Matrix menu appears on the screen. The dimensions of matrices *A*, *B*, and *C* are displayed (Fig. 9.1). You may pick the dimensions of matrix *A*, written [*A*], and of *B*, written [*B*]. Matrix *C*, written [*C*], is the result obtained from arithmetic operations on matrices [*A*] and [*B*]. Matrix [*C*] can be interchanged with either [*A*] or [*B*] to do successive calculations.

Figure 9.1

NOTE for the 6300 and 7000: These models do not have matrix capabilities.

Example: Enter the 2×2 matrix

$$A = \begin{bmatrix} 4 & -3 \\ 2 & 6 \end{bmatrix}.$$

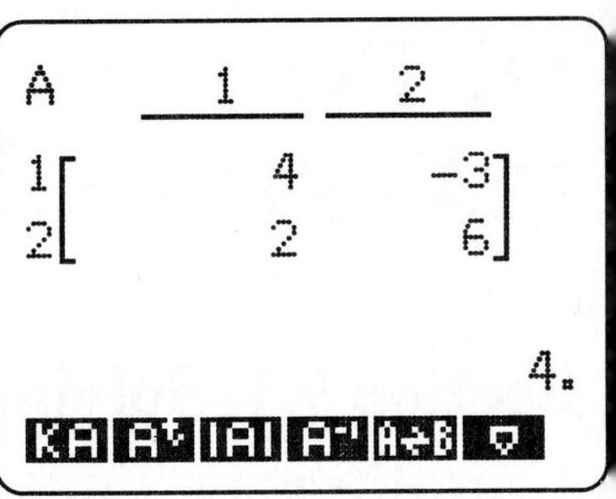

Figure 9.2

Change to Matrix mode by pressing [MODE] **0** and read the dimension of matrix A. If it is a 2×2, press [F1:A] and enter the matrix. Press

4 [EXE] [(−)] **3** [EXE] **2** [EXE] **6** [EXE]

(Fig. 9.2). If A is not a 2×2 matrix, press

[F1:A] [F6:↓] [F1:DIM] **2** [EXE] **2** [EXE].

Then enter the matrix as previously shown. Press [PRE] returns to the Matrix menu and clears the screen.

Adding and Subtracting Matrices

Example: Find

$[A] + [B]$ and $[A] - [B]$.

You have already entered matrix A. You must enter matrix

$$B = \begin{bmatrix} 2 & 5 \\ 1 & 2 \end{bmatrix}.$$

Start from the Matrix menu by pressing [PRE] and then press [F2:B]. If $[B]$ has dimensions 2×2 enter matrix B. Press

2 [EXE] **5** [EXE] **1** [EXE] **2** [EXE].

If not, press

[F6:↓] [F1:DIM] **2** [EXE] **2** [EXE].

Then enter matrix B. The answer is 2 (Fig. 9.3). Pressing [PRE] returns to the Matrix menu.

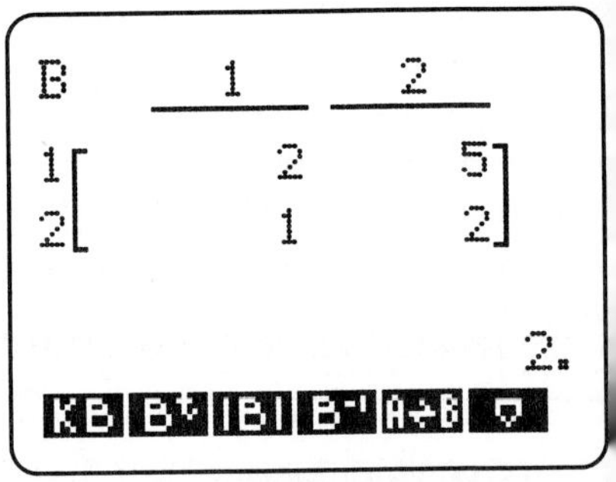

Figure 9.3

As [A] and [B] are now stored, press [F3:+]. Then matrix [C] =[A]+[B] is displayed

$$[A]+[B]=\begin{bmatrix}6 & 2\\ 3 & 8\end{bmatrix}$$

(Fig. 9.4). Press

PRE [F4:–].

This shows

$$[C]=[A]-[B]=\begin{bmatrix}2 & -8\\ 1 & 4\end{bmatrix}$$

(Fig. 9.5).

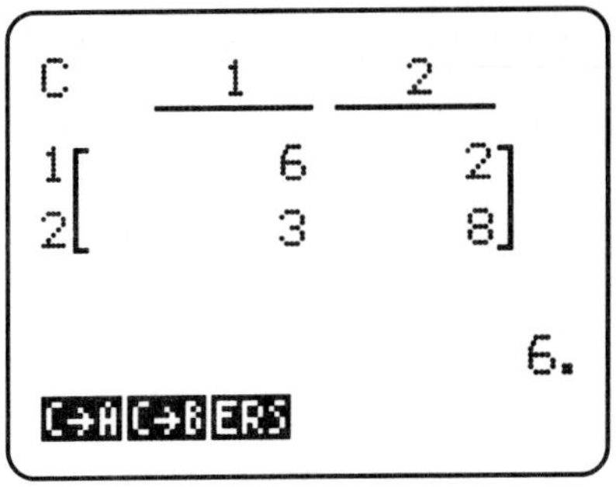

Figure 9.4

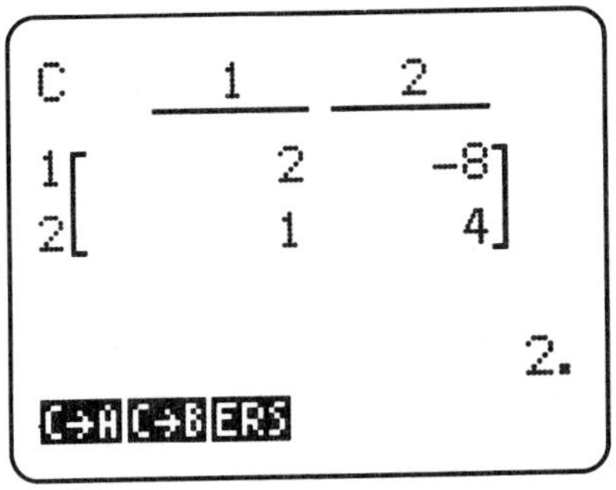

Figure 9.5

Editing Matrices

Matrix entries can be edited by moving to the entry and entering the new number. If the new entry is negative use the negative key by pressing SHIFT (–). Otherwise the difference in the entries will be entered.

Example: Change [A] to $\begin{bmatrix}1 & 2\\ -3 & 4\end{bmatrix}$.

From the Matrix window press

PRE [F1:A] **1** EXE **2** EXE SHIFT (–) **3** EXE **4** EXE.

Fractions in Matrices

The Casio fx-7700G will allow fractional entries in matrices. Although a decimal representation is shown for matrix entries, the fraction is stored and is shown above the Matrix menu on the right. These fractions can be checked by using the ◀ and ▶ keys.

Example: Enter $A=\begin{bmatrix}1/2 & 1/3\\ 1/4 & 5/6\end{bmatrix}$.

From the Matrix window press

[F1:A] **1** [a b/c] **2** [EXE] **1** [a b/c] **3** [EXE] **1** [a b/c] **4** [EXE] **5** [a b/c] **6** [EXE]

(Fig. 9.6).

If the entries of A and B are fractions, then the entries of C are fractions.

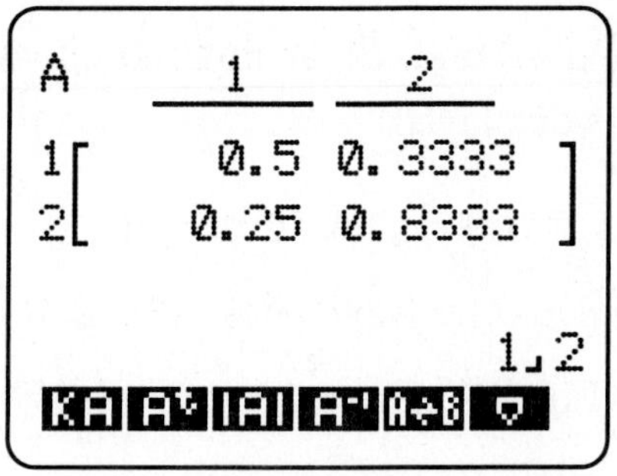

A 1 2
1 [0.5 0.3333]
2 [0.25 0.8333]
1⌟2
KA A^t |A| A^-1 A↔B ▽

Figure 9.6

Scalar Multiplication of Matrices

Matrices can be multiplied by a constant. This is called **scalar multiplication**.

Example: Find 3[A] if

$$[A] = \begin{bmatrix} 2 & 1 & 5 \\ 6 & -1 & 3 \end{bmatrix}.$$

Start from the matrix menu. Press

[PRE] [F1:A] [F6:↓] [F1:DIM] **2** [EXE] **3** [EXE]

to change the dimension. Then pressing

2 [EXE] **1** [EXE] **5** [EXE] **6** [EXE] [(-)] **1** [EXE] **3** [EXE]

enters the matrix (Fig. 9.7). Pressing

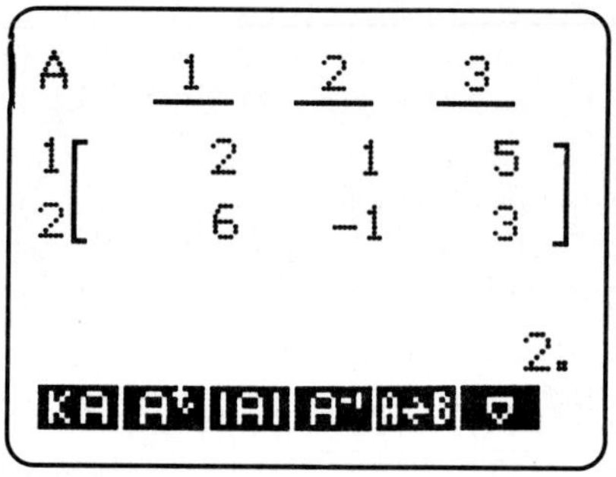

3 [F1:KA]

multiplies [A] by 3 and stores it in [C]. So you have

$$C = \begin{bmatrix} 6 & 3 & 15 \\ 18 & -3 & 9 \end{bmatrix}.$$

A 1 2 3
1 [2 1 5]
2 [6 -1 3]
2.
KA A^t |A| A^-1 A↔B ▽

Figure 9.7

Section 9.3 Multiplication of Matrices

Matrix Multiplication

Matrix multiplication is done from the Matrix menu. Check (M Disp) for the dimensions of matrix A and matrix B. If necessary, change them both to 2×2. Enter the matrices and multiply, by pressing [F5:×].

Example: Matrix $A = \begin{bmatrix} 3 & -4 \\ 2 & -8 \end{bmatrix}$ and matrix $B = \begin{bmatrix} 4 & -2 \\ -1 & 9 \end{bmatrix}$.

Now find $[A] \times [B]$ by pressing

(PRE) [F1:A] **3** (EXE) (SHIFT) ((-)) **4** (EXE) **2** (EXE) (SHIFT) ((-)) **8** (EXE)
(PRE) [F2:B] **4** (EXE) (SHIFT) ((-)) **2** (EXE) (SHIFT) ((-)) **1** (EXE) **9** (EXE)
(PRE) [F5:×].

The product is stored as matrix C (Fig. 9.8).

Figure 9.8

Identity Matrices

The **Identity** matrix is a square matrix with the number one everywhere on the main diagonal, and zero elsewhere. It can be entered into matrices A, B, or C. When a matrix is multiplied by its inverse matrix, the product is the identity.

Inverse Matrices

To find the inverse of a matrix A, use [F4:A^{-1}] in the Matrix submenu of A.

Example: Find the inverse matrix of $A = \begin{bmatrix} 5 & 1 \\ 4 & 1 \end{bmatrix}$.

Start from the Matrix menu by pressing

(PRE) [F1:A].

If A is not 2×2, change the dimension, then enter the matrix by pressing

5 EXE 1 EXE 4 EXE 1 EXE

(Fig. 9.9). Then press [F4:A-1]. The inverse is stored in C (Fig. 9.10).

If the inverse matrix contains fractions, the entries will show as decimals, but will be stored as fractions.

Figure 9.9

Example: Compute the inverse of

$$B = \begin{bmatrix} 4 & -2 \\ -1 & 9 \end{bmatrix}$$

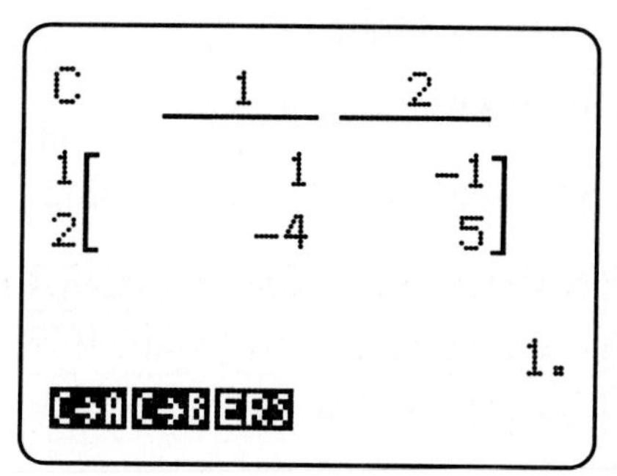

Figure 9.10

by pressing

PRE [F2:B] 4 EXE SHIFT (-) 2 EXE SHIFT (-) 1 EXE 9 EXE [F4:B-1].

The entries show as fractions on the right of the screen (Fig. 9.11).

Determinant of a Matrix

To compute the determinant of matrix B from the previous example, press

PRE [F2:B] [F3:|B|].

The answer is 34.

Some matrices do not have inverses.

Figure 9.11

Example: Find the inverse of

$$A = \begin{bmatrix} 3 & 1 \\ 3 & 1 \end{bmatrix}.$$

Press

PRE [F1:A]

and enter the matrix in [A], then press [F4:A-1]. An error message is displayed (Fig. 9.12). Press ▶ to delete the error message. Press [F3:|A|]. The answer is zero. Matrices with determinants equal to zero do not have inverses.

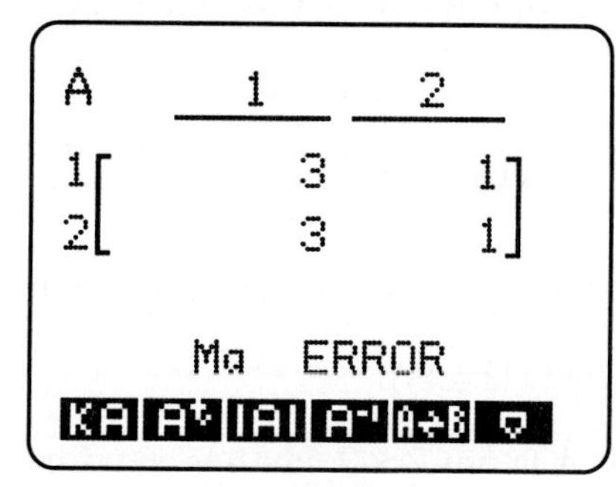

Figure 9.12

Section 9.4 Solving a System of Equations Using Matrices

Using an Inverse Matrix to Solve Systems of Equations

A system of linear equations can be solved with matrices. The system

$AX = B$

has the solution

$[X] = [A^{-1}]\,[B]$.

(Multiplication is not commutative, so the matrice must be entered in this order. Matrix $[A]$ is the coefficient matrix, matrix $[B]$ is the constant matrix, and matrix $[X]$ is the solution matrix.

Example: Enter

$$\begin{aligned} 4x - 3y &= -26 \\ 15x + 11y &= 36 \end{aligned}$$

with

$$A = \begin{bmatrix} 4 & -3 \\ 15 & 11 \end{bmatrix} \text{ and } B = \begin{bmatrix} -26 \\ 36 \end{bmatrix}.$$

Start from the Matrix menu. Be sure that you have dimensioned A as 2×2 and B as 2×1. Enter

PRE AC [F1:A] **4** EXE SHIFT (-) **3** EXE **15** EXE **11** EXE PRE [F2:B] SHIFT (-) **26** EXE **36** EXE PRE [F1:A] [F4:A^{-1}] [F1:C→A] PRE [F5:×].

The answer is stored in

$$[C] = \begin{bmatrix} -2 \\ 6 \end{bmatrix}.$$

So $x = -2$ and $y = 6$. Use

[MODE] [+]

to change back to Computational mode.

CHAPTER 10 CONIC SECTIONS

Conic Sections

Be sure the Mode screen is set as shown in Fig. 10.1. If not, reset to match.

```
RUN / COMP
G-type : REC/CON
angle  : Rad
display:Nrm2
```

Figure 10.1

The equations whose graphs are the conic sections are not functions. When solving for y, a complicated radical occurs. To simplify graphing, enter the radical into the [ⒻMEM] function. Then use it in the functions to graph.

Example: Graph

$$\frac{(x-1)^2}{4} + \frac{(y+2)^2}{9} = 1.$$

Solve this for y to get

$$y = -2 + 3\sqrt{4-(x-1)^2}.$$

Set the range to Default by pressing

[RANGE] [F1:INIT] [RANGE] [RANGE].

Enter

[√] [(] **4** [−] [(] [X/θ/T] [−] **1** [)] [SHIFT] [x²] [)] [SHIFT] [ⒻMEM] [F1:STO] **1** [AC] [GRAPH] [−] **2** [+] **3** [a b/c] **2** [F2:RCL] **1** [SHIFT] [↵] [GRAPH] [−] **2** [−] **3** [a b/c] **2** [F2:RCL] **1** [EXE].

Then press [▼] and press [▼] again (Fig. 10.2).

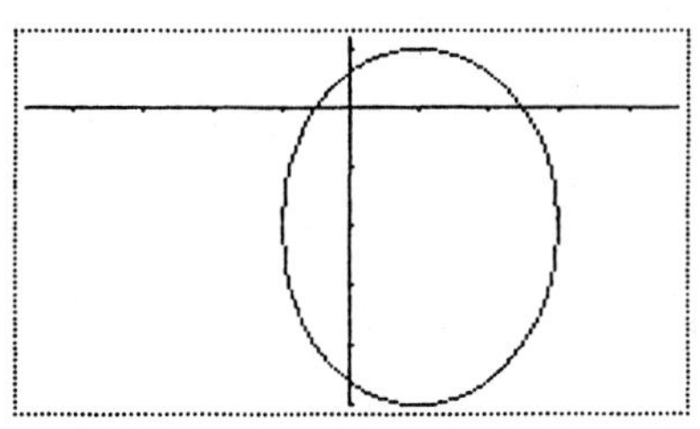
Figure 10.2

Section 10.1 Circles

Graphing a Circle

The conics need to be graphed in the Default range or a multiple of the Default range to show their true shape.

Example: Graph $x^2 + y^2 = 25$.

Set the range to Default by pressing

[RANGE] [F1:INIT] [RANGE] [RANGE].

Enter

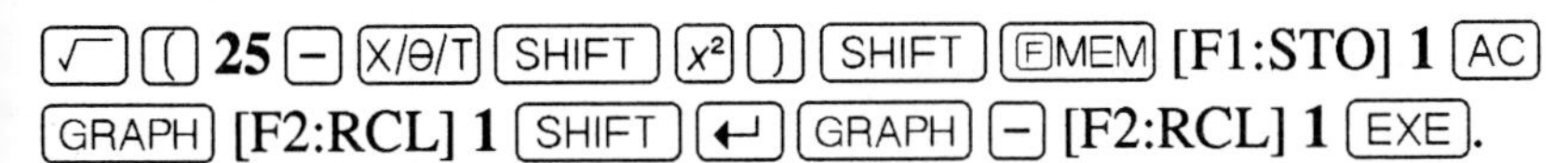
[√] [(] **25** [−] [X/θ/T] [SHIFT] [x^2] [)] [SHIFT] [MEM] [F1:STO] **1** [AC]
[GRAPH] [F2:RCL] **1** [SHIFT] [↵] [GRAPH] [−] [F2:RCL] **1** [EXE].

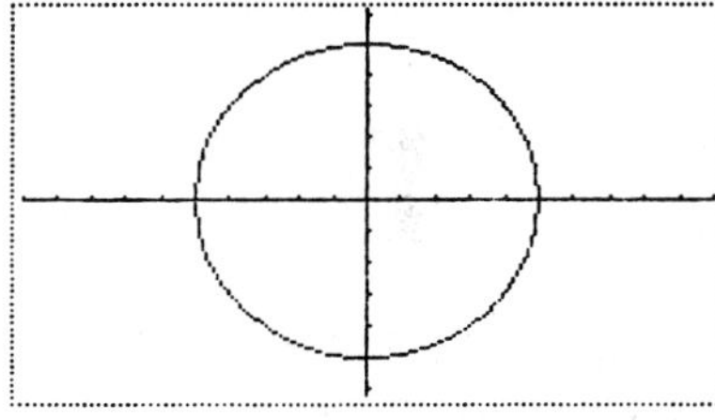
Figure 10.3

The circle does not fit in the window. Press

[F2] [F2:FCT] **2** [EXE] **2** [EXE] [F4:x1/f].

Now you can see the entire circle and it is round (Fig. 10.3). Change the range to [–10, 10] by [–10, 10]. Press

[RANGE] [RANGE] [EXE].

The circle is oval, which is not its true shape (Fig. 10.4).

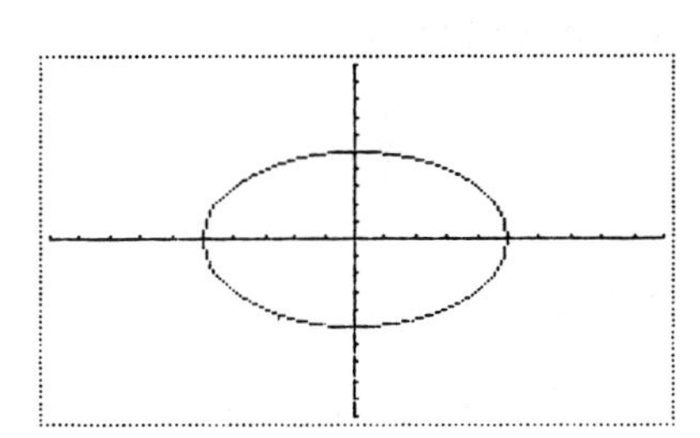
Figure 10.4

Section 10.5 Nonlinear Systems of Equations

Solving Nonlinear Systems Graphically

Review the material in Section 3.5 in this manual on solving systems of equations graphically. To ensure an error of at most 0.01, use *X*scl and *Y*scl as 0.01.

Example: Graphically solve the system

$x^2 + y^2 = 16$ and $y = 2x^2 - 2$.

Set the range to default by pressing

[RANGE] [F1:INIT].

Then change *X*scl and *Y*scl to 0.01. Then press

[RANGE] [RANGE] [AC] [√] [(] **16** [−] [X/θ/T] [SHIFT] [x^2] [)] [SHIFT] [F MEM] [F1:STO] **1** [AC] [GRAPH] [F2:RCL] **1** [SHIFT] [↵] [GRAPH] [−] [F2:RCL] **1** [SHIFT] [↵] [GRAPH] **2** [X/θ/T] [SHIFT] [x^2] [−] **2** [EXE].

Then press [▼] (Fig. 10.5). Press

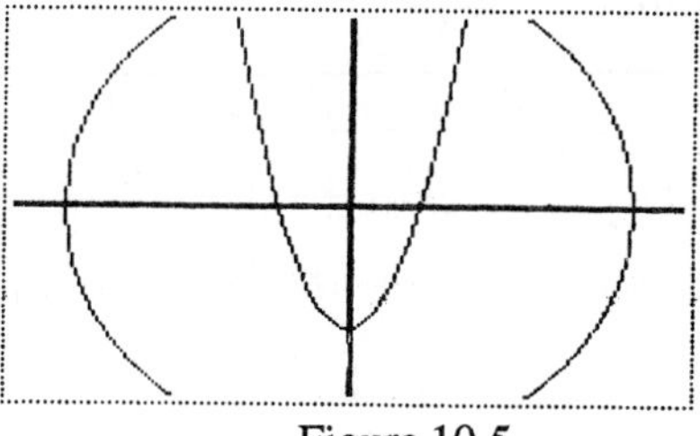
Figure 10.5

[F2] [F2:FCT] **5** [EXE] **5** [EXE]

to set the zoom factor to 5. Press

[PRE] [F1:TRC]

and trace to the left intersection. Use [▶] and [▼] to move along the circle. Pressing

[F2] [F3:xf]

zooms in by a factor of 5.

Continue tracing and zooming until you can read the scale marks (Fig. 10.6). The range is [–1.83, –1.53] by [3.53, 3.73]. The solution is approximated by (–1.678, 3.631). The other solution is (1.678, 3.631).

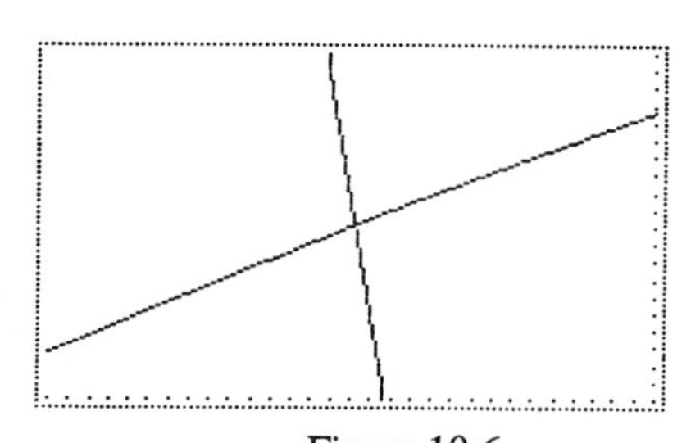
Figure 10.6

CHAPTER 11 EXPONENTIAL AND LOGARITHMIC FUNCTIONS

Section 11.3 Logarithmic Function

The [LOG] and [LN] Keys

Logarithms to the base 10 and base e can be found easily using your calculator. The key [LOG] is used for base 10, and [LN] is used for base $e = 2.71828\ldots$.

Example: Evaluate log 125 and ln 12.3 (Fig. 11.1). Press

[LOG] **125** [EXE] [LN] **12** [.] **3** [EXE].

```
log 125
      2.096910013
ln 12.3
      2.509599262
```

Figure 11.1

Section 11.5 Solving Logarithmic Equations

Revisiting Tables

Review Section 1.3 of this manual on evaluating expressions and building tables using [F MEM].

Example: Generate a table of values for $y = P(1.07)^x$, where $P = 1000$. Use this table to find the value of x when $y = 3000$. Start with $x = 5, 10, 15, 20$. Set the display at **Fix 2** by pressing

[SHIFT] [DISP] [F1:Fix] **2** [EXE] [PRE] **5** [→] [X/θ/T] [SHIFT] [↵] **1000** [(] **1** [.] **07** [)] [x^y] [X/θ/T] [EXE].

Use ▶ to change the value of x, then press EXE. Press

▶ **1** SHIFT INS **0** EXE.

x	5	10	15	20
y	1402.55	1967.15	2759.03	3869.68

Now try	16.1	16.2	16.3
	2972.21	2992.38	3012.70

Now try	16.22	16.23	16.24	16.25
	2996.44	2998.46	3000.49	3002.52

The answer to two decimal places is 16.24.

It is also possible to evaluate x from a graph. Set the range [0, 20], Xscl = 5 by [1000, 4000], Yscl = 500. Press

GRAPH **1000** (**1** . **07**) x^y X/θ/T EXE

(Fig. 11.2). Then press F1 and use the ▶ key to trace the point close to $y = 3000$. Zoom and trace until you get the required accuracy.

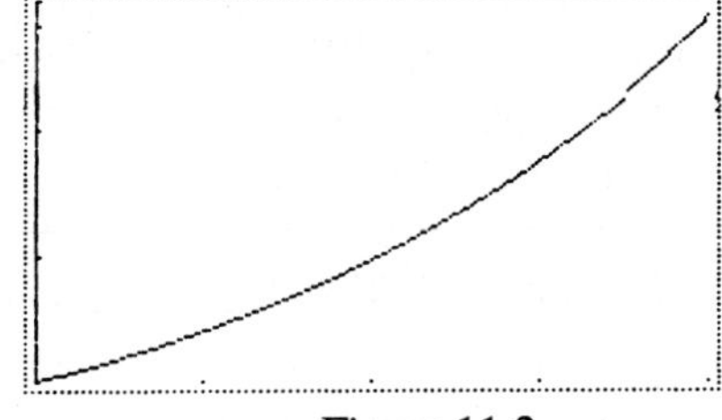

Figure 11.2

CHAPTER 12 SEQUENCES, SERIES, AND THE BINOMIAL THEOREM

Section 12.1 Arithmetic Sequences and Series

Generating Sequences

Sequences can be generated with the **Store** option on the Casio. Use → to store numbers.

Example: Generate the sequence defined by $A_1 = 3$ and $d = 2$. First switch back to the floating decimal display, and press

[SHIFT] [DISP] [F3:Nrm] [EXE].

Press

3 [→] [ALPHA] [A] [EXE] [ALPHA] [A] [+] **2** [→] [ALPHA] [A] [EXE] [EXE] [EXE] [EXE]

(Fig. 12.1).

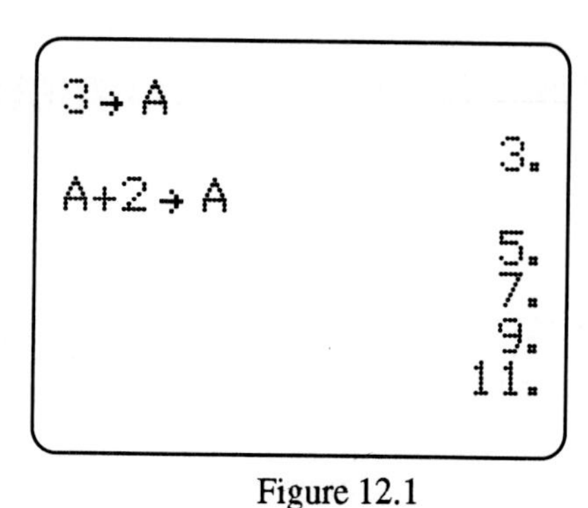

Figure 12.1

Displaying a Graph of a Sequence

The graph of a sequence can be displayed in the integer window with non-zero values for x and y. This is the window [0, 94] by [0, 62] on a Casio.

NOTE for the 6300: The Range is [0, 76] by [0, 44].

Example: Display the graph of the sequence 23, 26, 29 . . . , and find the sixth term.

Change the calculator to **Dot** mode as a graphing option by pressing

[MODE] [SHIFT] **6** [EXE] [GRAPH] **23** [+] **3** [(] [X/θ/T] [−] **1** [)] [EXE]

(Fig. 12.2). Then press

[F1] [▶] [▶] [▶] [▶] [▶] [▶].

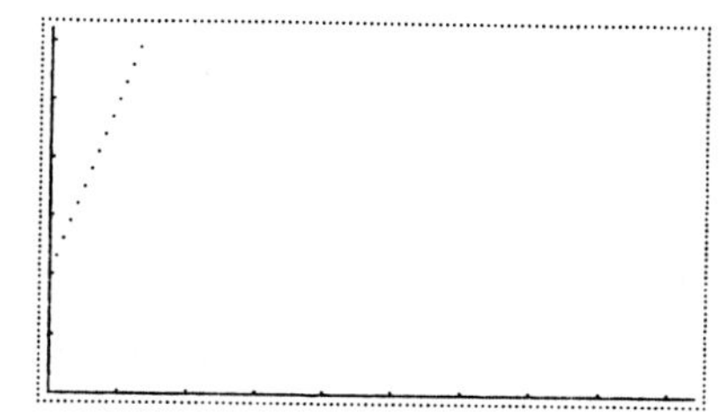

Figure 12.2

The sixth term is 38. Pressing

[MODE] [SHIFT] **5** [EXE]

changes back to Connected mode.

NOTE for the 6300 and 7000: These calculators do not have a Dot Graphing mode, so the points on the line are connected. Use [TRACE] to get the term number and [X↔Y] to get the value of the sixth term.

Section 12.2 Geometric Sequences and Series

Refer to Section 12.1 of this manual to review generating and graphing sequences.

Example: Generate the sequence defined by $A_1 = 3$ and $r = 2$. Press

3 [→] [ALPHA] [A] [EXE] [ALPHA] [A] [X] **2** [→] [ALPHA] [A] [EXE] [EXE] [EXE] [EXE].

Section 12.3 Binomial Expansion

The Factorial Key

The **Factorial** key is found on the Casio fx 7700G in the Math menu and the Probability submenu. It is on the Screen menu on the left. To evaluate a factorial, press the number, then

[F1:x!] [EXE].

Example: Evaluate 5! and 8!5!/(4!3!). Press

[SHIFT] [MATH] [F2:PRB] **5** [F1:x!] [EXE].

The answer is 120 (Fig. 12.3). Press

8 [F1:x!] [×] **5** [F1:x!] [÷] [(] **4** [F1:x!] [×] **3** [F1:x!] [)] [EXE].

The answer is 33600 (Fig. 12.3).

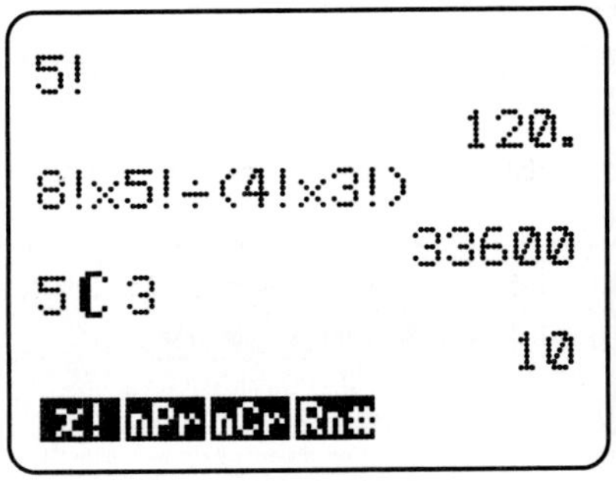

Figure 12.3

Evaluating a Binomial Coefficient on the Casio

The **binomial coefficient** $\binom{n}{r}$ can be evaluated directly on the Casio fx-7700. It is written as ${}_nC_r$ and is also in the Probability submenu in the Math menu. You enter n, then press [${}_nC_r$], then enter r, and execute.

Example: Evaluate $\binom{5}{3}$.

On the calculator this is written as ${}_5C_3$. If necessary, return to the PRB submenu. Enter

5 [F3:nCr] **3** [EXE].

The result is 10 (Fig. 12.3). Pressing [PRE] deletes a submenu.

NOTE for the 6300 and 7000: ${}_nC_r$ is not an option. Use the factorial definition ${}_nC_r = n! \div (r!(n-r)!)$.

Section 12.4 Counting Principle, Permutations, and Combinations

Evaluating Permutations and Combinations

Permutations and combinations are in the Probability submenu of the Math menu. Press

[SHIFT] [MATH] [F2:PRB].

First, you enter n, then press the operation key, and enter r.

Example: Evaluate $26P6$ and $4C3$.

If necessary, return to the PRB submenu. Press

26 [F2:nPr] **6** [EXE].

The answer is 165,765,600 (Fig. 12.4). Press

4 [F3:nCr] **3** [EXE].

The answer is 4 (Fig. 12.4).

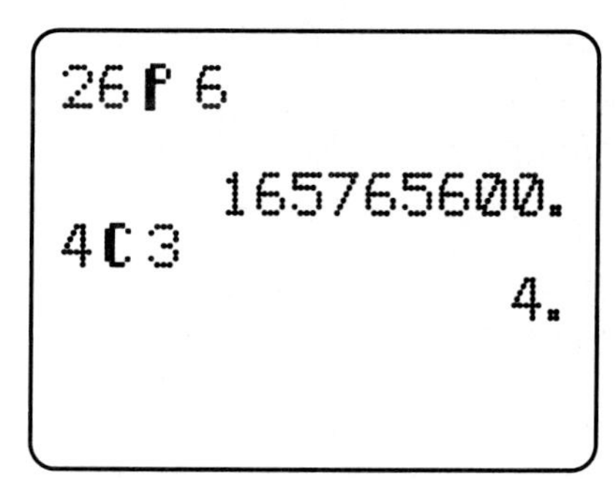

Figure 12.4

Preliminary Chapter

Getting Started
Using the EL-9200C/9300C

The Sharp EL-9200C and EL-9300C calculators are the same except the EL-9300C has a Solver mode, more memory, a backup battery to protect against data loss, and a communications port for unit-to-unit transfer and/or printing.

The Keyboard

The Sharp EL-9300C keyboard is shown in Figure 1.

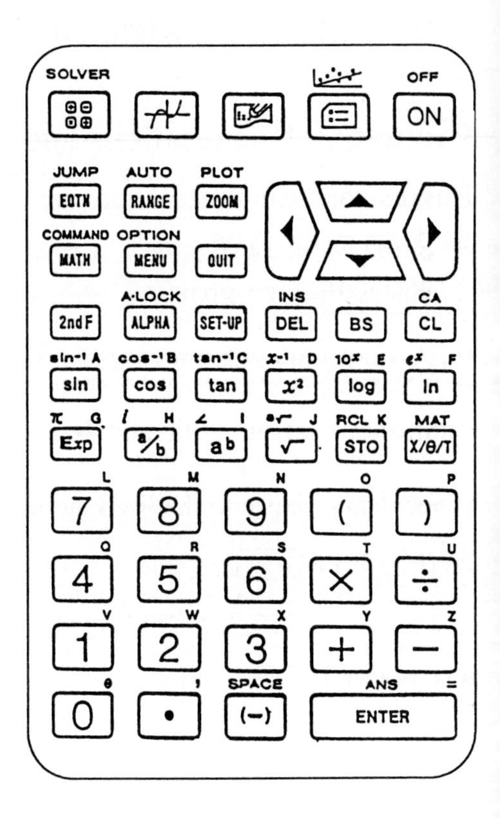

The top row of keys is the operation mode keys. The modes are (from left to right) Calculation mode, Graphing mode, Programming mode, and Statistics mode. The secondary modes (EL-9300C only) are the Solver mode and Statistical Graphing mode.

The second row consists of features for graphing and the third row contains the menu keys. To the right of these keys are the arrow keys. The fourth row of keys is for control and editing. The fifth and sixth rows are the function keys, with the arithmetic keys in the remaining rows.

The ON and OFF Keys

Press the ON key (the right key on the first row) to turn the calculator on. To turn the calculator off, press 2ndF OFF. Notice that 'OFF' is written in yellow above the ON key. The yellow selections are accessed by first pressing the 2ndF key followed by the desired selection. If you forget to turn the calculator off, don't worry, the calculator will automatically shut off after five minutes of inactivity.

To reset the calculator, press the reset button located on the back of the calculator with an ink pen. Press [CL] and all the memories are erased.

Adjusting Contrast

To darken or lighten the display screen, first press

[2ndF] [OPTION],

then hold the [+] key to darken the screen or the [−] key to lighten. Press [QUIT] to exit the Option screen.

The Home Screen

Press the Calculation mode key [⊞] followed by the [CL] key. Your Home screen should look like Figure 2. The Home screen is used for calculations.

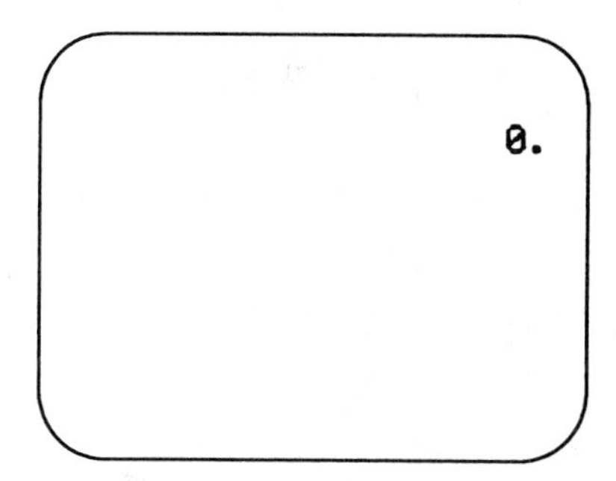

Figure 2

The Set Up Menu

Press [SET UP] to display the current configuration of the calculator. To change this configuration, press the [▼] key (or the letter to the left of the choices) to move to the desired submenu, then press the number (to the left of the choices) of your selection.

Put the calculator in the Equation Editor mode by pressing

[F:EDIT] [1:EQUATION]

(do this).
The 'F' is written in blue above the [ln] key and is usually accessed by pressing

[ALPHA] [V].

However, when you're in a menu, you don't have to press the [ALPHA] key. In addition, put the calculator in Decimal mode by pressing

[G:ANS] [1:Decimal]

(do this also).

Press [SET UP] again to exit the screen and save the configuration.

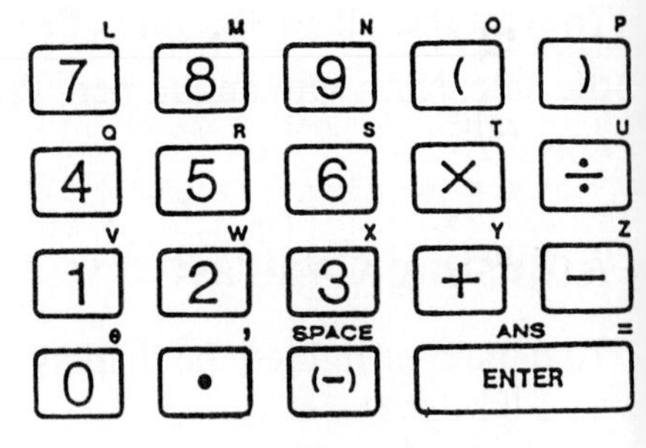

Figure 3

Arithmetic Keys

The arithmetic keys consist of the bottom rows displayed in Figure 3.

Enter an expression to be evaluated by "typing it as you see it." Try evaluating 18 × 145 by pressing

1 8 [×] **1 4 5** [ENTER].

You will press [ENTER] at the end of each expression to evaluate it. Also evaluate 8 ÷ 156 by pressing

8 [÷] **1 5 6** [ENTER].

Note the × is represented as an * and the ÷ is represented as a / on the display screen (Fig. 4).

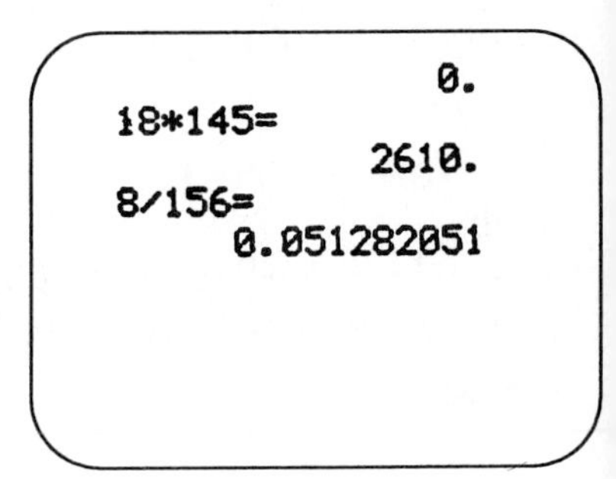

Figure 4

When reading about the Set Up menu described above, you should have placed the calculator in Equation Editor mode. Sharp's Equation Editor will allow you to enter the 8 ÷ 156 as a fraction. To do this, press

8 [a/b] **1 5 6** [ENTER]

(Fig. 5).

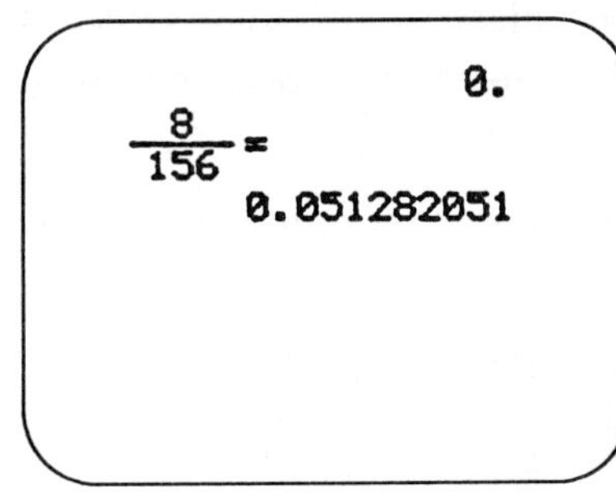

Figure 5

The [(−)] and [−] Keys

The [(−)] key is for entering a negative expression (the minus sign), whereas the [−] key is for entering the subtraction symbol. For example, to evaluate –3 + 5 press

[(−)] **3** [+] **5** [ENTER].

Now, to evaluate 5 – 3, press

5 [−] **3** [ENTER]

(Fig. 6).

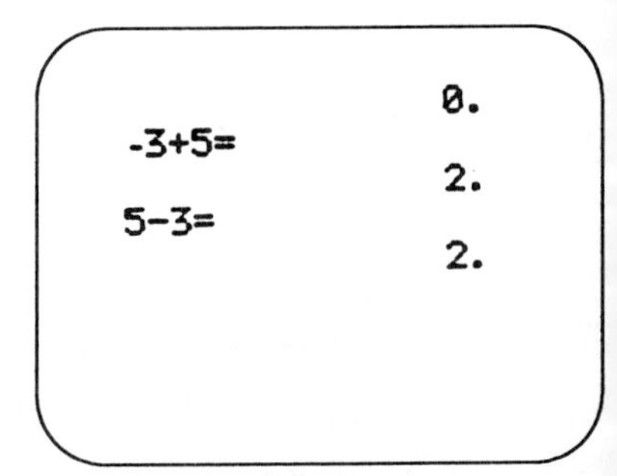

Figure 6

Using the [x²], [x⁻¹], and [aᵇ] Keys

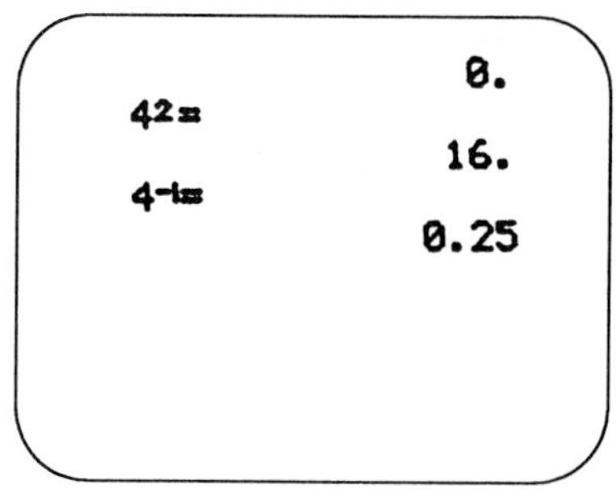

Figure 7

To evaluate a base to a power, you must enter the base first, followed by the power (remember, type it as you see it). Therefore, to find 4^2, you must enter the 4 first followed by the square. To evaluate 4^2, press

4 [x²] [ENTER].

To evaluate 4^{-1}, press

4 [2ndF] [x⁻¹] [ENTER]

(Fig. 7).

Using Sharp's Equation Editor, you can enter 4^3 or 4^{1+2} by pressing

4 [aᵇ] **3** [ENTER] or **4** [aᵇ] **1** [+] **2** [ENTER]

(Fig. 8).

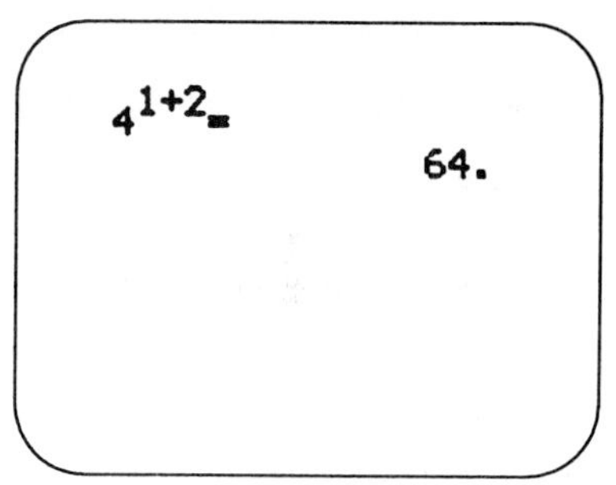

Figure 8

The [CL] Key and Expression Recall

To clear the screen of all expressions displayed, press [CL]. To recall the previous expression evaluated, press [▲]. You can recall older expressions by pressing [2ndF] [▲] repeatedly until the older expression appears. The grapher will store the last one hundred keystrokes.

Editing Keys

To edit an expression you are entering or an older expression you have recalled, you can use the [DEL] (delete) and [BS] (backspace) keys.

To delete the '1+' in the exponent of 4^{1+2}, you can recall the expression and move the cursor to the 1 and press [DEL] twice or move the cursor to the 2 and press [BS] twice.

The calculator operates in Insert mode. To insert something in the expression, simply move the cursor to the character before which you wish to insert, and enter the desired items.

For example, to edit the previous expression 4^{1+2} and make it 4^{1+12}, recall the expression 4^{1+2} and notice the blinking cursor over the '=' (Fig. 9). Press ◄ twice to move the blinking cursor over the 2. Press **1** to insert the 1 in front of the 2 to form 12. Press ENTER to evaluate the new expression (Fig. 10).

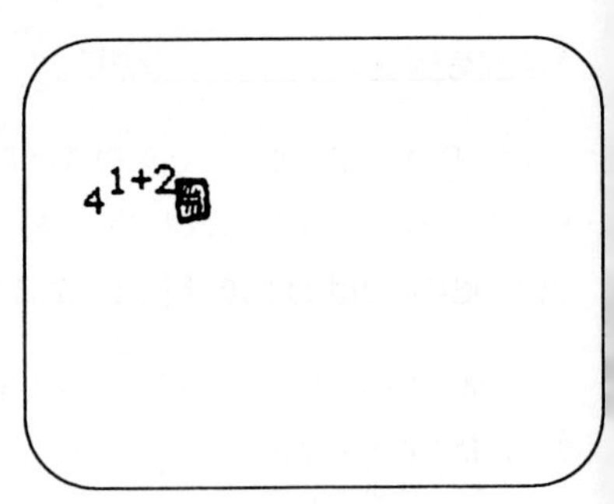

Figure 9

Error Messages

When you receive an error statement, press the ◄ or ► and a blinking cursor will appear near the error. Compare the displayed expression with the expression you want to evaluate, and make corrections with the editing features discussed above.

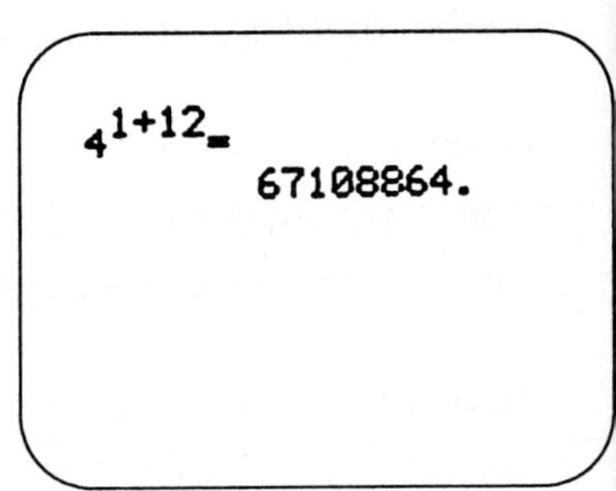

Figure 10

For example, enter 4^{1+} and press ENTER (Fig. 11). An error statement will be given. Press ► and the blinking cursor will appear on the plus sign '+.' This indicates that the '+' was entered in error or that the second term needed for addition was left off.

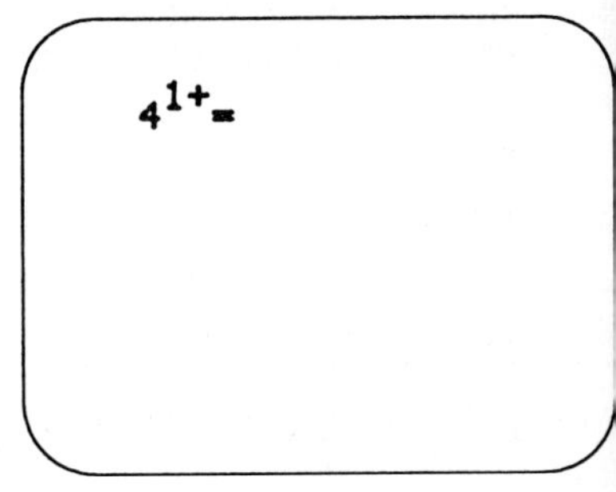

Figure 11

Chapter 1 Numerical Mathematics and the Graphing Calculator

Section 1.1 Real Numbers and a Graphing Calculator

The Sharp calculators do not change decimal representation to fractional representation. However, they will allow you to work with and display mixed numerals ($2^3/4$) and improper fractions (5/4). Unless told to do otherwise, set the calculator to give you decimal answers by pressing

[SET UP] [G:ANS] [1:Decimal] [QUIT].

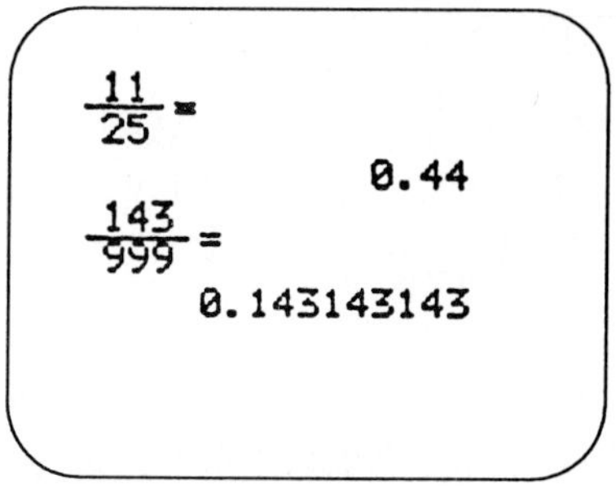

Repeating and Terminating Decimals

An example of a terminating decimal is 11/25, which can be evaluated by pressing

1 1 [a/b] **2 5** [ENTER].

Note the exactness of the decimal representation in Figure 1.1.

An example of a repeating decimal is 143/999. Note the repetition of 143 in the decimal expansion. It is impossible to display all the digits so these are considered approximations (Fig. 1.1).

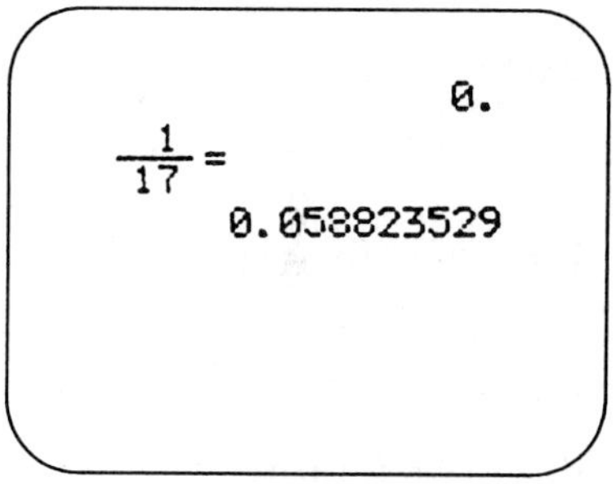

An example of a repeating decimal that cannot be determined using the grapher is 1/17. No repetition is observed in the number of decimal places displayed (Fig. 1.2).

$\frac{11}{25}$ =
0.44
$\frac{143}{999}$ =
0.143143143

Figure 1.1

0.
$\frac{1}{17}$ =
0.058823529

Figure 1.2

Floating Decimal Versus Fixed-Point Display

Previously, 11/25 and 143/999 were displayed with a floating decimal display. Notice that only two decimal places were needed to express 11/25 as a decimal and that all nine decimal places available were used to display 143/999. This describes the floating decimal display: only the needed number of decimal places is shown from zero to nine places.

You may wish to fix the number of decimal places shown. For example, when working with money problems, only two decimal places are needed, and when working with a science problem, teachers prefer to work with four decimal places.

To fix the number of decimal places to two digits, press

[SET UP] [C:FSE] [2:Fix]

to fix the number of decimal places, then press

[D:TAB] **2**

to set the number of decimal places to two. Press [QUIT] to return to the Home screen and then find the decimal equivalent for 143/999 again (Fig. 1.3).

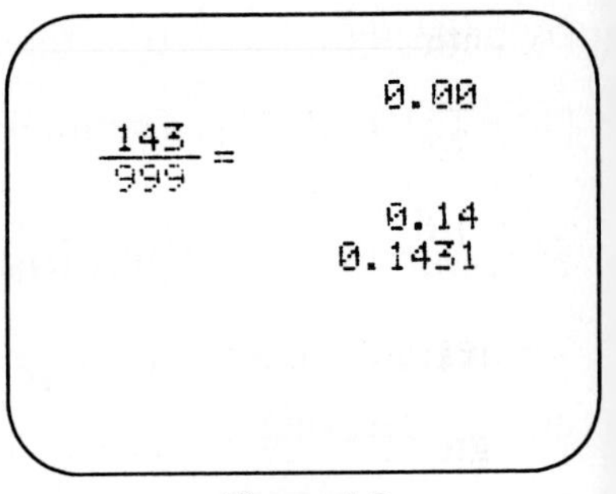

Figure 1.3

Fix the number of decimal places to four by repeating the instructions given previously, but replace

[D:TAB] **2** by [D:TAB] **4**.

When you press [QUIT], you will return to the Home screen with the four-decimal equivalent for 143/999 shown in Figure 1.3.

Return the calculator to the floating decimal display by pressing

[SET UP] [C:FSE] [1:Float Pt] [QUIT].

0.
$\sqrt{25}$ =
5.
$\sqrt{2}$ =
1.414213562

Figure 1.4

Square Roots

To find the $\sqrt{25}$, press

[$\sqrt{\ }$] **2 5** [ENTER].

Note in Figure 1.4 that the radical extends over the 25 like it should. Find the $\sqrt{2}$ by repeating the keystrokes given above (Fig. 1.4).

Exponentiation

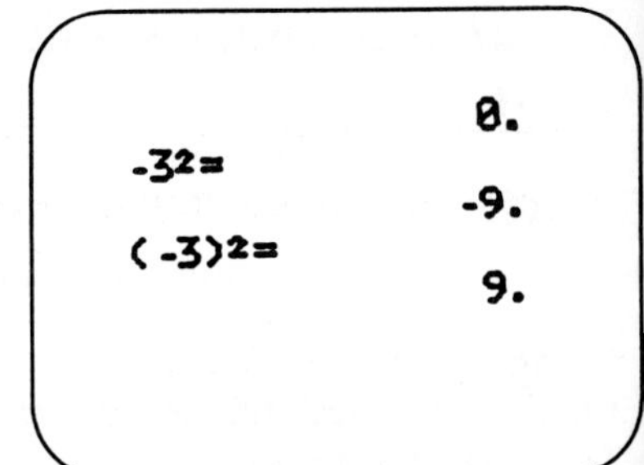

Figure 1.5

Parentheses are important with exponents involving negatives. For example, -3^2 is distinctly different from $(-3)^2$. Compare these two expressions using the grapher by pressing

[(-)] **3** [x^2] [ENTER]

followed by

[(] [(-)] **3** [)] [x^2] [ENTER]

(Fig. 1.5).

Parentheses are important with exponents involving fractions. Parentheses around a fraction, followed by an exponent, will raise the entire fraction to the power. To evaluate $(2/5)^3$, press

(**2** a/b **5** ▶) a^b **3** ENTER

(Fig. 1.6).

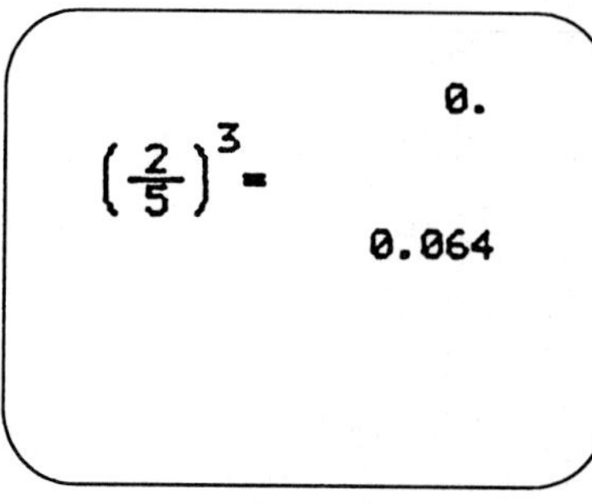

Figure 1.6

Order of Operations

The grapher will perform calculations in the proper arithmetic order. Expressions in parentheses are calculated first, followed by operations in fractions, powers, multiplications and divisions, and then additions and subtractions.

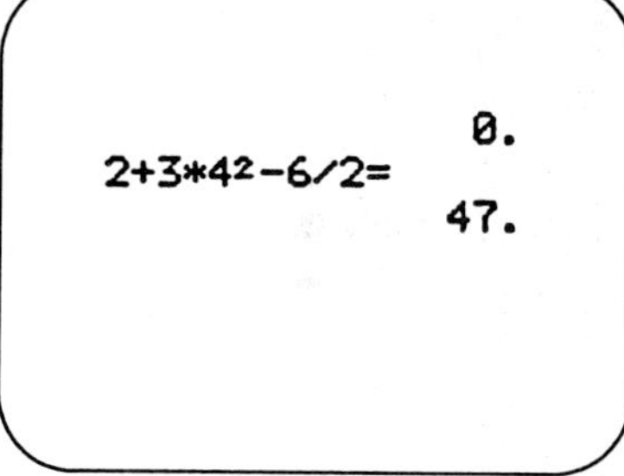

Figure 1.7

Evaluate

$2 + 3 \times 4^2 - 6 \div 2$

on the grapher by pressing

2 + **3** × **4** x² − **6** ÷ **2** ENTER.

Note this is the same value as in the text (Fig. 1.7).

Section 1.2 Properties of Real Numbers and the Basic Rules of Operations

Properties of Opposites

The differences between the negative or minus key and the subtraction key are covered in the Getting Started chapter.

Absolute Value

The absolute value function is found within the Math menu. Evaluate $-|-7|$ by pressing

(−) MATH [1:abs] (−) **7** ENTER.

Note in Figure 1.8 that the –7 appropriately appears within the absolute value symbols.

0.
-|-7|=
-7.
|5.2-17.8|=
12.6

Figure 1.8

Evaluate |5.2 – 17.8| by pressing

[MATH] [1:abs] **5** [.] **2** [−] **17** [.] **8** [ENTER]

(Fig. 1.8).

Section 1.3 Algebraic Expressions and Problem Situations

Evaluate an Expression

To store a value within a variable, you will use the [STO] key. Store 25 in the variable X by pressing

25 [STO] [X] or [X/θ/T].

Note the calculator switches into Alpha mode for you to enter the X variable (the Alpha indicator appears on the top of the screen).

Now that the value 25 has been stored in the X variable, you can evaluate an expression containing X for $X = 25$. Evaluate the expression

$X^3 + 17$ for $X = 25$

by pressing [X/θ/T] or [ALPHA] [X] to display the X variable, and then pressing

[a^b] **3** [▶] [+] **17** [ENTER]

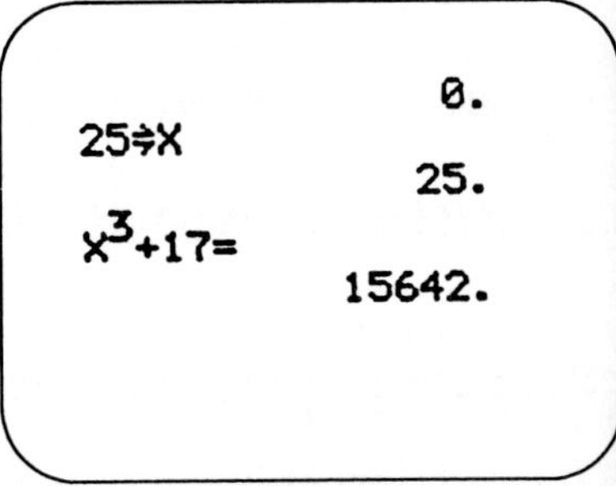

Figure 1.9

(Fig. 1.9). Note the [▶] moves the cursor out of the exponent to allow you to continue entering the expression properly.

To evaluate the expression for $X = 39$, press

39 [STO] [X]

to store 39 in the variable X. Now recall the $X^3 + 17$ and press [ENTER] (Fig. 1.10).

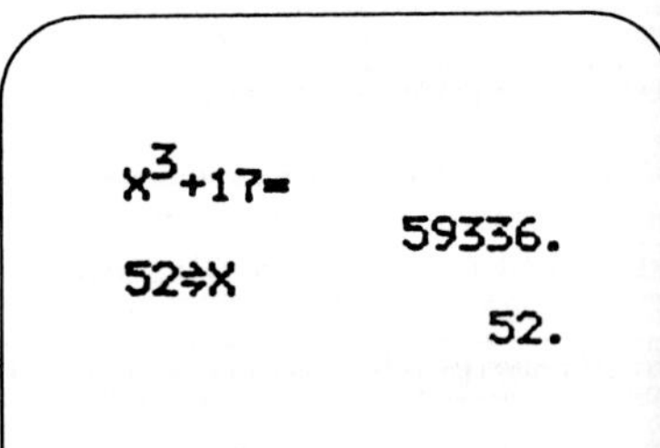

Figure 1.10

Evaluate $X^3 + 17$ for $X = 52$ by repeating the instructions above (Figs. 1.10 and 1.11).

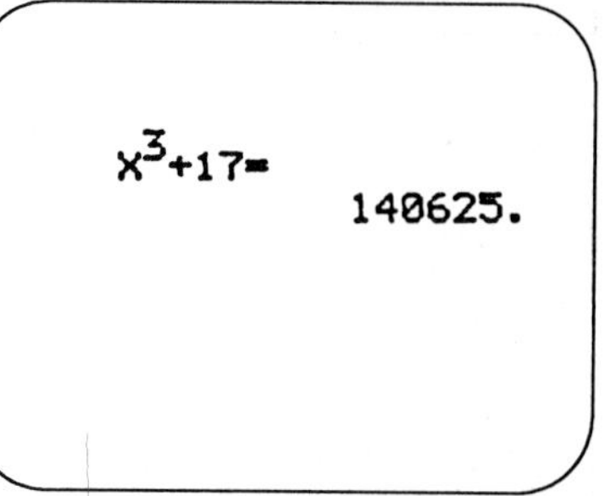

Figure 1.11

To evaluate a multivariable expression, first store all the values in the variables, then enter and evaluate the expression. For example, evaluate $(4T - 3S^3)/5R$ for $T = 3.28$, $S = -23.7$, and $R = 13.5$ by pressing

3 [.] **2 8** [STO] [T] [(-)] **2 3** [.] **7** [STO] [S] **1 3** [.] **5** [STO] [R] [(] **4** [ALPHA] [T] [−] **3** [ALPHA] [S] [a^b] **3** [▶] [)] [a/b] **5** [ALPHA] [R] [ENTER]

(Figs. 1.12 and 1.13).

3.28⇒T 3.28
-23.7⇒S -23.7
13.5⇒R 13.5

Figure 1.12

(4T−3S³)/5R = 591.8411704

Figure 1.13

Concatenation

The Sharp grapher does not link expressions together using ':'.

Implied Multiplication

Implied multiplication will sometimes perform computations in an order that you do not desire. For example, to evaluate $^5/_9(41 - 32)$ you may need to put parentheses around the $^5/_9$ or the grapher will evaluate it as $5/[9(41 - 32)]$.

Wrong:

5 [÷] **9** [(] **4 1** [−] **3 2** [)] [ENTER]

(Fig. 1.14).

5/9(41−32)= 0.061728395

Figure 1.14

Correct:

[(] **5** [÷] **9** [)] [(] **4 1** [−] **3 2** [)] [ENTER]

(Fig. 1.15).

Instead of using extra parentheses and the [÷] key, you can enter the equation as you see it using the [a/b] key. Enter the expression by pressing

5 [a/b] **9** [▶] [(] **4 1** [−] **3 2** [)] [ENTER]

(Fig. 1.16).

Earlier, we worked with

$(4T - 3S^3)/5R$.

If the parentheses are not used, our expression would look like $4T - 3S^3/5R$ and only the $3S^3$ would be divided by the $5R$. Our result would be distinctly different and incorrect. See the result in Figure 1.17.

```
(5/9)(41-32)=
                5.
```

Figure 1.15

Section 1.5 Numerical Representations of Problem Situations

Making Tables

It is not possible to create tables on the Sharp grapher.

```
                0.
5/9 (41-32)=
                5.
```

Figure 1.16

Finding Percents

You can use the Sharp grapher to calculate percents in Computation mode using a decimal representation of the percent. Once again, it is not possible to create tables on the Sharp grapher.

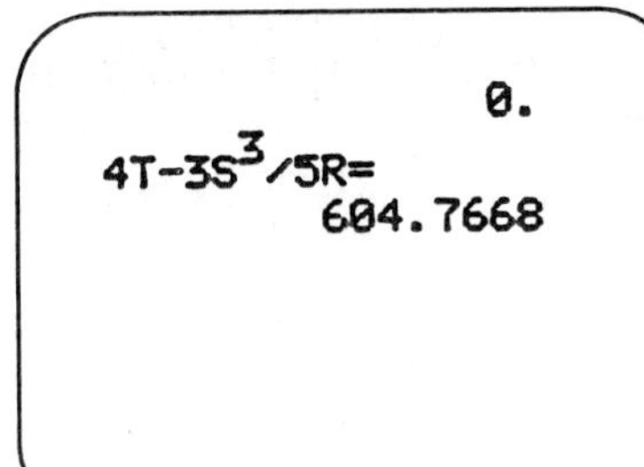

Figure 1.17

Section 1.6 Algebraic Representations of Problem Situations

Parametric Simulation, Parametric Mode, and Simultaneous Mode

Change the grapher to Parametric Graphing mode by pressing the [mode icon] mode key followed by

[SET UP] [E:COORD] [3:XYT] [QUIT].

Define the WINDOW by pressing

[RANGE] [ENTER] 5 [.] 5 [ENTER].

```
T RANGE
Tmin=
                0.
Tmax=
                5.5
Tstp=
  5.85106383E-02
```

Figure 1.18

Note that the *T*stp (*T*step) is automatically defined for you by the grapher at $5.85106383 \times 10^{-2}$ or approximately .0585 (Fig. 1.18). This is a bigger step than is defined in the textbook. Do not try to change the *T*stp to .01, for the grapher will automatically change the *T*max to an incorrect value.

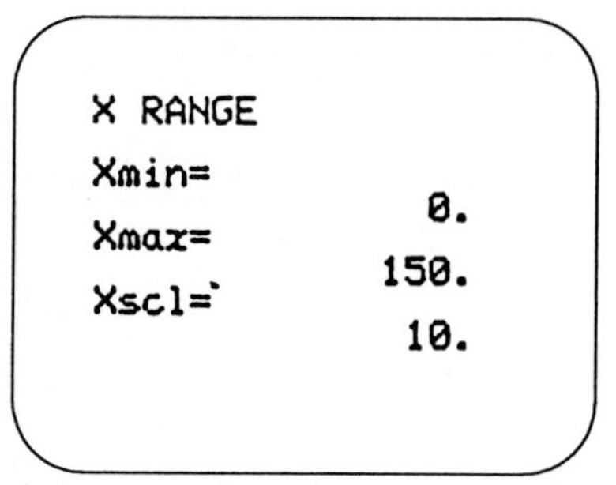

Figure 1.19

Continue to define the window, setting *X*min = 0, *X*max = 150, *X*scl = 10 (scale for *x*-axis) (Fig. 1.19), *Y*min = 0, *Y*max = 8, and *Y*scl = 1 (scale for *y*-axis) (Fig. 1.20), by pressing

[ENTER] **0** [ENTER] **1 5 0** [ENTER] **1 0** [ENTER] **0** [ENTER] **8** [ENTER] **1** [ENTER].

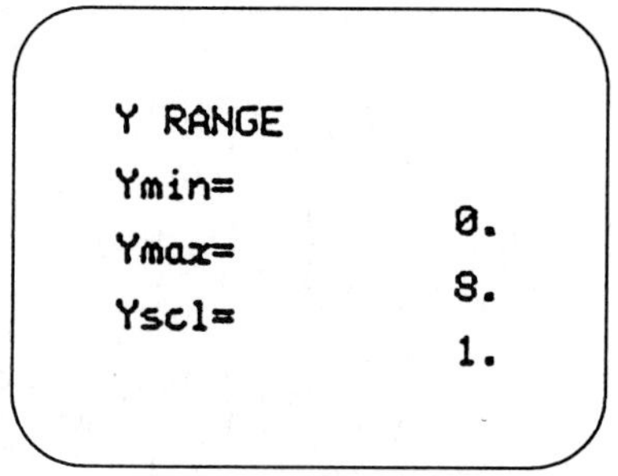

Figure 1.20

Your Range screens should look like those in Figs 1.19 and 1.20. You can check them by pressing [▲] repeatedly to move back through the values.

Press [EQTN] to exit the Range screens and enter the equations by pressing

3 0 [X/θ/T] [ENTER] **5** [ENTER] **2 8** [X/θ/T] [+] **1 0** [ENTER] **2** [ENTER].

Press the [graph key] to view the graph. Press [▶] to engage the grapher's trace feature. Pressing [▶] repeatedly will allow you to move along the line. Notice the coordinates at the bottom of the screen. The *T* represents the time in seconds that has passed. The *X* represents the running position in feet. Pressing the [▲] or [▼] will move you from one runner's line to another's line. Notice the *T* remains the same, but the position on the runner's line changes. You need to move right until the running position (the *X* value) is virtually the same and then read the time it takes for Juan (Fig. 1.21) to catch Carlos (Fig. 1.22).

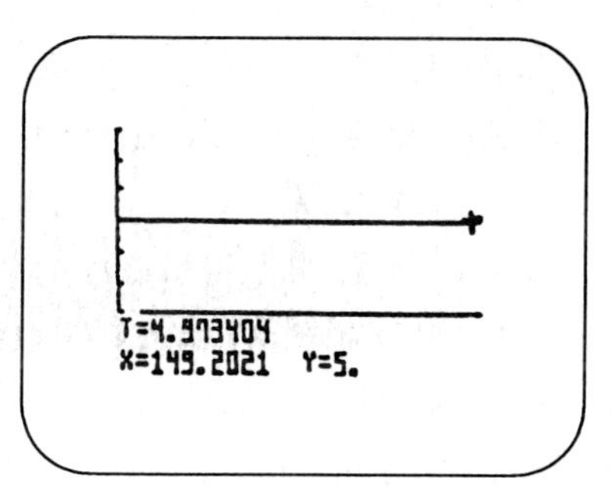

Figure 1.21

Change the grapher back to its normal operating mode (rectangular) by pressing

[SET UP] [E:COORD] [1:XY] [QUIT].

You can return to the Home screen by pressing the [mode key] mode key followed by [CL].

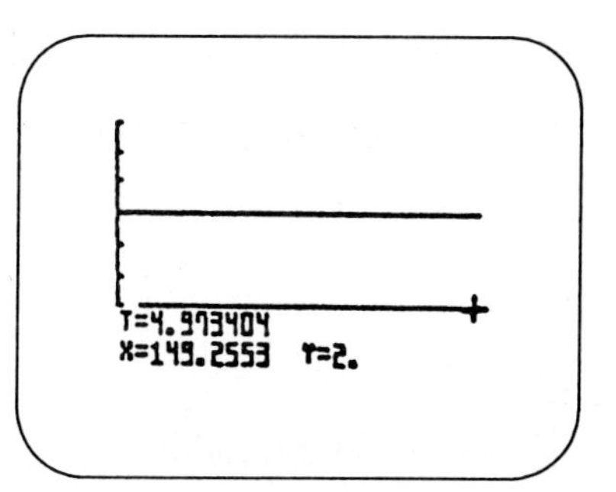

Figure 1.22

Chapter 2 Graphing Equations and Functions

Section 2.1 Rectangular Coordinate Plane

Representing a Coordinate Plane on a Grapher

The Sharp EL-9200C/9300C can be changed into Graph mode by pressing the [Graph]. A 'Y1=_' will appear on the screen. The viewing window for the grapher is defined using the Range screens (as we did in Section 1.6) or using the built-in ranges. To access these nineteen built-in ranges, press

.

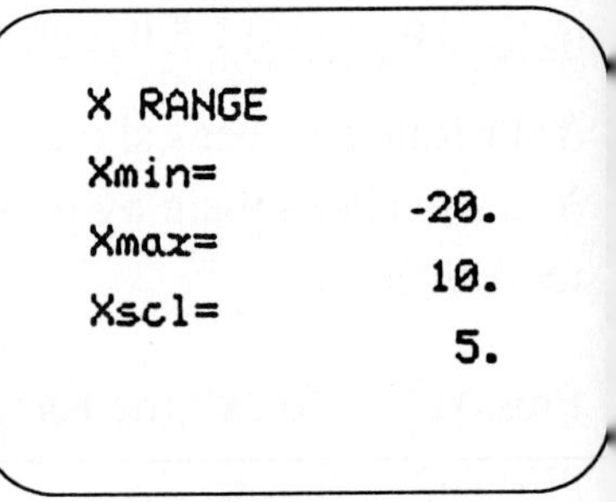

Figure 2.1

The Window

The default window range is now displayed on the screen. The additional eighteen ranges can be chosen by pressing [▼] and the number to the left of the desired choice. Press [QUIT] until you return to the 'Y1=_' screen.

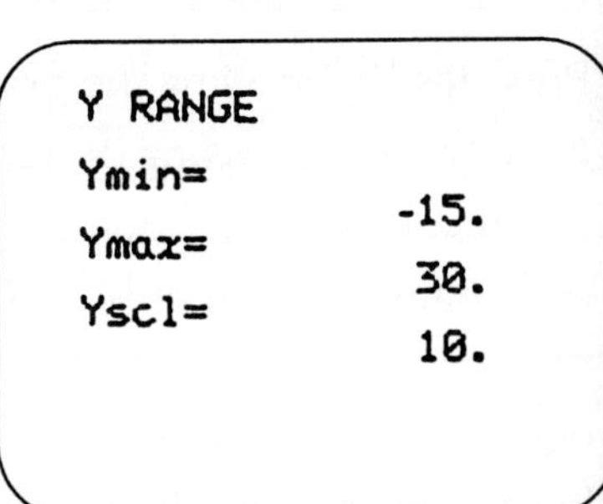

Figure 2.2

The Range

To enter a specific range for a viewing window, we follow a similar path as in Section 1.6. Set the *X* range to [–20, 10] with a scale of 5 (tick marks on the coordinate axis) and the *Y* range to [–15, 30] with a scale of 10, by pressing

[RANGE] [(-)] **2 0** [ENTER] **1 0** [ENTER] **5** [ENTER] (Fig. 2.1) [(-)] **1 5** [ENTER] **3 0** [ENTER] **1 0** [ENTER] (Fig. 2.2).

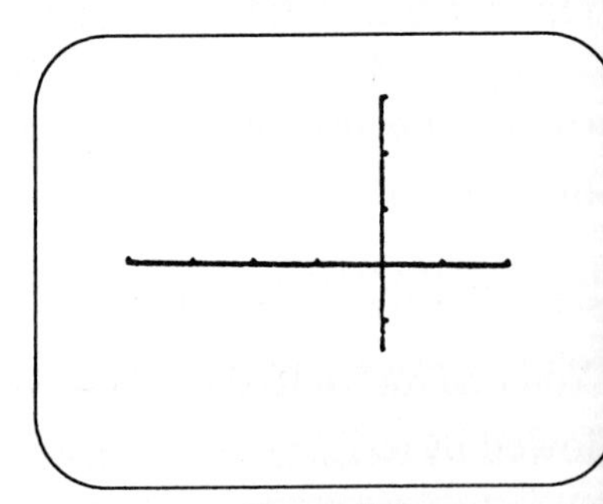

Figure 2.3

Pixels

Press the Graph mode key [Graph] to see the viewing window (Fig. 2.3). The Sharp grapher does not have a free-moving cursor engaged by pressing an arrow key. However, it does have a free-moving cursor within its point-plotting feature. To engage this cursor press

[2ndF] [PLOT] [B:PLOT] [2:FREE].

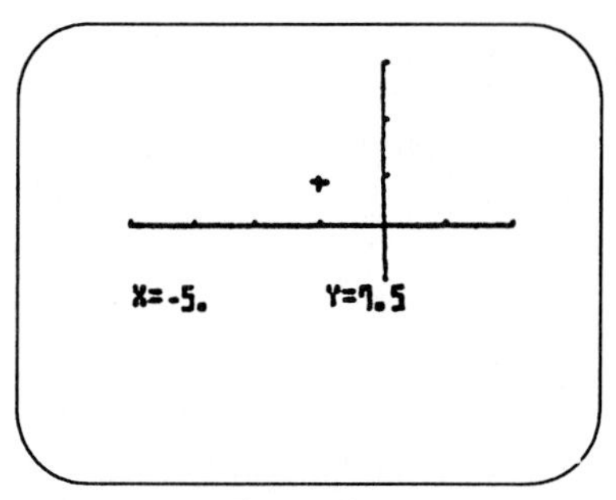

Figure 2.4

A blinking cursor will appear in the center of the viewing window (Fig. 2.4). You can move it around by pressing the arrow keys. Note that the cursor coordinates are displayed at the bottom of the screen. These coordinates are often a little distorted by our window. For example, move the cursor to the origin and observe that the coordinates displayed are not zero, but are close to zero (Fig. 2.5). This is because the cursor moves from "pixel" to "pixel" on the grapher's display screen. A pixel is a block of the screen that is either blackened in or left alone to display a graph or expression.

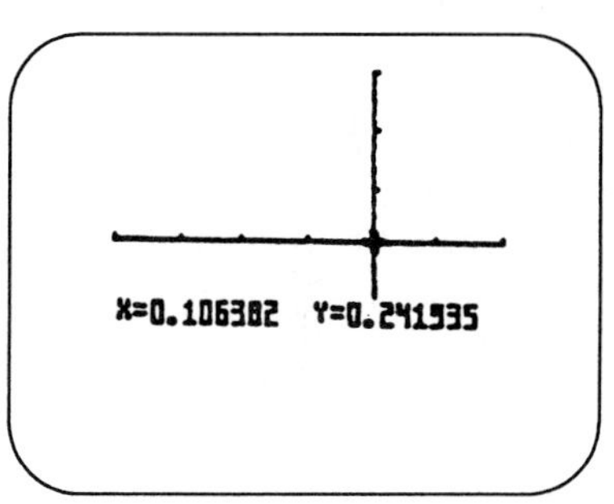

Figure 2.5

The Integer Window

The integer window is a viewing window, centered at the origin, where the moving cursor will display integers only. The Range values for this window are shown in Figures 2.6 and 2.7.

Enter these Range values by pressing

[RANGE] [(-)] 4 7 [ENTER] 4 7 [ENTER] 1 0 [ENTER] [(-)] 3 1 [ENTER] 3 1 [ENTER] 1 0 [ENTER].

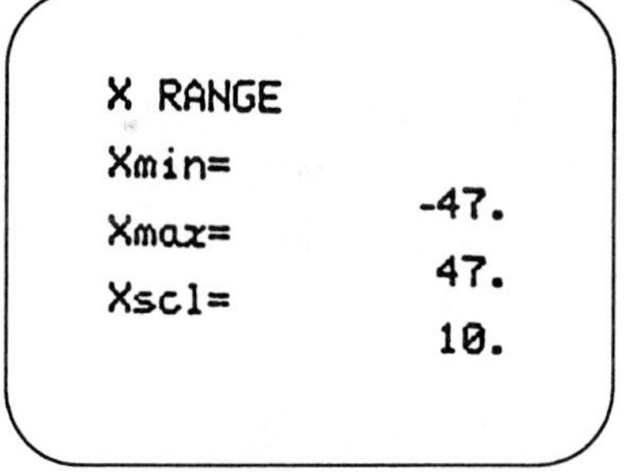

Figure 2.6

Press the Graph mode key [graph] to view the window. Use the Plot cursor (described previously) to move around the window. Notice how the cursor coordinates remain integers as you move around. Move over to the origin and notice the coordinates displayed are X=0 and Y=0.

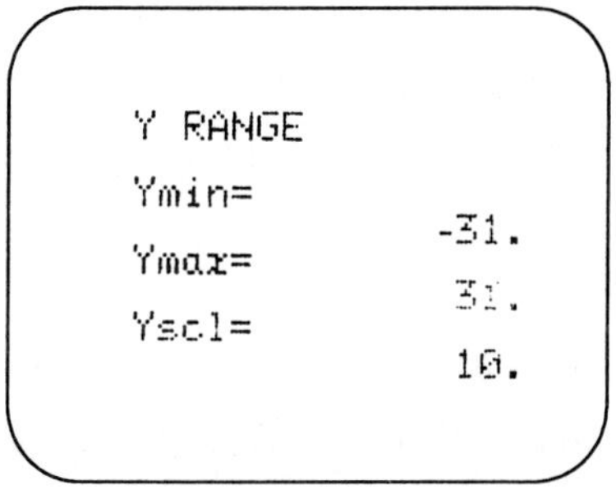

Figure 2.7

The 0.1 Window

For a cleaner and more accurate window than an individually specified window, use the default range values. To set these default values, press

[RANGE] [MENU] [A:DEFLT] [ENTER].

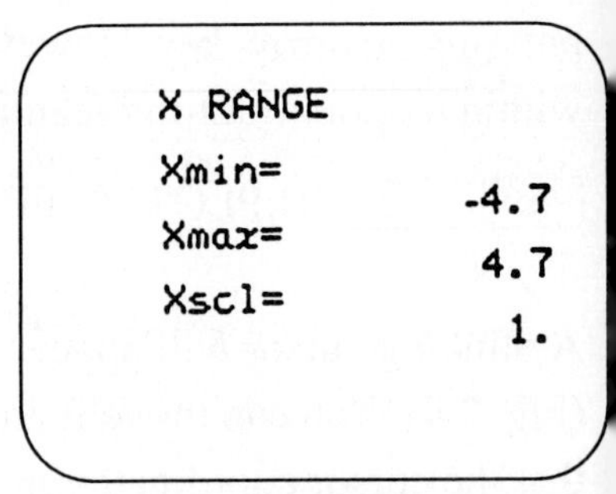

Figure 2.8

These range values are displayed in Figures 2.8 and 2.9.

Press the Graph mode key [graph] to view the window. Use the Plot cursor (described previously) to move around the window. Notice how the cursor coordinates only show one decimal place. This is considered a clean window and is referred to as the decimal or 0.1 window. Move over to the origin and notice the coordinates displayed are *X*=0 and *Y*=0.

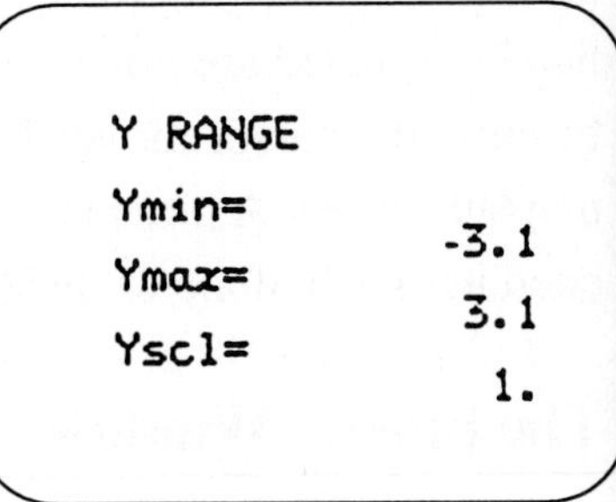

Figure 2.9

The 0.2 Window

To create the 0.2 window, begin with the default range values (the decimal or 0.1 window) and multiply the *X*min, *X*max, *X*scl, *Y*min, *Y*max, and *Y*scl by 2. You can do this by pressing

[×] **2** [ENTER]

on each line of the *X* Range and *Y* Range screens. The resulting values should be *X*min = –9.4, *X*max = 9.4, *X*scl = 2, *Y*min = –6.2, *Y*max = 6.2, and *Y*scl = 2.

Within this window, tracing the graph will result in increments of 0.2 in the x value displayed at the bottom of the screen. This window will be used in Chapters 6 and beyond.

Other Windows

The integer window described previously is not the only integer window available. Another window might be $0 < x < 94$ and $0 < y < 62$. You can create other integer windows by letting the difference between *X*min and *X*max be 94. In addition, changing the *Y*min and *Y*max will not change the type of window you are in. For example, a decimal window is established by setting $-4.7 < x < 4.7$. The *Y* values can be $-3.1 < y < 3.1$ or $-10 < y < 10$ and you will still have a decimal window, where the x values will increment by 0.1 as you trace a curve.

Section 2.2 Equations in Two Variables and Their Graphs

Using the Statistics and Statistical Graphing Mode to Graph Solution Pairs

Press [STAT] to enter Statistics mode. If a data set is present, a data card will appear instead of the menu desired. Press

[MENU] [D:DEL] [2:ALL DATA] [ENTER]

to remove the data set and view the menu. Press 3 to select the two-variable data format. Now, enter the x- and y-coordinates of the points you would like to plot. For example, to enter the points (1,2) and (3,4), press

1 [ENTER] 2 [ENTER] 3 [ENTER] 4 [ENTER].

Press

[2ndF] [STAT GRAPH]

to enter Statistical Graphing mode, and press [E: S.D.] to plot the points entered in the Statistics mode. You can adjust the Range of the viewing window by pressing

[RANGE] [▶]

and entering the desired values. For the previous example, set the viewing window by pressing

0 [ENTER] **5** [ENTER] **1** [ENTER] **0** [ENTER] **5** [ENTER] **1** [ENTER].

Press [QUIT] to exit the range menu, and press [ENTER] to view the points.

Press [GRAPH] to exit Statistical Graphing mode and return to Graphing mode.

Plotting Points

You have used the point-plotting cursor as a free-moving cursor in Section 2.1. You can plot points with the cursor or you can plot them directly by entering the coordinates for the point. In the following example, you will directly plot the solution points for an equation.

Plot the solution points (–15, 13.5), (–10, 11), (0, 6), and (15, –1.5) for $14X + 28Y = 168$ in a viewing window of Xmin = –20, Xmax = 20, Ymin = –15, Ymax = 15. Set the viewing window by pressing

[RANGE] [(–)] **2 0** [ENTER] **2 0** [ENTER] **5** [ENTER] [(–)] **1 5** [ENTER] **1 5** [ENTER] **5** [ENTER].

Next, press

[2ndF] [PLOT] [B:PLOT] [1:DIRECT]

and enter the coordinates of the first solution point. The point will appear in the viewing window. Repeat the

[2ndF] [PLOT] [B:PLOT] [1:DIRECT]

keystrokes and enter the coordinates for the remaining solution points. The resulting plot is displayed in Figure 2.10.

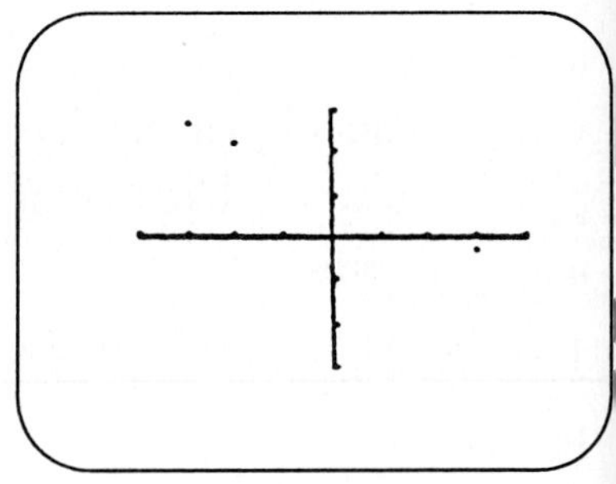

Figure 2.10

Now, you can plot an additional point using the free-moving cursor by pressing

[2ndF] [PLOT] [B:PLOT] [2:FREE].

A blinking cursor will appear at the origin. To view the cursor better, move it right and up by pressing [▶] and [▲], respectively. Now, move the cursor to the desired location for the point, and press [ENTER]. Move the cursor to view the point. Pressing [ENTER] again will plot another point.

How to Graph an Equation

To graph a function, you will enter the function, adjust the range, and then graph thc function.

Function-Defining Menu

To enter a function for graphing, press [graph] [EQTN] and a 'Y1=' will appear. Enter your function at this time. For example, enter $Y1 = 2X - 8$ by pressing

2 [X/θ/T] [−] **8**.

Pressing [ENTER] at the end of the line will allow you to enter a second function for graphing. You can graph up to four functions at any time.

To adjust the range, press [RANGE] and enter the values desired or select from one of the built-in ranges by pressing [MENU]. Enter the values for the integer window discussed previously in Section 2.1 (Xmin = −47, Xmax = 47, Xscl = 10, Ymin = −31, Ymax = 31, and Yscl = 10).

Now, graph the function by pressing the Graph mode key [graph] (Fig. 2.11).

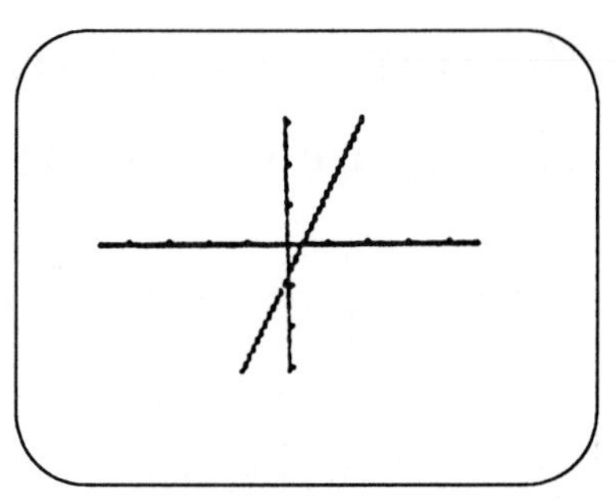

Figure 2.11

Tracing in an Integer Window

Press [▶] to engage the grapher's trace feature. Pressing [▶] repeatedly will allow you to move along the line. Notice the coordinates at the bottom of the screen.

Press [EQTN] to enter a new function for graphing. If a 'Y2=_' is displayed, press [2ndF] [▲] to display the $Y1$ function. Press [CL] to remove the old function and then enter the new function.

For example, enter the function $Y1 = {}^3/_4\, X - 6$ by pressing

[EQTN] **3** [a/b] **4** [▶] [X/θ/T] [−] **6**,

use the integer range, and then graph the function by pressing [graph]. Press [▶] to engage the trace feature. Check to see if (−9, −12.75) is on the graph (Fig. 2.12).

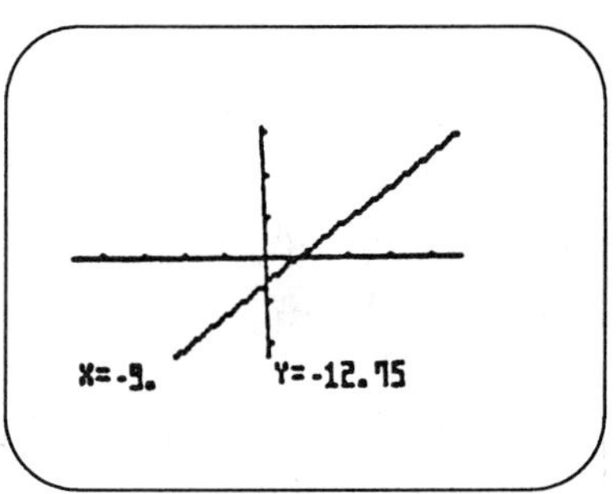

Figure 2.12

Further, find the y-coordinate for points on the graph at $X = -12$ (Fig. 2.13), $X = -11$ (Fig. 2.14), and $X = 3$.

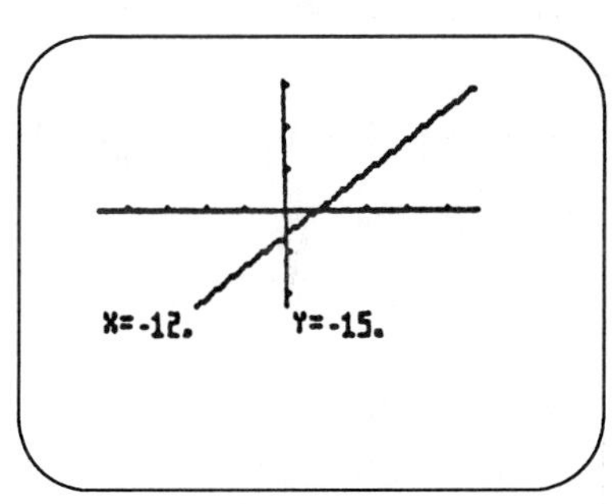

Figure 2.13

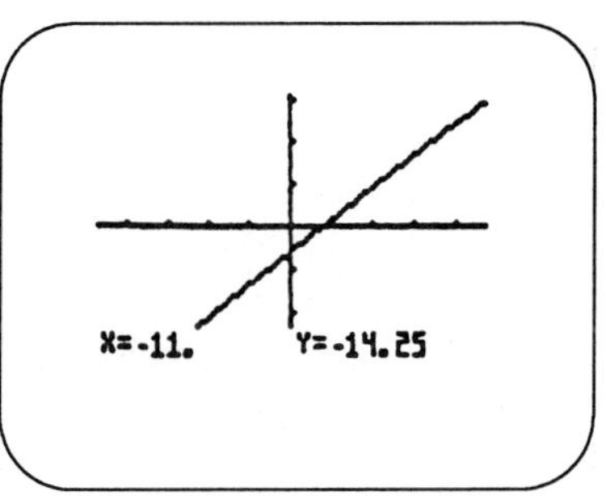

Figure 2.14

Section 2.3 Linear Equations and the Slope of a Line

Graphing a Vertical Line

You can graph a line segment in one of two ways, directly or with the free-moving cursor. To graph the vertical line segment directly, first clear the window by pressing

[2ndF] [PLOT] [A:CLEAR] [ENTER]

and then press

[2ndF] [PLOT] [C:LINE] [1:DIRECT]

and enter the coordinates for the endpoints of the line you desire to graph. Remember, the x-coordinates for vertical lines will be the same, whereas the y-coordinates for horizontal lines will be the same.

For example, to graph the vertical line $X = 4$ using an integer range, you would set the X range by pressing [–47, 47] with the Xscl = 10, and the Y range [–31, 31] with the Yscl at 10, then you enter the coordinates

($X1$, $Y1$) = (4, Ymax) = (4, 31) and ($X2$, $Y2$) = (4, Ymin) = (4, –31)

by pressing

[2ndF] [PLOT] [C:LINE] [1:DIRECT] and **4** [ENTER] **3 1** [ENTER] **4** [ENTER] [(–)] **3 1** [ENTER]

(Fig. 2.15).

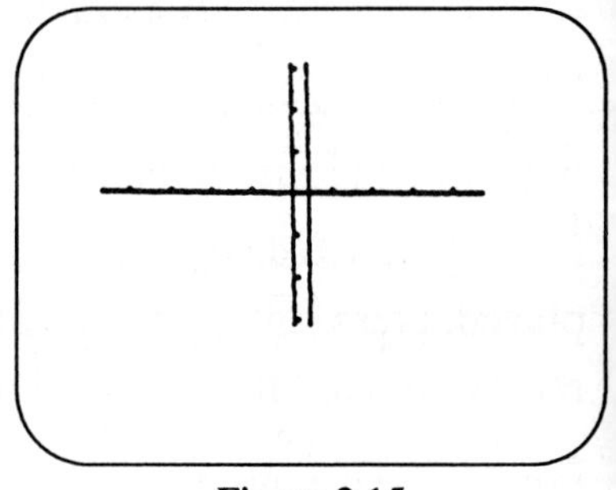

Figure 2.15

To graph the vertical line, $X = 10$, using the free-moving cursor, press

[2ndF] [PLOT] [C:LINE] [2:FREE]

and use the [▶] and [▲] keys to move the cursor to (10, 31) and press [ENTER]. Next, use the [▼] key to move the cursor down to (10, –31) and press `ENTER` again (Fig. 2.16). The vertical line X=10 will appear (Fig. 2.17).

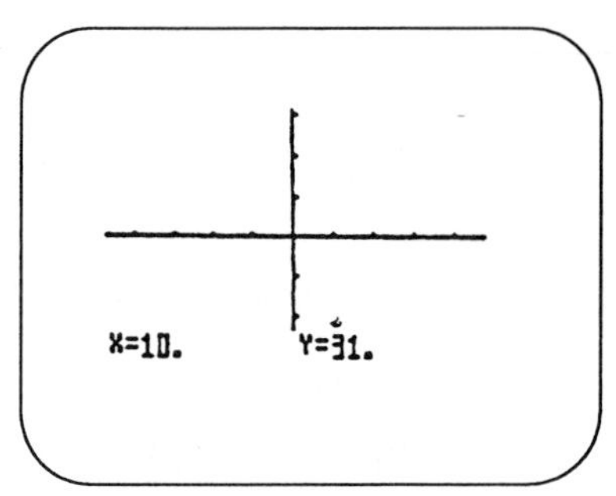

Figure 2.16

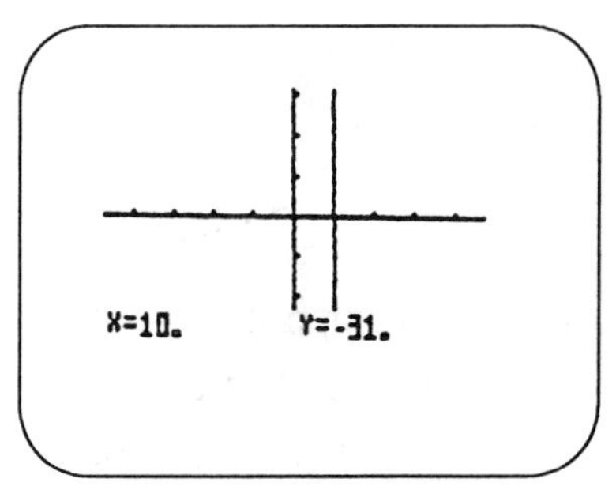

Figure 2.17

Section 2.4 The Slope-Intercept Form and Point-Slope Form of Linear Equations

Squaring a Window

The display screen on the grapher is a rectangle. Hence, using a range of –10 to 10 [–10, 10] [–10, 10] on both the x- and y-axes is not appropriate. It will lead to some distortion in the graph, since the distance between the tick marks on the x-axis is not equal to the distance between the tick marks on the y-axis. Notice this difference in Figure 2.18.

The *cleanest* windows are the integer and default ranges. They are considered square, which means the distance between tick marks on the axes is equivalent. Another range that appears square is $-15 < X < 15$ and $-10 < Y < 10$.

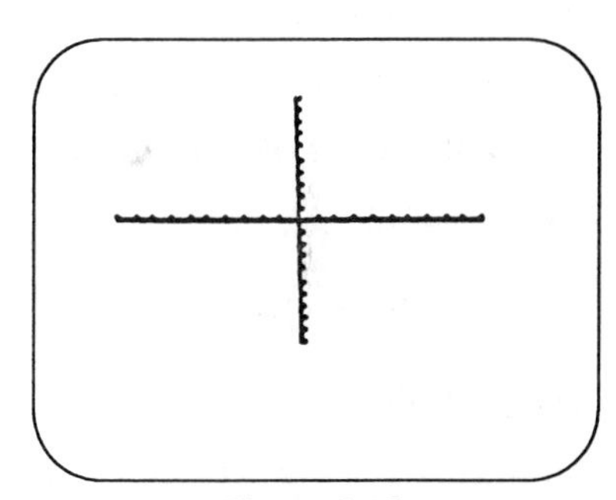

Figure 2.18

Graph distortion is important. For example, perpendicular lines will not appear perpendicular in a non-square range. Enter the graphs of $Y1 = .5X + 2$ and $Y2 = -2X - 3$ by pressing

`EQTN` `MENU` [1:Y1] `CL` `.` **5** `X/θ/T` `+` **2** `ENTER` `CL` `(-)` **2** `X/θ/T` `−` **3**.

Set the non-square range to –10 to 10 for both axes by pressing

`RANGE` `(-)` **1 0** `ENTER` **1 0** `ENTER` **1** `ENTER` `(-)` **1 0** `ENTER` **1 0** `ENTER` **1** `ENTER`.

Press `⩘` to view the graph displayed in Figure 2.19. Notice the lines do not appear perpendicular.

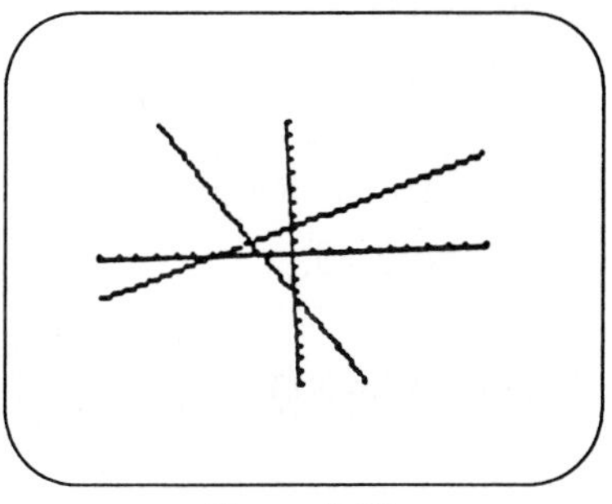

Figure 2.19

Now, set a square range with $-15 < X < 15$ and $-10 < Y < 10$ by pressing

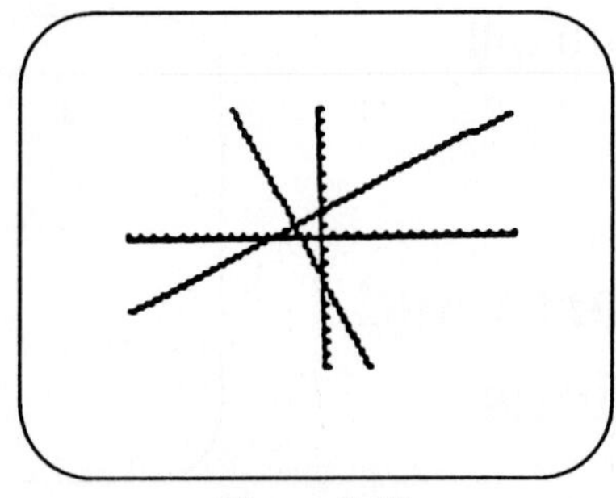

Figure 2.20

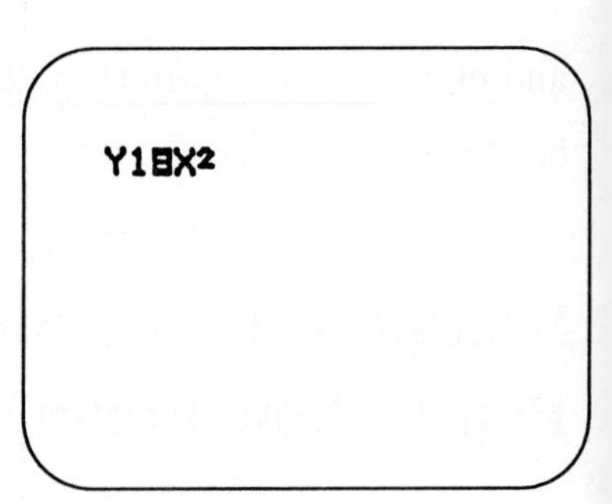

Figure 2.21

[RANGE] [(-)] **1 5** [ENTER] **1 5** [ENTER] **1** [ENTER] [(-)] **1 0** [ENTER] **1 0** [ENTER] **1** [ENTER]

Press [graph] to view the graph displayed in Figure 2.20. Notice the lines now appear perpendicular.

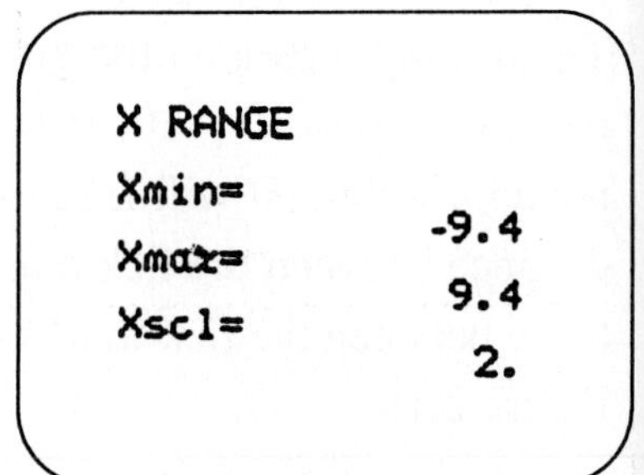

Figure 2.22

Section 2.5 Functions and Graphs of Functions

Graphing Functions

To graph the function $f(x) = x^2$, you will use $Y1 = X^2$, with the built-in x^2 power range. Press

[EQTN] and [2ndF] [▲]

to move up to $Y1$. Press [CL] to clear any $Y2$ and $Y1$ equations previously stored in the grapher. Enter the function in 'Y1=' by pressing [X/θ/T] [x^2] (Fig. 2.21). Enter the built-in range by pressing

[RANGE] [MENU] [B:POWER] [2:x^2].

The 'Y1=', X Range, and Y Range screens are displayed in Figs. 2.22 and 2.23.

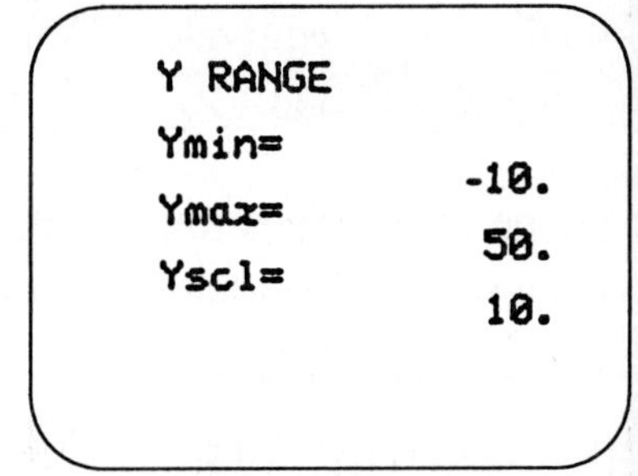

Figure 2.23

Press the Graph mode key [graph] to display the graph (Fig. 2.24).

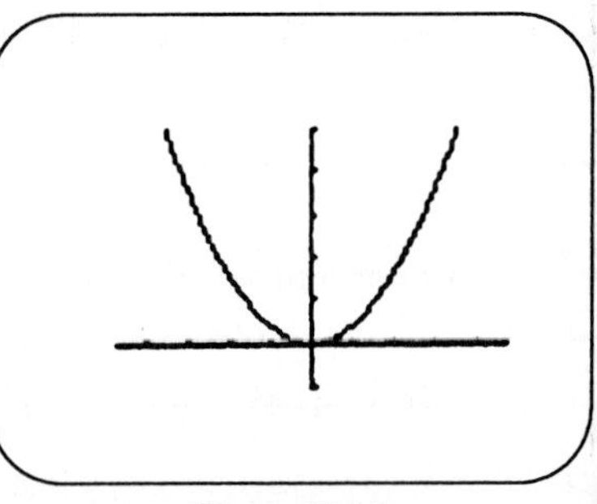

Figure 2.24

Graph

$Y1 = \sqrt{x}$

with its built-in range. Enter the function by pressing

[EQTN] [CL] [√] [X/θ/T].

Select the built-in range by pressing

[RANGE] [MENU] [B:POWER] [3: √].

The range should be Xmin = –1, Xmax = 8.4, and Xscl = 1. Press the [GRAPH] to display the graph (Fig. 2.25).

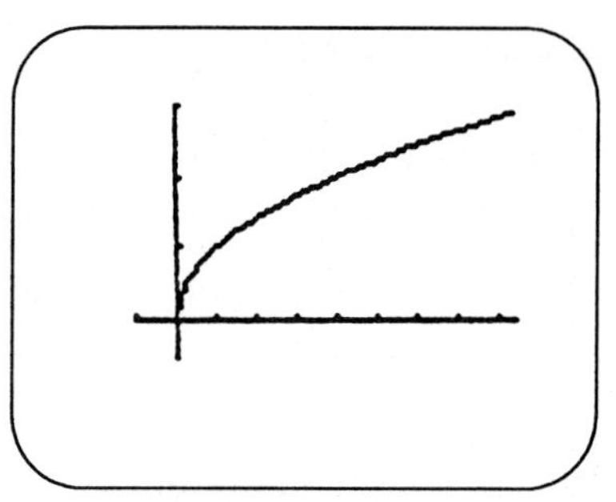
Figure 2.25

Graph $Y1 = X^{-1}$ with its built-in range. Enter the function by pressing

[EQTN] [CL] [X/θ/T] [2ndF] [x^{-1}].

Select the built-in range by pressing

[RANGE] [MENU] [B:POWER] [1:X^{-1}].

The range should be Xmin = –4.7, Xmax = 4.7, Xscl = 1, Ymin = –4, Ymax = 4, and Yscl = 1. Press [GRAPH] to display the graph (Fig. 2.26).

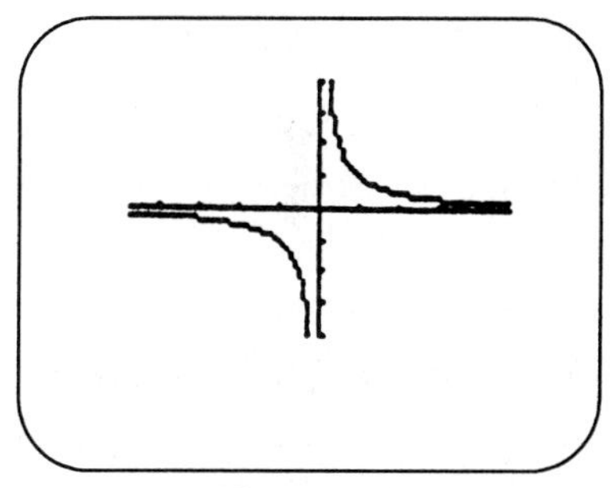
Figure 2.26

Graph $Y1 = -X^2 - 3X$ with a default range. Enter the function by pressing

[EQTN] [CL] [(–)] [X/θ/T] [x^2] [–] **3** [X/θ/T].

Select the standard or default range by pressing

[RANGE] [MENU] [A:DEFLT] [ENTER].

The range should be Xmin = –4.7, Xmax = 4.7, Xscl = 1, Ymin = –3.1, Ymax = 3.1, and Yscl = 1. Press [GRAPH] to display the graph (Fig. 2.27).

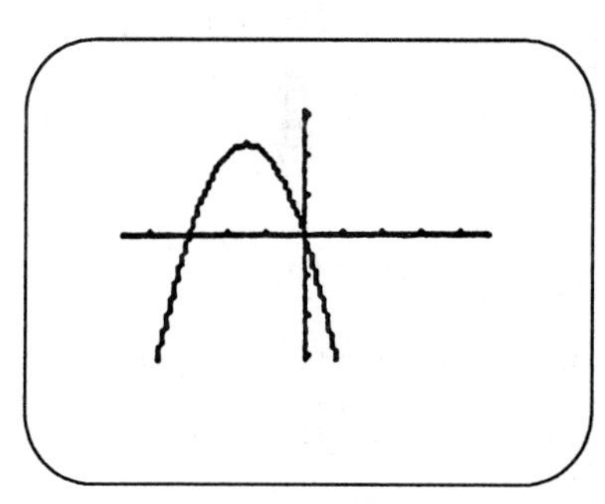
Figure 2.27

The Standard Window

The Standard Window is $-10 < X < 10$ [–10, 10] and $-10 < Y < 10$ [–10, 10]. This window is not one of the built-in Ranges available within the Sharp grapher's Range menu. However, it can be entered as any other range by typing the values in within the X Range and Y Range screens.

Using the Trace Feature to Find Domain and Range

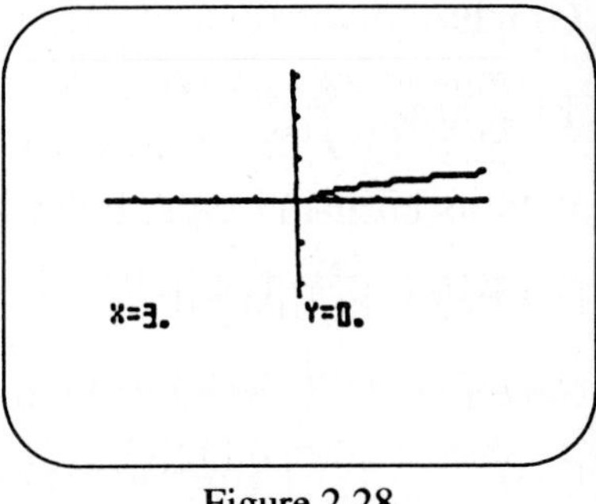

Figure 2.28

You can use the trace feature to determine the domain and range of a function. For example, to find the domain and range of $f(x) = \sqrt{x} - 3$, enter the function by pressing

[EQTN] [CL] [√] [X/θ/T] [▶] [−] **3**,

set the Range to the Integer Window, and press [graph] to graph the function (Fig. 2.28). Press [▶] to engage the trace feature. The cursor appears at the leftmost point of the function, where the graph ends. This point is (3,0) and it implies the domain is $[3, \infty)$. Also, notice that the graph lies entirely above the x-axis, which implies the range is $[0, \infty)$. Press [▶] repeatedly to trace right to confirm the domain and range. Note that you can press [2ndF] [▶] to jump the trace cursor to the right-hand side of the display. Pressing [▶] again will scroll the display to see the cursor. Scrolling occurs when you move the trace cursor outside of the viewing window, and the grapher adjusts the viewing window so that you can view the cursor again. Scroll several times to confirm that the graph is continuing to rise, with the x- and y-coordinates increasing (Fig. 2.29).

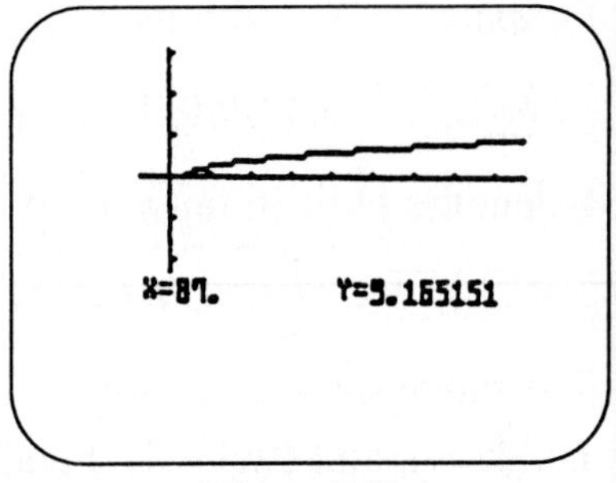

Figure 2.29

Determine the range of

$f(x) = x^4 - 3x^2 + 5x - 6$.

Enter the function by pressing

[EQTN] [CL] [X/θ/T] [a^b] **4** [▶] [−] **3** [X/θ/T] [x^2] [+] **5** [X/θ/T] [−] **6**.

Set the range to

$-5 < X < 5$ and $-10 < Y < 10$,

with scales of one, and then graphing the function (Fig. 2.30). Note that the graph is incomplete. Repeatedly press [▶] and trace to find the bottom of the graph. The grapher will scroll the display several times. Watch the y values decrease and then increase. Find the minimum y value.

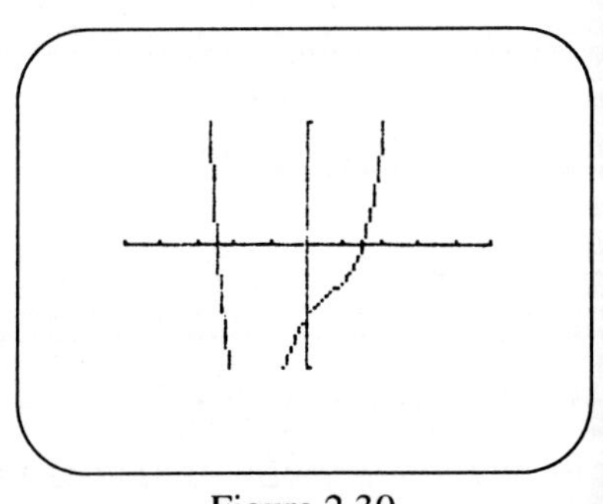
Figure 2.30

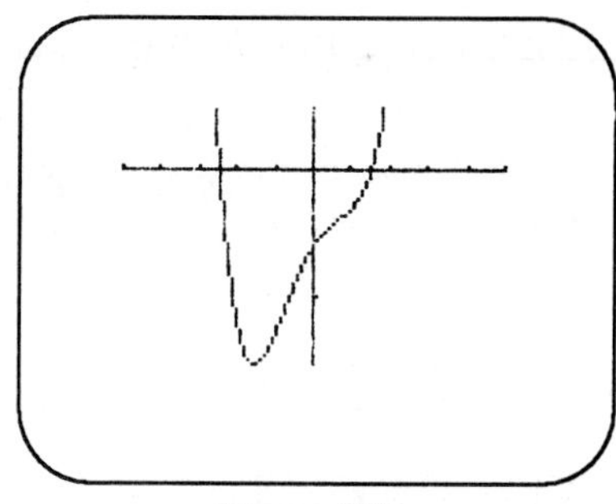

Figure 2.31

The minimum y value appears to be approximately –15.2. Press [CL] to remove the coordinates from the screen and view the bottom of the graph (Fig. 2.31). The range appears to be the interval $[-15.2, \infty)$.

Graphing a Relation That Is Not in Function Form

To graph a relation, solve the equation for the y variable and then graph the corresponding equations generated. For example, $y^2 = x$ is a relation. Solving for y you will get $y = \pm\sqrt{x}$. Now, graph

$Y1 = \sqrt{x}$ and $Y2 = -\sqrt{x}$

simultaneously using the default range. Set the default range by pressing

[RANGE] [MENU] [A:DEFLT] [ENTER].

Then press [graph key] and the result is the graph of the relation (Fig. 2.32).

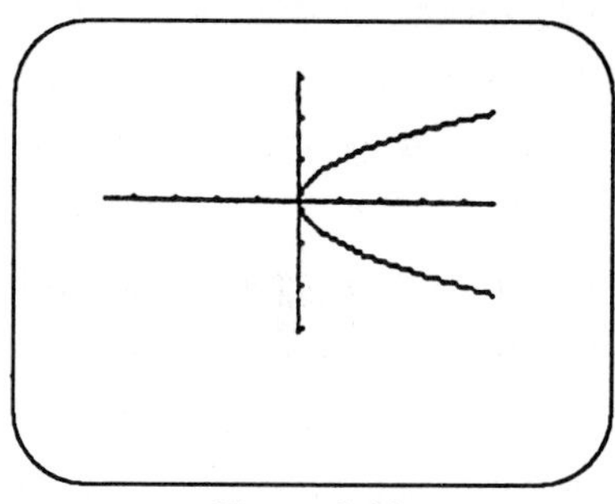

Figure 2.32

Chapter 3 Solving Equations and Systems of Equations

Section 3.1 Finding Graphical Solutions

Features for Viewing Multiple Graphs

The features for viewing multiple graphs include the ability to select only those graphs you wish to view and the ability to have those graphs drawn simultaneously (at the same time) or sequentially (in order of entry).

Enter the two graphs $Y1 = 2X + 5$ and $Y2 = 21$ and graph them in an integer window (Fig. 3.1). You can graph one or the other function alone or both together. To graph one alone, you will need to deselect

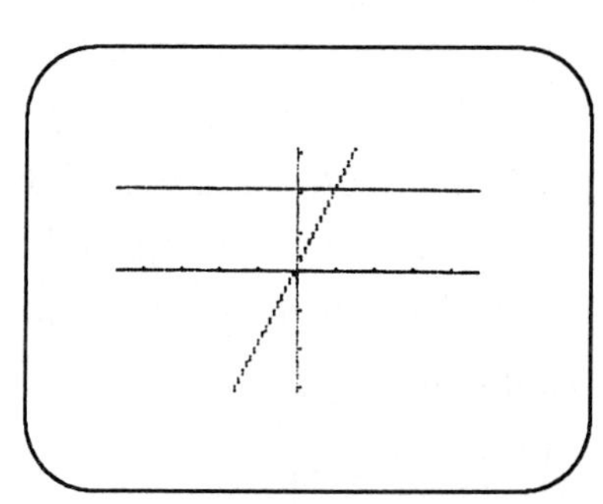

Figure 3.1

the other graph. You do this by moving to the equation you desire to deselect, and pressing [◄] followed by [ENTER]. This will remove the fixed-highlighted block over the equals sign. This equation is deselected and will not graph.

To select a deselected graph, you repeat the above procedure and the fixed-highlighted block will appear over the equals sign, indicating that it is selected.

To set the grapher to either sequential or simultaneous graphing, you press [MENU] [D:PLOT2] followed by the number to the left of your choice (**1** for sequential, **2** for simultaneous).

Multigraph Method for Solving Linear Equations

To use the multigraph method for solving a linear equation, you first enter the left side of the equation for $Y1$ and the right side of the equation for $Y2$.

$17 - 1.5X = -10$

By pressing

'Y1=' **17** [−] **1** [.] **5** [X/θ/T] and 'Y2=' [(−)] **10**.

Set the viewing window to the integer range (–47, 47, 10, –31, 31, 10) and graph the equations. Press [►] and trace the line to the intersection. Confirm that you have found the intersection by pressing [▲] and [▼] to toggle between the two lines, observing the same coordinates at the bottom of the display. Pressing the arrow keys moves the tracer from one graph to the other. Unlike some graphing calculators, the Sharp grapher does not have an indicator (a 1 for $Y1$, a 2 for $Y2$) in the upper right-hand corner to tell you which curve you are tracing. Use this method to solve the linear equation

$17 - 1.5X = -10.$

Your solution is X=18 (Fig. 3.2).

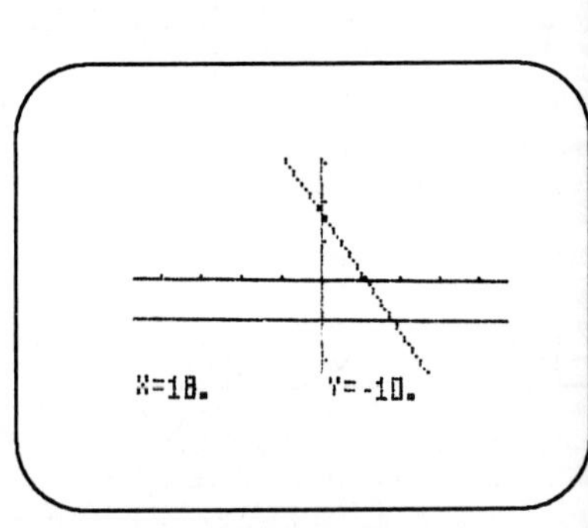

Figure 3.2

Section 3.2 Solving Equations: Using Zoom In

Finding a Graphical Solution Using a 0.1 Window

The decimal (0.1 or default) window is discussed in Section 2.1 of this manual.

Zooming

Zooming is the process of changing the viewing window in order to obtain information desired about the graph. You may want to zoom out to see more of a graph, or zoom in to get a closer look at an x-intercept, an intersection of two graphs, etc.

Press [ZOOM] to view a menu of zoom features. These features include the zoom box, zoom in, zoom out, an ability to change the zoom factors, and an automatic or auto zoom.

The zoom factors for the Sharp grapher are set at 1.5 and 1.2 for the x- and y-axes, respectively. Increasing this number will allow you to enlarge (zoom in) or 'back off' (zoom out) the graph more rapidly. Decreasing the number will slow down the process. Change the factors to 10 for both axes by pressing

[ZOOM] [4:FACTOR] **1 0** [ENTER] **1 0** [ENTER] [ZOOM].

To zoom in on an x-intercept, you will graph the function, trace near the intercept, and press

[ZOOM] [2:IN].

For example, to find a solution to

$2.7X + 3.28 = 0,$

enter the function for $Y1$ by pressing

[⁄⁄] [MENU] [A:SCRN] **1** [CL] **2** [.] **7** [X/θ/T] [+] **3** [.] **2 8** [ENTER] (remember to clear $Y2$, $Y3$, $Y4$, if you used the equations before this exercise), and use the default range by pressing

RANGE MENU [A:DEFLT] ENTER.

Set the Xscl and Yscl to 0.01. Finding the intercept between two tick marks on the axis will satisfy the textbook's error requirement. Now, graph the function, trace near the intercept, and press

ZOOM [2:IN]

to zoom in. Repeat until accuracy is achieved. See Fig. 3.3.

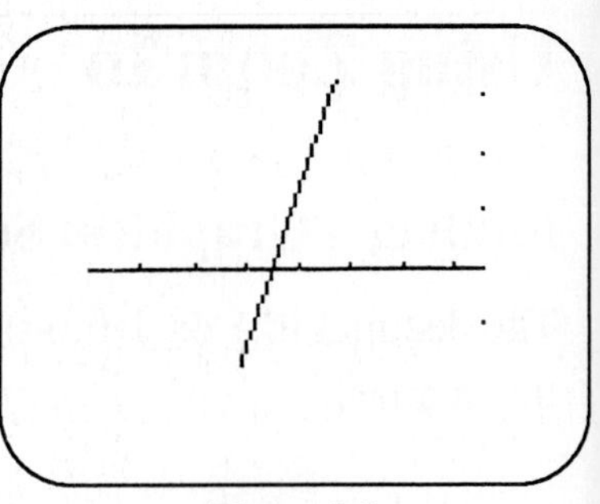

Figure 3.3

Problems with Zooming In

Continue zooming in until the tick marks appear on the x-axis and the graph passes between two of them. Use the trace to find the value of the x-intercept lying between the two .01 tick marks. Note that the tick marks are also displayed to the right even though the y-axis is not visible (Fig. 3.4). The answer is approximately –1.215.

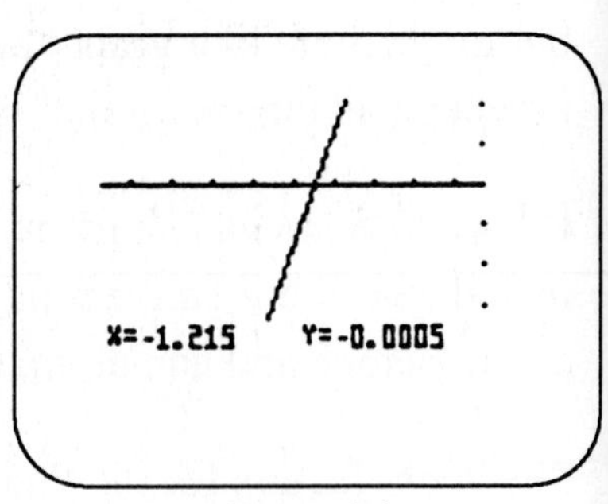

Figure 3.4

Accuracy of a Solution on the Grapher

Graphers are used to approximate answers, not find them exactly. For example, the solution to $3X - 2 = 0$ in the default range is $^2/_3$ algebraically. Graphically, no matter how many times you zoom in, you can only approximate $^2/_3$ as .66 (Fig. 3.5).

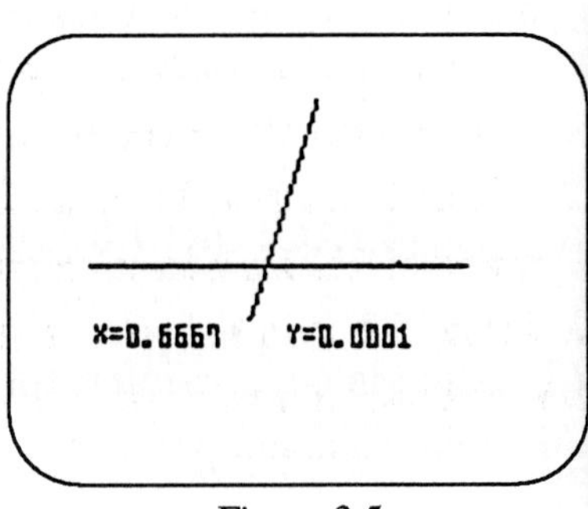

Figure 3.5

How to Determine an Appropriate Window for a Graph

When you don't know an appropriate viewing window for the graph, start with the default range and press

ZOOM [5:AUTO]

to automatically set the Y Range for the default X Range. For example, enter $Y1 = 32000 + 960X$, first set the default range by pressing

RANGE MENU [A:DEFLT] ENTER,

and then press

[ZOOM] [5:AUTO].

The Y Range is automatically set for you to view the graph (Fig. 3.6).

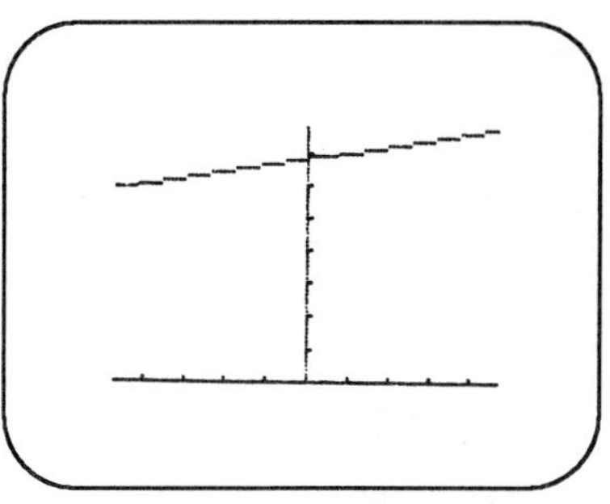

Figure 3.6

Section 3.3 Some Non-Linear Equations

Using Zoom with the Multigraph Method

You need to be careful when zooming in on an intersection for a multigraph method solution. For example, solve

$.7\sqrt{(X-3)} = .3$

using the multigraph method. Graph

$Y1 = .7\sqrt{(X-3)}$ and $Y2 = .3$

(press [EQTN] [2ndF] [▲] to move to $Y1$) in the default window (–4.7, 4.7, 1, –3.1, 3.1, 1). The .3 barely appears on the screen (Fig. 3.7). To zoom in, trace along the $Y2$ to the intersection, since it remains constant throughout the window. Then, press

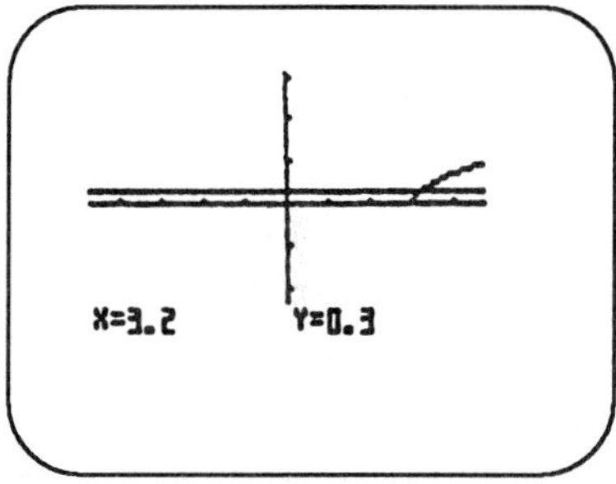

Figure 3.7

[ZOOM] [2:IN]

to zoom in on the intersection. Repeat this three times with factors of 5 until you receive the accuracy you need (your graph will be similar to Fig. 3.8).

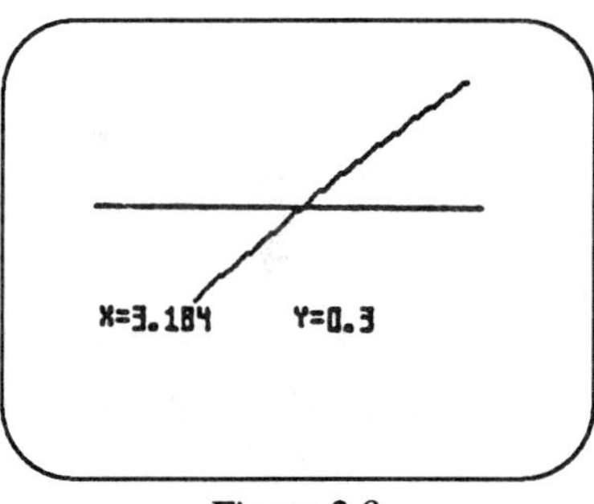

Figure 3.8

Section 3.4 Graphical Representation of Problem Situations

Motion Simulation

Example 3 in Section 3.4 of your main text refers to a parametric solution to a baseball problem. In Section 1.6 of this manual, you can find the information on changing to Parametric mode. Once you have changed to Parametric mode, press [EQTN] and enter the $X1T =$ 3 (fixes the graph in the middle of our display) and

$Y1T = -16T^2 + 88T$ (the height of the ball at time T).

Enter $Y1T$ by pressing

[(-)] **1 6** [X/θ/T] [x^2] [+] **8 8** [X/θ/T].

Set the viewing window by pressing [RANGE] and entering Tmin = 0, Tmax = 5.5 (Tstp is automatically set), Xmin = 0, Xmax = 5.5, Xscl = 1, Ymin = 0, Ymax = 150, and Yscl = 10. Press [graph] to sketch the line (Fig. 3.9).

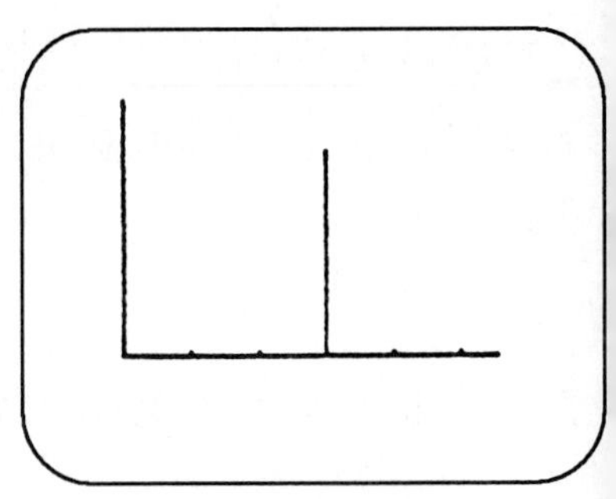

Figure 3.9

Trace the line. At the bottom of the display you will see the height of the ball (y-coordinate) for the time T. Note the ball rises to a height of 121 feet in 2.75 seconds and then starts falling (Fig. 3.10).

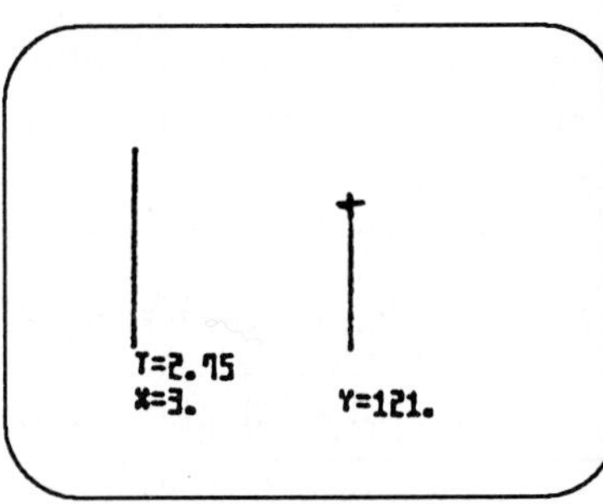

Figure 3.10

Remember to change your grapher back to XY (rectangular) mode by pressing

[SET UP] [E:COORD] [1:XY] [QUIT].

Drawing

Basic point plotting was discussed in Section 2.2 of this manual. You can take the numeric representations of a problem and plot these on the grapher. Examine your data to determine the viewing window you will need to plot the points. In the text's Section 3.4, Example 4, the data span from 10 to 50 for average speed (x-axis) and from 10 to 50 for time of trip (y-axis). Therefore, to see all the data points and the axes, you will need to go beyond the data limitations. For example, press [RANGE] and set Xmin = −10, Xmax = 60, Xscl = 10, Ymin = −10, Ymax = 60, and Yscl = 10. Press [EQTN] and [CL] to clear any expressions for $Y1$ through $Y4$, then press [graph] to view the window. Now, plot all the data points for Example 4 by pressing

[2ndF] [PLOT] [B:PLOT] [1:DIRECT]

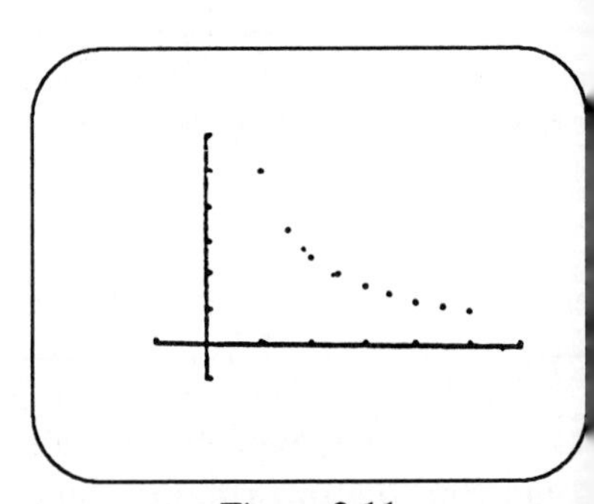

Figure 3.11

and entering the average speed for X and time of trip for Y each time.

The plotted data appear in Figure 3.11.

X	Y
10	50
15	33.33
20	25
25	20
30	16.67
35	14.29
40	12.5
45	11.11
50	10

To clear off the data points, press

[2ndF] [PLOT] [A:CLEAR] [ENTER].

The graphical representation of the problem is a plot of the algebraic representation found in the text. This representation is $y = 500/x$. To plot this function press [EQTN], enter the representation, and press [graph] to view it (Fig. 3.12).

Figure 3.12

Custom Integer Window

A custom integer window can be found by using [0, Xmax] and [0, Ymax], where Xmax is 94 or multiples of 94, and Ymax is 62 or multiples of 62. These values were presented at the end of Section 2.1.

Section 3.5 Solving Systems of Equations by Graphing

Tracing Two Functions

The ability to trace more than one function was discussed in the multigraph method in Section 3.1 of this manual. Tracing can be used to find the solution (intersection) of a system of equations. For example, graph

$Y1 = 3X - 2$

and

$Y2 = -.5X + 5$

with an integer range (Fig. 3.13). Now, trace over to the intersection. Verify that you have the intersection by toggling up and down between the two graphs and observing the same coordinates at the bottom of the display screen.

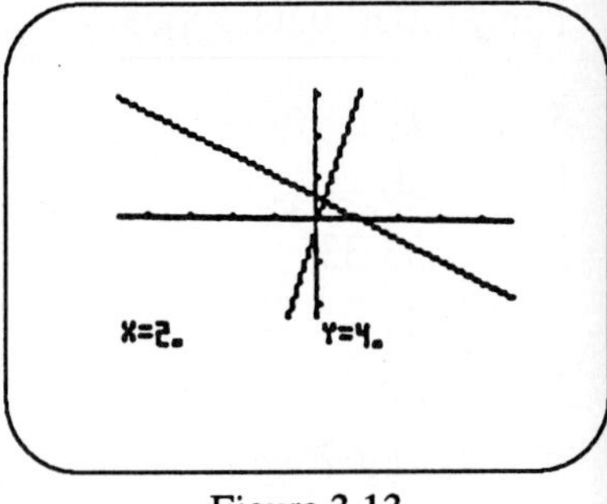

Figure 3.13

Solving a System by Zooming

Zooming, in some cases, can be used to locate an intersection and then to find its coordinates. For example, graph

$Y1 = -2X + 175$

and

$Y2 = 3X - 320$

with an integer range. Note that the graphs do not appear on the screen. You need to zoom out to find the intersection. When zooming within the integer range, it is convenient to set the zoom factors at 5 or 10. Set them to 5 by pressing

[ZOOM] [4:FACTOR]

and entering the 5s. Now, zoom out to find the intersection by pressing

[ZOOM] [3:OUT]

(Fig. 3.14).

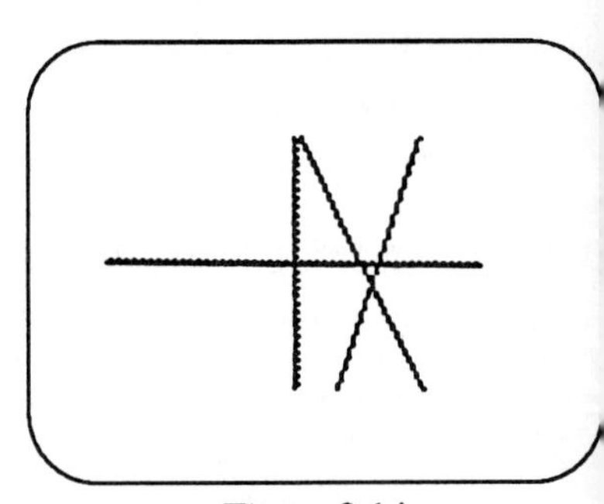
Figure 3.14

Trace over to the intersection and zoom in to find the solution. Remember to toggle back and forth between the lines to ensure you have found the intersection (the coordinates should not change). After zooming in on the intersection, your graph should look similar to Figure 3.15.

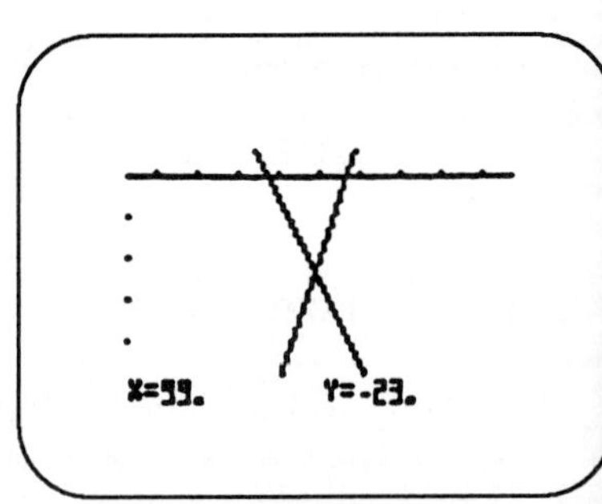

Figure 3.15

Convenient Zoom Factors

Using zoom factors of 5 and 10 are convenient for keeping the scale values clean on display. We like scales of 1, 5, 10, etc.

Chapter 4 Linear Inequalities and Systems of Linear Inequalities

Section 4.1 Solving Linear Inequalities Algebraically

Please refer to Section 1.3 of this lab manual for a discussion of evaluating an expression with the Recall or Replay feature and the Editing features.

Solving Inequalities Graphically

To solve the inequality

$.8X - 4 < 0$,

enter the function $Y1 = .8X - 4$ and graph it with a default range (remember to clear off $Y2$). Next, change your grapher to the Dot mode by pressing

[MENU] [C:PLOT1] [2:Dot].

Zoom out (zoom factors of 5) to see more of the graph and the x-intercept. Trace the curve to find the x-intercept (where $y = 0$) (Fig. 4.1). The function is equal to 0 at $X = 5$. The function is negative (less than 0) to the left of 5 (X less than 5). Therefore, the solution is all $X < 5$.

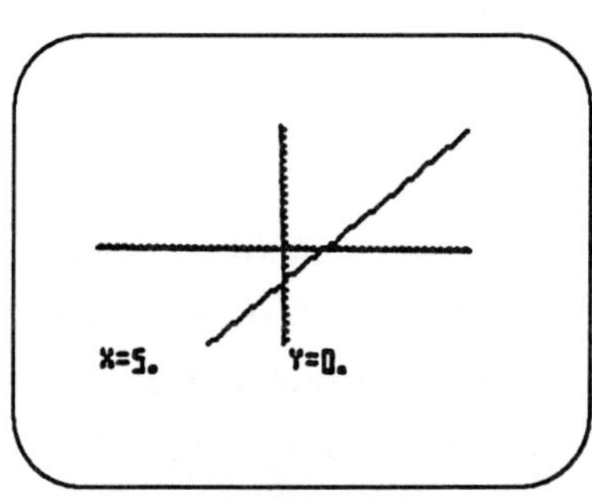

Figure 4.1

Change the grapher back to Connected mode by pressing

[MENU] [C:PLOT1] [1:Connect].

Section 4.4 Solving Absolute Value Equations and Inequalities

Please refer to Section 1.5 of this manual for a discussion of generating grapher-produced tables.

Section 4.5 Linear Inequalities in Two Variables

Shading

To graph the inequality $Y > 4X - 5$, first enter the function as $Y1 = 4X - 5$ and then graph it with a default range (Fig. 4.2).

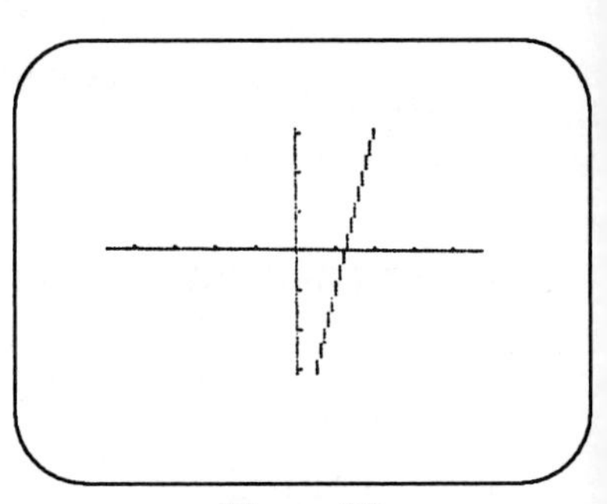

Figure 4.2

When the graph does not pass through the origin, test the inequality by plugging in $X = 0$ and $Y = 0$. If the inequality is 'true' for the origin, then you want to shade toward the origin. If it is 'false,' then you want to shade away from the origin. In your example,

$0 > 0 - 5$

is 'true.' You want to shade toward the origin or, in the grapher's case, above the graph of $Y1$. To shade the solution set for the graph (above $Y1$), press

[MENU] [A:SCRN] [5:FILL]

to view the shading menu. Press [▼] to leave the 'fill below' option as 'none' and press [▶] to highlight $Y1$ to be filled (shaded) above. Press the [graph key] to see the shaded solution set (Fig. 4.3).

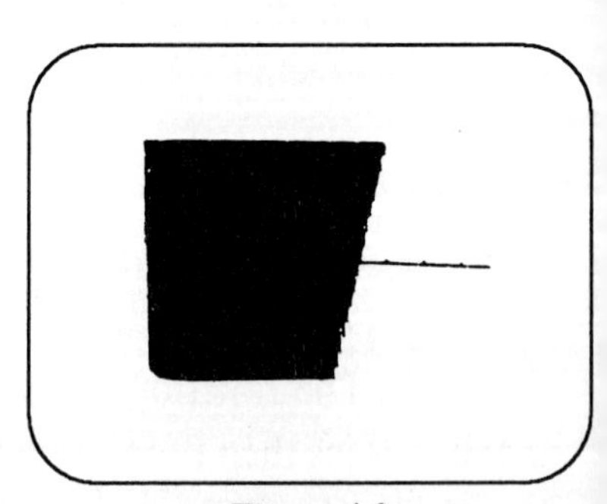

Figure 4.3

On the Sharp grapher, you cannot change the resolution (light to dark) of the shading.

Section 4.6 Solving Systems of Inequalities

Graphing Systems of Inequalities Using Shading

The Sharp grapher works best with a system of inequalities where one is a '<' and one is a '>.' For example, to find the solution region for

$y < -x + 9$ and $y > 3x - 15$,

you will graph $Y1 = -X + 9$ (enter by pressing [graph] [MENU] [A:SCRN] [1:Y1] [CL] [(-)] [X/θ/T] [+] **9** [ENTER])
and $Y2 = 3X - 15$ in a default range

([RANGE] [MENU] [A:DEFLT] [ENTER]).

You will need to zoom out (set zoom factors at 5 by pressing [ZOOM] [4:FACTOR] **5** [ENTER] **5** [ENTER] [ZOOM]) by pressing

[ZOOM] [3:OUT]

to see the intersection shown in Figure 4.4. Now, press

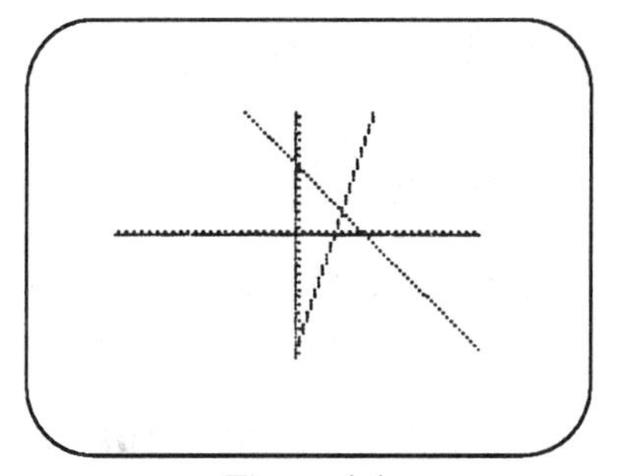
Figure 4.4

[MENU] [A:SCRN] [5:FILL]

and select to fill below $Y1$ and fill above $Y2$. Notice the inequalities to the right of the 'Y' on the display; they match the inequality structure in your original inequalities. After setting the shading, press [graph] to view the graph (Fig. 4.5).

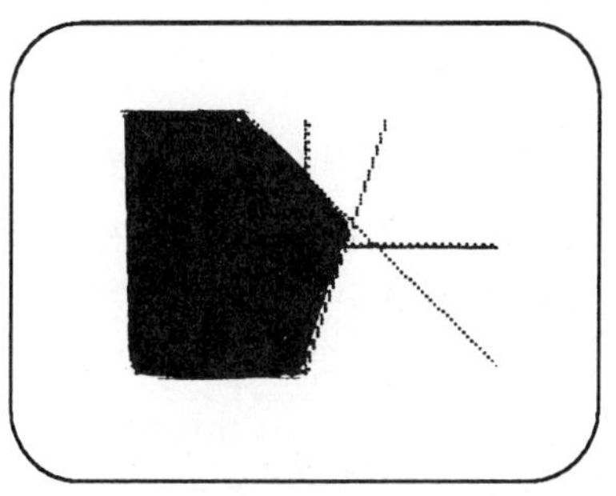
Figure 4.5

When a system has more than one '$y <$' or '$y >$,' you must investigate each one separately and then find their common shading. Turn off the shading by pressing

[MENU] [A:SCRN] [5:FILL]

and setting both the FILL BELOW and FILL ABOVE to none ('NON'). Press

[MENU] [A:SCRN] [1:Y1]

to return to the $Y1$ prompt.

Chapter 5 Polynomials

Section 5.1 Exponents and Their Properties

Using the Trace Feature to Show Two Graphs Are Identical

To verify that

$(2x^3)(3x^2) = 6x^5$,

graph

$Y1 = (2x^3)(3x^2)$

(enter by pressing

[⇄] [MENU] [A:SCRN] [1:Y1] [CL] [(] **2** [X/θ/T] [a^b] **3** [▶] [)] [(] **3** [X/θ/T] [x^2] [)] [ENTER])

and

$Y2 = 6x^5$

by pressing

6 [X/θ/T] [a^b] **5**

in the default viewing window, and see if the two graphs are identical (Fig. 5.1). The appearance of virtually one curve indicates the two expressions are identical. Trace the curve by pressing [▶] (Fig. 5.2). Press [▲] and [▼] to toggle between $Y1$ and $Y2$. Please refer to Section 3.1 of this manual for a discussion of switching back and forth between two graphs. Notice the values for both $Y1$ and $Y2$ are the same for each cursor location.

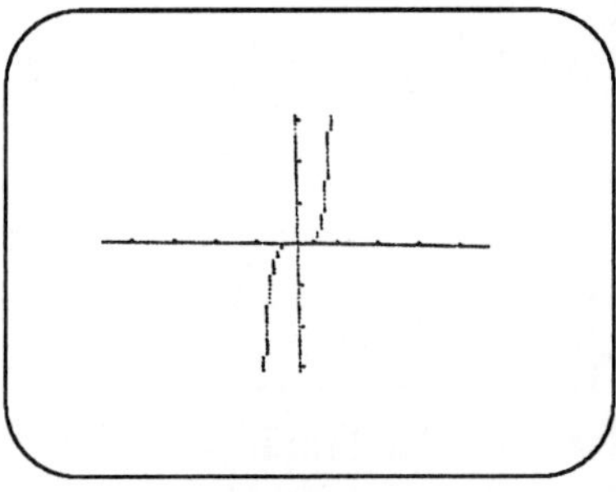

Figure 5.1

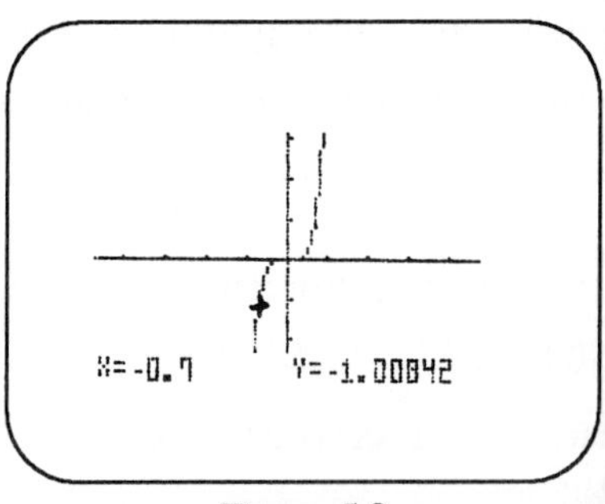

Figure 5.2

Scientific Notation

To enter a number in scientific notation, you will use the [EXP] key. Make sure your grapher is in Computational mode and set up with a floating decimal point, a tab of four decimal places, and with decimal answers by pressing

SET UP [C:FSE] [1:Float Pt] [D:TAB] [4:4] [G:ANS] [1:Decimal] QUIT

Enter 1.23×10^3 by pressing

1 . **2 3** EXP **3** and ENTER.

Note, the calculator translates it to decimal form (Fig. 5.3).

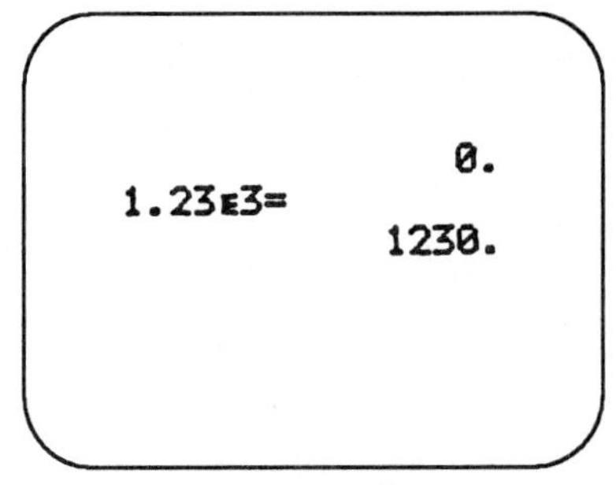

Figure 5.3

This results because the calculator is operating with a floating decimal point. If you want your answers in scientific notation, press

SET UP [C:FSE] [3:SCI] QUIT

to change the operation. Now, add

3.87×10^{-2} and 4.329×10^2.

The E near the end of the answer indicates the number is in scientific notation (Fig. 5.4).

0.
7E-2+4.329E2=
4.3294E 02

Figure 5.4

Change your grapher back to a floating decimal point by pressing

SET UP [C:FSE] [1:Float Pt] QUIT.

Large Numbers on the Grapher

When the grapher is operating in a normal floating decimal point mode, a number that is larger than ten digits will be represented in scientific notation. For example, 2^{33} is a ten-digit number and it is represented regularly. However, 2^{34} is an eleven-digit number and it is represented in scientific notation (Fig. 5.5).

0.
2^{33}=
8589934592.
2^{34}=
1.717986918E 10

Figure 5.5

If the grapher was operating in the Scientific Notation mode, the 2^{33} would be represented as 8.5899E 09 (if the grapher has the tab set at four).

Numbers Close to Zero on the Grapher

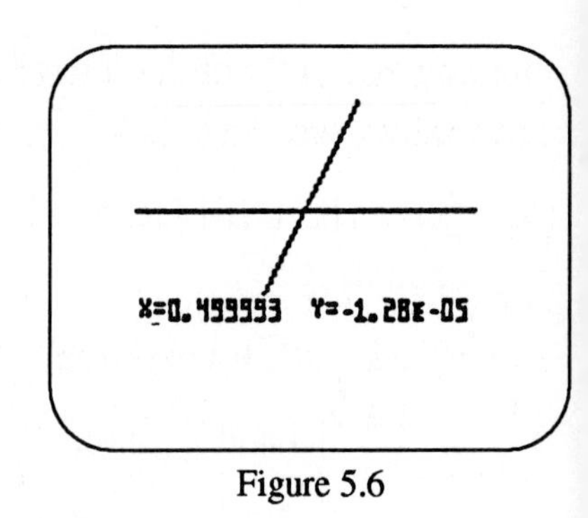

Figure 5.6

Numbers close to zero are often represented in scientific notation. For example, 0.000000000123 would be represented as 1.23 E –10. These numbers are often observed when finding the x-intercepts of a graph in Graphing mode. For example, graph $Y1= 2X - 1$ with a default range, then trace to the x-intercept and zoom in repeatedly. After zooming in several times, notice that the y values close to the x-intercept are represented in scientific notation and are very small (close to zero) (Fig. 5.6).

Section 5.2 Polynomial Expressions

Evaluating a Polynomial Expression on a Grapher

The Sharp grapher will not evaluate a polynomial expression using the $Y1$ from graphing mode. You cannot access the variables $Y1$ through $Y4$ within the Computational mode of the Sharp grapher. Therefore, you will use the method of storing a value and recalling an expression discussed in Section 1.3.

The Method of Entering the Expression in the Function-Defining Menu

You will not enter functions within the Graphing mode as $Y1$, $Y2$, $Y3$, or $Y4$, since the Sharp grapher will not evalute these in the Computational mode.

Table Building for Evaluating Polynomial Expressions

The Sharp grapher does not have a table feature as found on the TI-82.

Entering the Sum of Two Polynomials into the Grapher

To enter the sum of two polynomials and graphically check the answer of adding polynomial expressions, you must first enter the two original polynomial expressions in $Y1$ and $Y2$. Then enter the expression $Y1 + Y2$ in $Y3$. To access the variables $Y1$ and $Y2$, press

MATH [E:VAR].

Now, enter the sum (derived algebraically) in $Y4$. Deselect $Y1$, $Y2$, and $Y3$ (discussed in Section 3.1 of this manual), and graph $Y3$ and $Y4$ in the window. If the algebraic sum was derived correctly, the two graphs will appear identical. Trace the curve and toggle between $Y3$ and $Y4$ to verify that the values are the same.

For example, find the sum of the polynomials

$6x^4 + 2x^3 + 2x^2 + 10x + 1$

and

$-5x^4 - 7x^3 + 3x^2 - 3x - 8.$

You would enter$Y2$

$Y1 = 6x^4 + 2x^3 + 2x^2 + 10x + 1$

$Y2 = -5x^4 - 7x^3 + 3x^2 - 3x - 8$

$Y3 = Y1+Y2$

$Y4 = x^4 - 5x^3 + 5x^2 + 7x - 7$

where $Y4$ is the algebraic answer.

The graphs of the sums are shown in Figure 5.7 within a decimal window.

Figure 5.7

Evaluating Expressions Having Two or More Variables

You can evaluate expressions having two or more variables by first storing values into the variables and entering or recalling an expression. This method was discussed in Section 1.3 of this manual.

For example, find the volume of a tennis ball container

$(V = \pi x^2 y + (2\pi/3)x^3)$

for $x = 3.75$ and $y = 16.25$. Set your grapher to the Calculation mode. You would first store 3.75 into x by pressing

3 [.] **7 5** [STO] [X]

and then store 16.25 into y by pressing

1 6 [.] **2 5** [STO] [Y]

(Fig. 5.8). Now, enter the expression for evaluation by pressing

[2ndF] [π] [X/θ/T] [a^b] **2** [▶] [ALPHA] [Y] [+] **2** [2ndF] [π] [a/b] **3** [▶] [X/θ/T] [a^b] **3**.

Press [ENTER] to evaluate the expression at the stored values (Fig. 5.9).

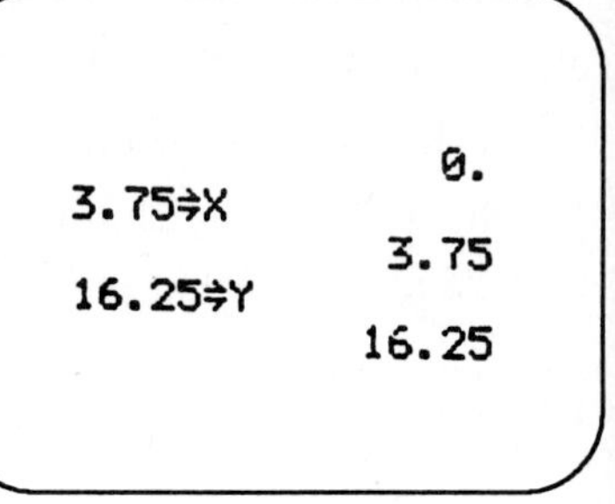

Figure 5.8

$\pi X^2 Y + \frac{2\pi}{3} X^3 =$ 828.3496255

Figure 5.9

Section 5.4 Polynomial Equations and Factoring

Please refer to Section 2.1 of this manual for a discussion of the 0.1 window.

Chapter 6 Rational Expressions and Equations

Section 6.1 Simplifying Rational Expressions

Using Parentheses in Rational Expressions

The Sharp grapher's Equation Editor allows you to enter rational expressions as you see them in the text. Therefore, fewer mistakes are made in entering and evaluating expressions. For example, without an Equation Editor mode, you would evaluate $(x - 2)/x$ at $x = 2$ by storing 2 for x and entering $(x - 2)/x$. Do this by pressing

[⊞] **2** [STO] [X/θ/T] [(] [X/θ/T] [−] **2** [)] [÷] [X/θ/T] [ENTER]

(Fig. 6.1). However, you might enter the expression incorrectly without parentheses as $(x - 2)/x$. Do this by pressing

[X/θ/T] [−] **2** [÷] [X/θ/T] [ENTER]

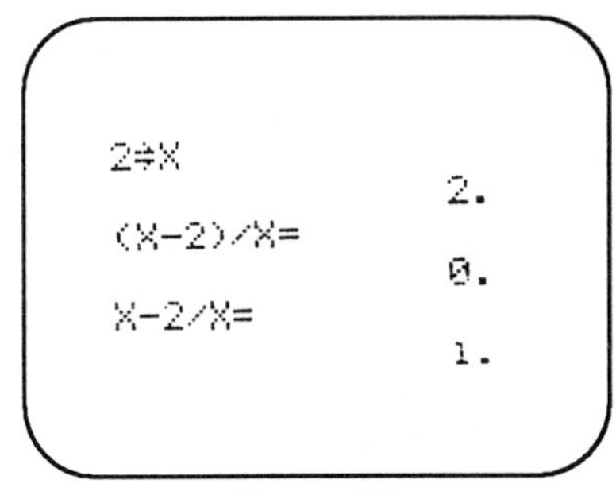

Figure 6.1

(Fig. 6.1). Without the parentheses, the denominator will only divide the 2 and not the complete numerator desired of $x - 2$. When working with the one-line editor, please remember your parentheses.

Error Messages in Rational Expressions

When working with rational expressions, you need to remember not to divide by zero. For example, store zero into X by pressing

0 [STO] [X/θ/T]

and then evaluate $1/X$ by pressing

1 [a/b] [X/θ/T] [ENTER].

You will receive a calculation error box, telling you that you did something wrong. Therefore, keep from evaluating rational expressions at points that make the denominator zero.

Section 6.2 Multiplying and Dividing Rational Expressions

Please refer to Section 3.1 of this manual for discussion of selecting and deselecting graphs.

Section 6.3 Adding and Subtracting Rational Expressions

Resolution of the Screen with Graphing Rational Expressions

The Sharp grapher is designed to display rational expressions accurately and not display the vertical lines present on other graphers.

The [x^{-1}] key can be used when working with rational expressions. For example, the graphs of

$y = (x - 2)^{-1}$ and $y = 1/(x - 2)$

are the same. Verify this on the grapher by entering

$Y1 = (X - 2)^{-1}$ and $Y2 = 1/(X - 2)$

by pressing:

[(] [X/θ/T] [−] **2** [)] [2ndF] [x^{-1}] [ENTER] and **1** [÷] [(] [X/θ/T] [−] **2** [)].

Make sure $Y3$ and $Y4$ are clear, and graph them in a default window (Fig. 6.2). Note a vertical line does not appear.

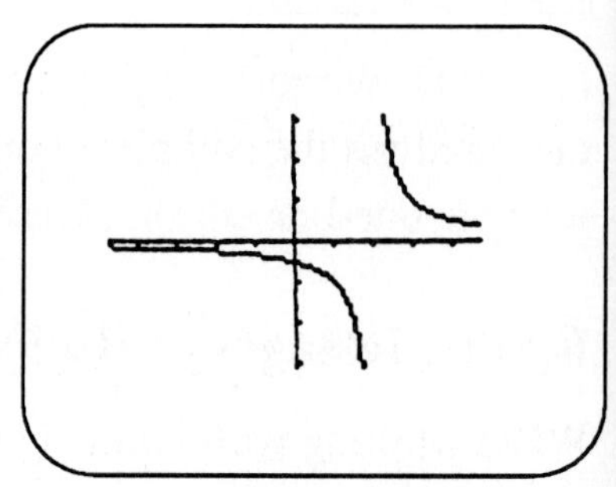

Figure 6.2

Section 6.4 Complex Fractions

Using Parentheses with Complex Fractions

Once again, with the Sharp grapher's Equation Editor (confirm your grapher's setup by pressing

[⊞] [SET UP] [F:EDIT] [1:EQUATION] [QUIT]),

you can enter complex fractions as you see them in the text. If you use the one-line editor, then you will add parentheses around the entire numerator and the entire denominator. For example, enter the complex fraction

$$\frac{\frac{1}{2}+\frac{2}{3}}{\frac{5}{6}+\frac{3}{4}}$$

by pressing

[a/b] [▲] [BS] **1** [a/b] **2** [▶] [+] **2** [a/b] **3** [▼] **5** [a/b] **6** [▶] [+] **3** [a/b] **4**.

Press [ENTER] to evaluate the expression (Fig. 6.3).

$$\frac{\frac{1}{2}+\frac{2}{3}}{\frac{5}{6}+\frac{3}{4}} =$$

0.736842105

Figure 6.3

Section 6.6 Solving Equations Containing Rational Expressions

Please refer to Section 2.1 of this manual for discussion of the 0.2 and integer windows.

Chapter 7 Rational Exponents, Radicals, and Complex Numbers

Section 7.1 Roots and Radicals

Error Messages with Grapher Limitations of Evaluating Square Roots of Negative Numbers

When evaluating even-powered roots, such as the square root, you need to avoid evaluating the square root of a negative number. For example, in Calculation mode, find the square root of –3 by pressing

[√] [(-)] **3** [ENTER].

You will receive a calculation error box. This simply means the answer is not a real number, since no real number squared is –3. This number is a complex number, and they will be discussed later in Chapter 7 of this manual.

Evaluating Expressions with Fractional Exponents

To evaluate an expression with fractional exponents, you will enter the expression as you see it using the [a^b] key. For example, to find $36^{1/2}$ you would press

3 6 [a^b] **1** [a/b] **2** [ENTER] or **3 6** [a^b] **1** [÷] **2** [ENTER].

The answer 6 will appear.

Finding Roots with Your Grapher

To find roots other than the square root, you can use the [$\sqrt[x]{\ }$] key. For example, to find $125^{1/3} = \sqrt[3]{125}$, press

3 [2ndF] [$\sqrt[x]{\ }$] **1 2 5** [ENTER]

(Fig. 7.1). You can find $\sqrt[4]{8}$ by pressing

4 [2ndF] [$\sqrt[x]{\ }$] **8** [ENTER]

(Fig. 7.1).

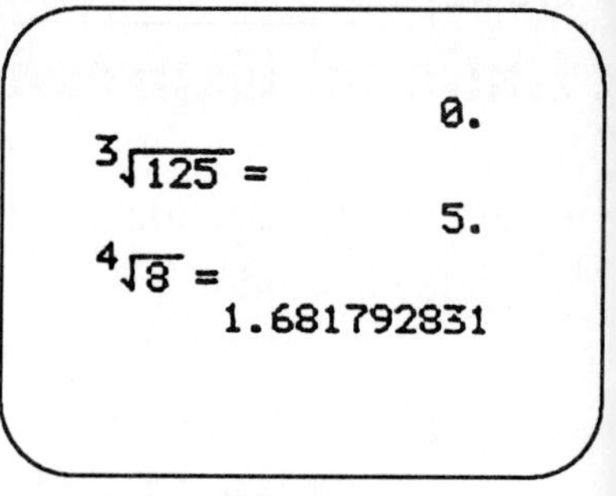

Figure 7.1

Adjusting the Number of Decimal Places Displayed on Your Grapher

Sometimes, you may want to limit the number of decimal places displayed by the grapher. To limit the number of decimal places to two, press

[SET UP] [C:FSE] [2:Fix]

to fix the number of decimal places, and press

[D:TAB] [2:2]

to set the number of places to two. Press [SET UP] to exit the menu. Now, recall the $\sqrt[4]{8}$ by pressing [▲] and press [ENTER] to evaluate the expression again (Fig. 7.2).

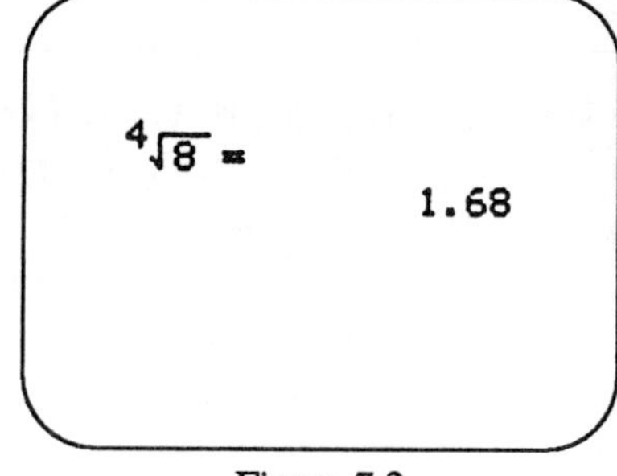

Figure 7.2

Press

[SET UP] [C:FSE] [1:Float Pt] [SET UP]

to change the grapher back to a floating decimal point.

Section 7.2 Rational Exponents and Radicals

Please refer to Section 7.1 of this manual for a discussion of evaluating an expression with rational exponents on the Sharp grapher.

Section 7.6 Complex Numbers

Using the Grapher to Work with Complex Numbers

To work with complex numbers, you must change to Complex mode by pressing

[MENU] [4:COMPLEX].

Note the Complex mode statement at the top of the screen (Fig. 7.3). Make sure your grapher is working with the complex form $a + bi$ by pressing

[SET UP] [H:CMPLX] [1:X + Y*i*] [SET UP].

Now, evaluate the $\sqrt{-3}$ to view its complex equivalent (Fig. 7.4).

To enter a complex number, you will use the [*i*] key (above [a/b]). For example, to find

$(1 - i)(8 + 7i)$,

press

[(] **1** [−] [2ndF] [*i*] [)] [(] **8** [+] **7** [2ndF] [*i*] [)] [ENTER]

(Fig. 7.5). To evaluate

$3 \div (2 + 3i)$,

press

3 [a/b] **2** [+] **3** [2ndF] [*i*] [ENTER]

(Fig. 7.5).

Change your grapher back to real mode by pressing

[MENU] [1:REAL].

COMPLEX MODE
0.

Figure 7.3

COMPLEX MODE
0.
√-3 =
1.7320508i

Figure 7.4

(1-i)(8+7i)=
15.-i
3/(2+3i) =
0.4615384+
-0.6923076i

Figure 7.5

Chapter 8 Quadratic Functions

Section 8.4 Sketching Graphs of Quadratic Functions

Please refer to Section 2.1 of this manual for a discussion of the 0.2 window.

Section 8.6 Vertex and Symmetry of a Parabola

Please refer to Section 1.1 of this manual for a discussion of fractions.

Chapter 9 Higher-Order Systems of Equations and Matrices

Section 9.1 Solving Systems of Equations in Three Variables

Equations in Three Variables That Cannot Be Graphed

Equations in three variables cannot be graphed on the Sharp grapher. You can only find the solution of a system of equations in three variables numerically. A graphical solution is not possible.

Section 9.2 Matrix Algebra

Entering a Matrix into the Grapher

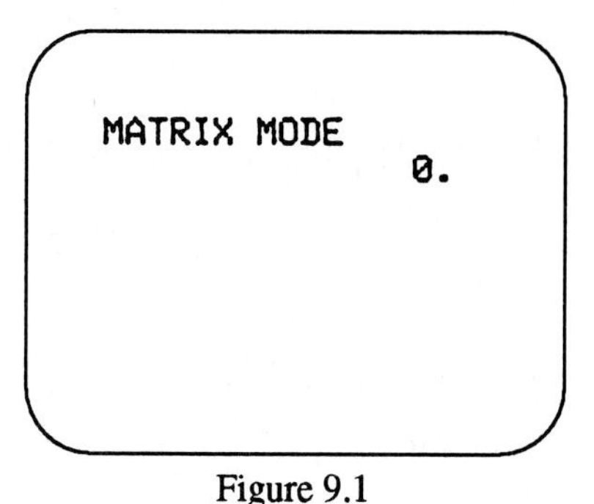

Figure 9.1

To enter a matrix into the grapher, you must change the grapher to Matrix mode. While in Calculation mode, press

[MENU] [3:MATRIX]

to enter Matrix mode. The words Matrix mode should appear at the top of the screen (Fig. 9.1). Press

[MENU] [C:DIM]

to select the matrix to be entered. You can select from twenty-six different matrices. Press

0 1

to select matrix *A*.

You now enter the size or dimension of the matrix. We will enter a 2×2 matrix

$$\begin{bmatrix} 4 & -3 \\ 2 & 6 \end{bmatrix},$$

so press

2 [ENTER] **2** [ENTER]

A[1,1]=
0.
A[2,1]=
0.

Figure 9.2

to set the dimension of the matrix at two rows by two columns. The grapher will now prompt you for the elements of the first column of the matrix by position (Fig. 9.2). $A[1, 1]$ represents the element in the first row, first column position. $A[2, 1]$ represents the element in the second row, first column position. Enter the elements of the column by pressing

4 [ENTER] **2** [ENTER].

The grapher will now prompt you for the elements of the second column of the matrix by position. $A[1, 2]$ represents the element in the

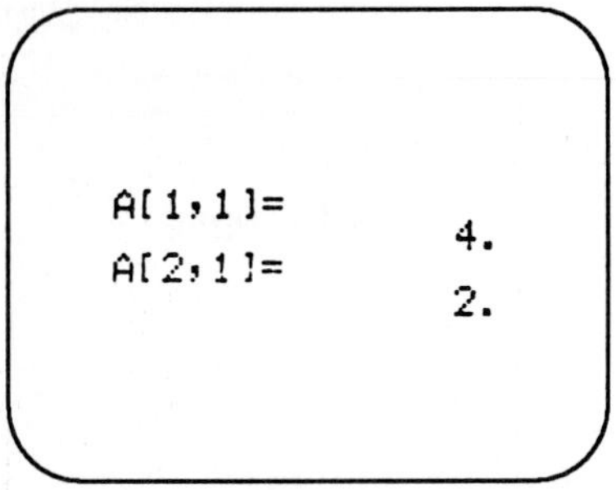

A[1,1]= 4.
A[2,1]= 2.

Figure 9.3

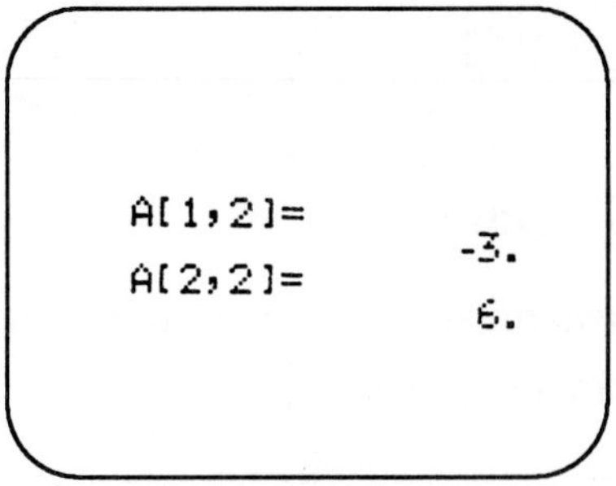

A[1,2]= -3.
A[2,2]= 6.

Figure 9.4

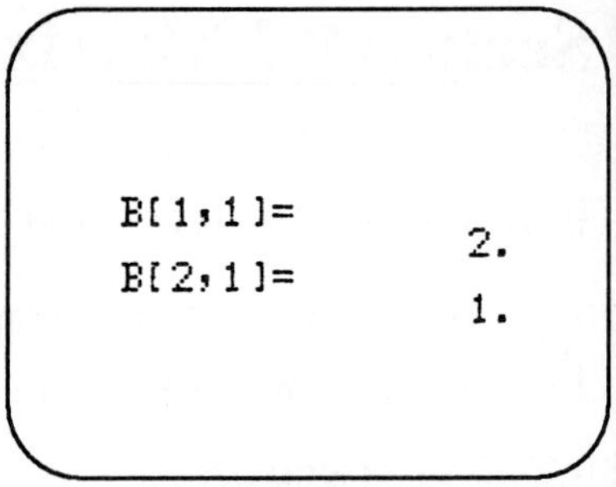

B[1,1]= 2.
B[2,1]= 1.

Figure 9.5

B[1,2]= 5.
B[2,2]= 2.

Figure 9.6

first row, second column position. Enter the elements of the second column by pressing

[(-)] **3** [ENTER] **6** [ENTER].

Use the arrow keys to move around the matrix to check each entry (Figs. 9.3 and 9.4).

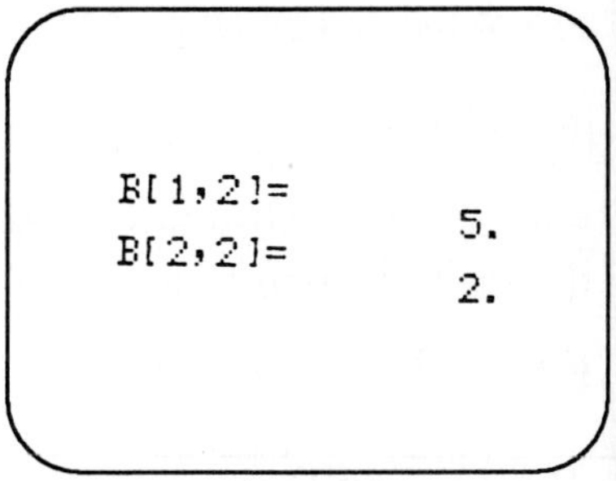

Repeat the process to enter a 2×2 matrix

$$B = \begin{bmatrix} 2 & 5 \\ 1 & 2 \end{bmatrix}.$$

Matrix *B* is displayed in Figures 9.5 and 9.6. Press [QUIT] to exit the display of matrix *B*.

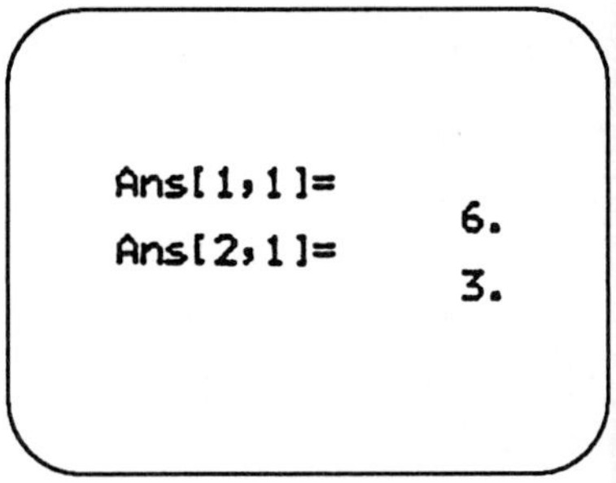

Ans[1,1]= 6.
Ans[2,1]= 3.

Figure 9.7

Adding and Subtracting Matrices

To add matrix *A* to matrix *B*, [*A*] + [*B*], press

[2ndF] [MAT] (above [X/θ/T]) [A] [+] [2ndF] [MAT] [B] [ENTER].

Figures 9.7 and 9.8 represent the resulting matrix. Remember to press [▶] to view the second column.

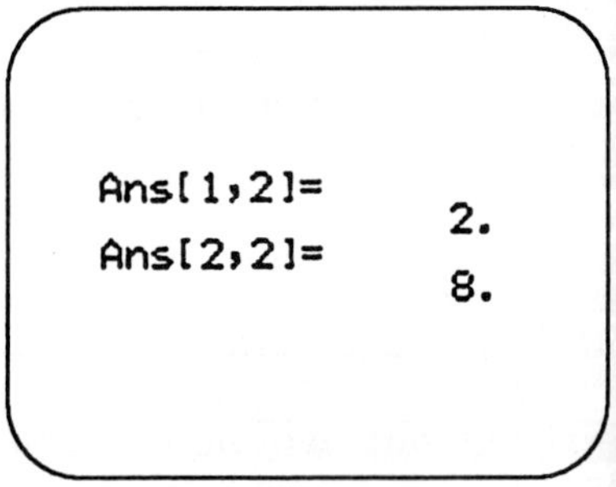

Ans[1,2]= 2.
Ans[2,2]= 8.

Figure 9.8

Subtract matrix *B* from matrix *A*, [*A*] – [*B*], by pressing

[2ndF] [MAT] [A] [−] [2ndF] [MAT] [B] [ENTER].

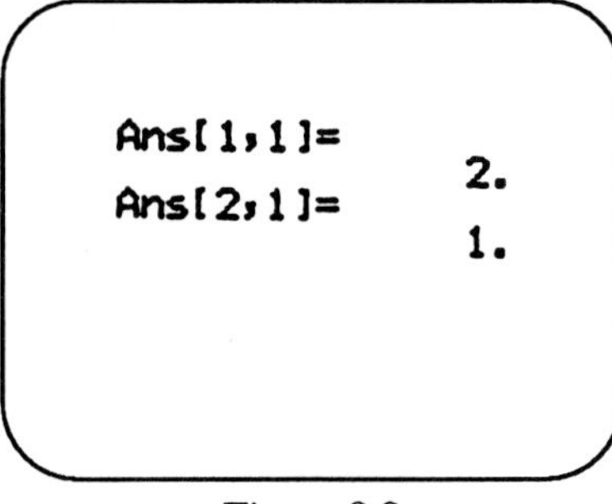

Figure 9.9

Ans[1,2]=
-8.
Ans[2,2]=
4.

Figure 9.10

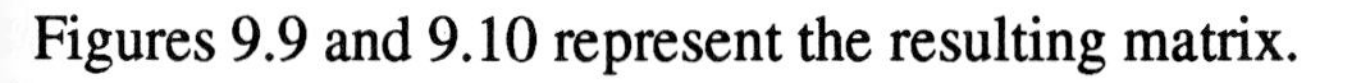
Figures 9.9 and 9.10 represent the resulting matrix.

To delete the matrix A, press

[MENU] [D:DEL] **0 1** [ENTER].

Now enter a new matrix

$$A = \begin{bmatrix} 2 & 1 & 5 \\ 6 & -1 & 3 \end{bmatrix}$$

by pressing

[B:EDIT] **0 1**,

entering the 2×3 dimension, and entering the elements of the new matrix. Remember to press [QUIT] to exit the display of matrix A.

Ans[1,1]=
6.
Ans[2,1]=
18.

Figure 9.11

Scalar Multiplication of Matrices

Scalar multiplication can be performed on matrices using the grapher. For example, to find 3[A], press [CL] or [QUIT] to clear the screen, and press

3 [2ndF] [MAT] [A] [ENTER]

to perform the scalar multiplication. Figures 9.11, 9.12, and 9.13 display the resulting matrix. Remember to press the arrow key to see the additional screens shown in Figures 9.12 and 9.13.

Ans[1,2]=
3.
Ans[2,2]=
-3.

Figure 9.12

Ans[1,3]=
15.
Ans[2,3]=
9.

Figure 9.13

Section 9.3 Multiplication of Matrices

Matrix Multiplication

Delete the old matrices [A] and [B] in the grapher

[MENU] [D:DEL] **01** [ENTER] [CL] [MENU] [D:DEL] **02** [ENTER] [CL].

Enter the new matrices

$$[A] = \begin{bmatrix} 3 & -4 \\ 2 & -8 \end{bmatrix}$$

and

$$[B] = \begin{bmatrix} 4 & -2 \\ -1 & 9 \end{bmatrix}.$$

Multiply the matrices A and B together by pressing [QUIT] to exit the Editing mode, and by pressing

[2ndF] [MAT] [A] [2ndF] [MAT] [B] [ENTER]

to perform the matrix multiplication. The resulting matrix is shown in Figures 9.14 and 9.15.

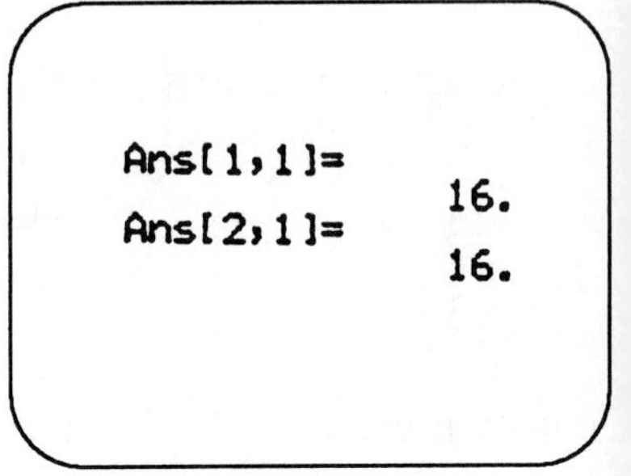

Figure 9.14

Ans[1,2]= -42.
Ans[2,2]= -76.

Figure 9.15

Identity Matrices on the Grapher

The Sharp grapher will not directly establish an identity matrix of a given size. However, you can easily enter the matrix by placing a 1 in all diagonal positions ([1, 1], [2, 2], [3, 3], etc.) and zeros elsewhere.

Inverse Matrices Using the [x^{-1}] Key

Find the inverse of the square matrix

$$A = \begin{bmatrix} 5 & 1 \\ 4 & 1 \end{bmatrix}$$

by entering the matrix A, pressing [QUIT] to exit the Editing menu, and pressing

[2ndF] [MAT] [A] [2ndF] [x^{-1}] [ENTER].

The resulting matrix is shown in Figures 9.16 and 9.17.

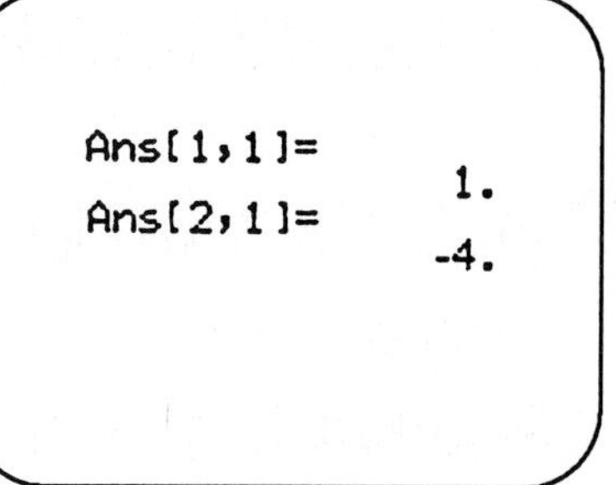

Figure 9.16

Ans[1,2]=
-1.
Ans[2,2]=
5.

Figure 9.17

Some square matrices generate error statements when calculating the inverse. For example, the matrix

$$A = \begin{bmatrix} 3 & 1 \\ 3 & 1 \end{bmatrix}$$

has no inverse, and an error statement is generated when you try to find it. Enter the matrix and try to find the inverse. Delete matrix A.

Determinants of a Square Matrix

To find the determinant of the square matrix

$$A = \begin{bmatrix} 3 & 1 \\ 3 & 1 \end{bmatrix},$$

with the matrix entered, press

[QUIT] [MATH] [E:MTRX] [6:det] [2ndF] [MAT] [A] [ENTER].

The result is shown in Figure 9.18.

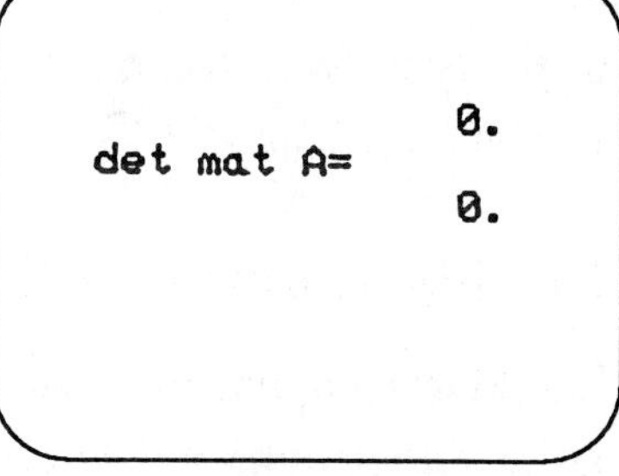

Figure 9.18

Section 9.4 Solving a System of Equations Using Matrices

Solving a System of Equations with Matrices on the Grapher

To solve the system of equations

$$\begin{aligned} 4x - 3y &= -26 \\ 15x + 11y &= 36 \end{aligned}$$

using matrices, first enter the coefficients into the matrix A (dimension 2×2),

$$A = \begin{bmatrix} 4 & -3 \\ 15 & 11 \end{bmatrix},$$

and the constants into the matrix B (dimension 2×1), where

$$B = \begin{bmatrix} -26 \\ 36 \end{bmatrix}.$$

Press [QUIT] to exit the display of the B matrix. The solution matrix X is found by multiplying $A^{-1} B$. This multiplication is derived from the equation

$$AX = B$$
$$A^{-1} A X = A^{-1} B \text{ (multiply both sides by } A^{-1}\text{)}$$
$$I X = A^{-1} B \;\; (A^{-1}A = I, \text{ identity matrix})$$
$$X = A^{-1} B \;\; (I X = X)$$

Note the multiplication is order sensitive, and you will not get the correct answer by multiplying $B A^{-1}$.

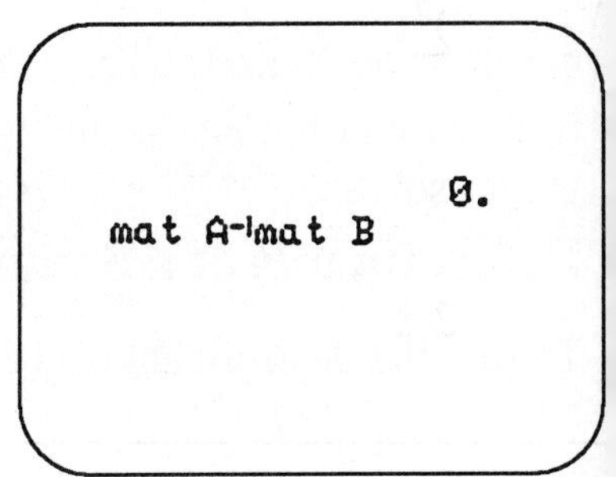

Figure 9.19

Multiply $A^{-1} B$ by pressing

[2ndF] [MAT] [A] [2ndF] [x^{-1}] [2ndF] [MAT] [B] [ENTER]

(Fig. 9.19), and the solution matrix (Fig. 9.20) will appear.

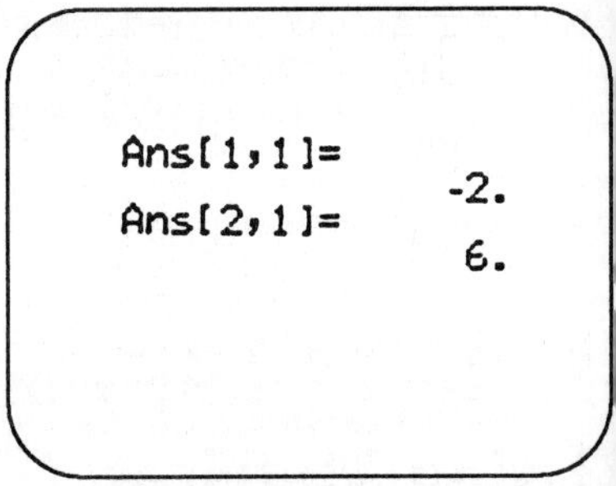

Ans[1,1]= -2.
Ans[2,1]= 6.

Figure 9.20

The [1, 1] element is the X-coordinate and the [2, 1] element the Y-coordinate of the solution point. Return the calculator to Real mode by pressing

[MENU] [A:MODE] [1:REAL].

Chapter 10 Conic Sections

Conic Sections

When graphing conic sections on the grapher, you must first solve the equation for y, and then enter the resulting positive and negative expressions as $Y1$ and $Y2$. These expressions are typically messy and involve a radical. For example, the ellipse

$$\frac{(x-1)^2}{4}+\frac{(y+2)^2}{9}=1$$

would solve to

$$y=-2\pm\frac{3\sqrt{4-(x-1)^2}}{2}.$$

Enter these as $Y1$ and $Y2$. Enter $Y1$ by pressing

[GRAPH] [MENU] [A:SCRN] [1:Y1] [CL] [(-)] 2 [+] [a/b] 3 [√] 4 [−] [(] [X/θ/T] [−] 1 [)] [a^b] 2 [▼] [▼] 2 [ENTER].

Enter $Y2$ by pressing

[CL] [(-)] 2 [−] [a/b] 3 [√] 4 [−] [(] [X/θ/T] [−] 1 [)] [a^b] 2 [▼] [▼] 2 [ENTER].

Set the viewing window to the 0.2 window. Please refer to Section 2.1 of this manual for a discussion of the 0.2 window, and refer to Section 3.1 for a discussion of selecting and deselecting graphs. Press [GRAPH] to view the graphs (Fig. 10.1).

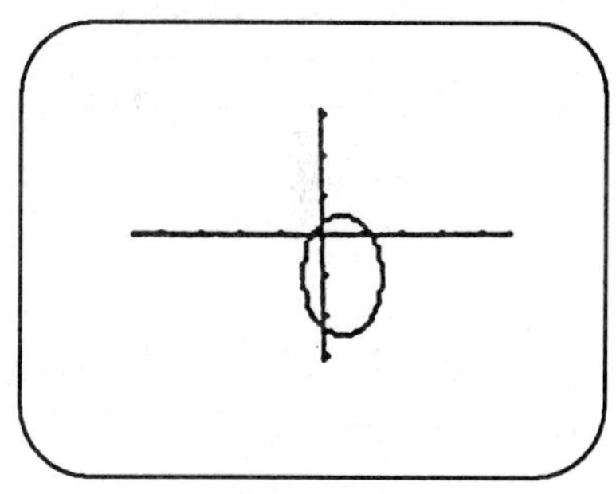

Figure 10.1

Section 10.1 Circles

Graph in the Square Window for a Circle

Circles will be graphed with a default range (the 0.1 window) or with a 0.2 window. These windows provide circles with low distortion and a true shape. To graph the circle

$x^2 + y^2 = 25,$

you must first solve for y

$\left(y = \pm\sqrt{25 - x^2}\right),$

and then enter the two equations as $Y1$ and $Y2$. Enter $Y1$ by pressing

[MENU] [A:SCRN] [1:Y1] [CL] [√] **2 5** [−] [X/θ/T] [a^b] **2** [ENTER].

Enter $Y2$ by pressing

[CL] [(−)] [√] **2 5** [−] [X/θ/T] [a^b] **2** [ENTER].

Set the display to a 0.2 window (–9.4, 9.4, 2, –6.2, 6.2, 2), and press [graph] to view the circle (Fig. 10.2).

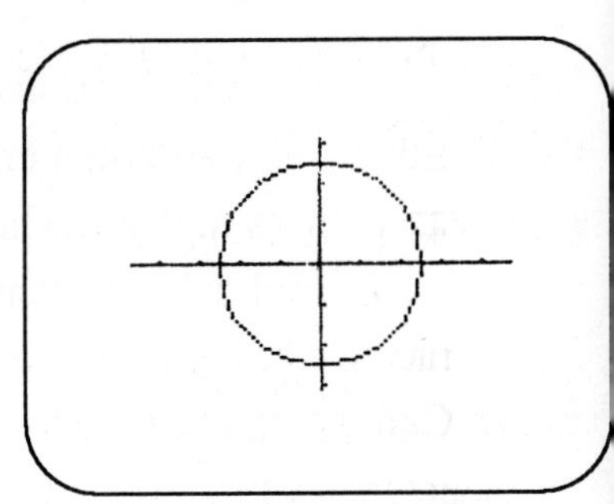

Figure 10.2

Section 10.5 Non-Linear Systems of Equations

A Review on Solving Systems of Equations Graphically

Solving systems of equations graphically is discussed in Chapter 3 of this manual. Setting the scale to 0.01 ensures the accuracy needed. To solve the system

$x^2 + y^2 = 16$ and $y = 2x^2 - 2,$

solve the first equation for y, which results in two equations

$y = \pm\sqrt{16 - x^2}.$

Enter these equations as $Y1$ and $Y2$. Enter $Y1$ by pressing

[MENU] [A:SCRN] [1:Y1] [CL] [√] **1 6** [−] [X/θ/T] [a^b] **2** [ENTER].

Enter $Y2$ by pressing

[CL] [(−)] [√] **1 6** [−] [X/θ/T] [a^b] **2** [ENTER].

Enter the system's second equation as $Y3$ by pressing

[CL] **2** [X/θ/T] [a^b] **2** [▶] [−] **2** [ENTER].

Set the range with a 0.2 window (–9.4, 9.4, 2, –6.2, 6.2, 2). Press [graph] to graph the three equations (Fig. 10.3).

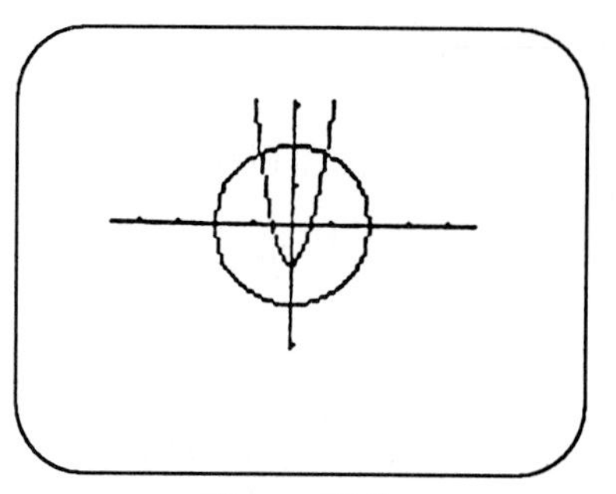
Figure 10.3

Note that *Y*3 does not intersect *Y*2. You can deselect *Y*2 by pressing [EQTN], moving to the *Y*2 equation line, and pressing

[◄] [ENTER].

Press [graph] to view the window with *Y*2 deselected (Fig. 10.4).

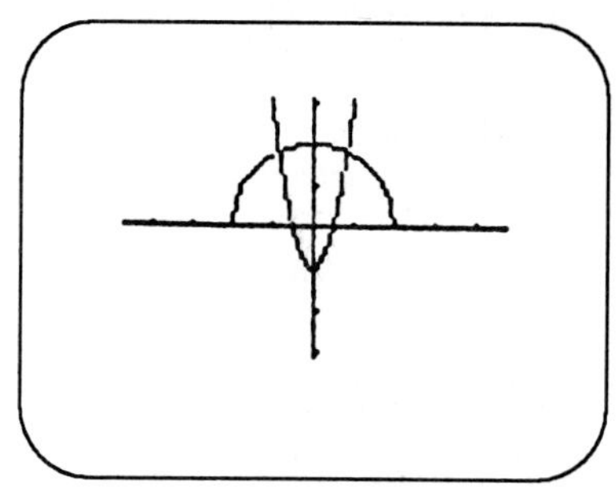
Figure 10.4

Now, trace to the right intersection and zoom in. Zoom in two more times. Change your *X*scl and *Y*scl to 0.01 to verify accuracy. Trace to the intersection and toggle between the curves to check that you in fact have the intersection. The values will change a little in the third or fourth decimal place (Fig. 10.5).

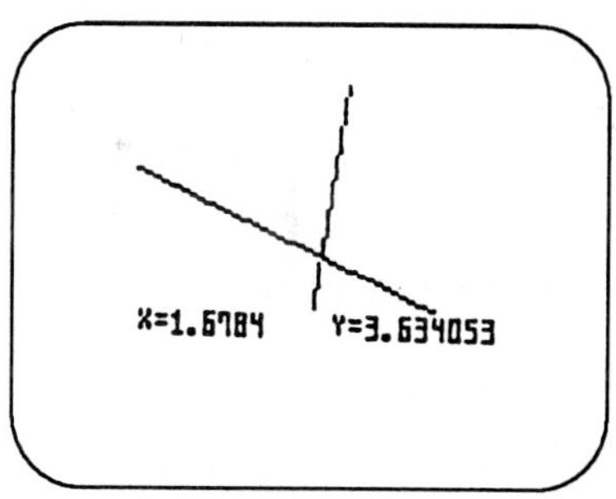

Figure 10.5

To find the left intersection point, return to the 0.2 window (–9.4, 9.4, 2, –6.2, 6.2, 2), by pressing

[RANGE] [MENU] [A:DEFLT] [ENTER]

and then multiplying each entry in the *X*Range and *Y*Range by 2. Press [graph] to view the graph. Now trace and zoom in repeatedly to find the left intersection. Remember to toggle between the curves to ensure you have the intersection (Fig. 10.6).

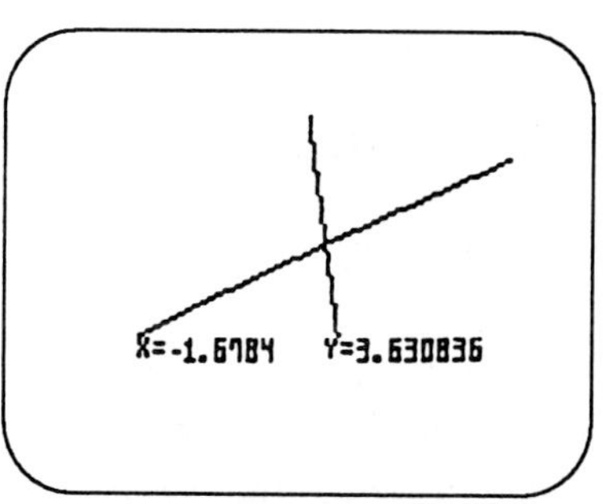

Figure 10.6

Chapter 11 Exponential and Logarithmic Functions

Section 11.3 Logarithmic Functions

The [log] and [ln] Keys

The logarithm keys on the grapher are [log], representing $\log_{10}$, and [ln], representing $\log_e$. To find log 125, press [computation] to enter Computation mode, and then press

[log] 1 2 5 [ENTER]

(Fig. 11.1). To find ln 12.3, press

[ln] 1 2 [.] 3 [ENTER]

(Fig. 11.1).

```
log 125=
     2.096910013
ln 12.3=
     2.509599262
```

Figure 11.1

Section 11.5 Solving Logarithmic Equations

Evaluating *Y*1 at 3 by Using the Notation *Y*1(3)

The Sharp grapher cannot evaluate *Y*1 at 3 using the notation *Y*1(3). Therefore, refer to Section 5.2 to review how to evaluate an expression on your grapher.

Revisiting the Grapher-Generated Table

Please remember that the Sharp grapher will not generate tables.

Chapter 12 Sequences and Series and the Binomial Theorem

Section 12.1 Arithmetic Sequences and Series

Generating Sequences Using a Command Sequence

The Sharp grapher does not have the capability of generating sequences.

Displaying a Graph of a Sequence

To display a graph of a sequence, press [graph key] to change the grapher to Graphing mode, and then change from Connected mode to Dot mode by pressing

[MENU] [C:PLOT1] [2:Dot].

All graphs of sequences should be displayed in a custom integer window with

$0 < x < 94$ and $0 < y < 62$.

Refer to Section 2.1 of this manual to review adjusting the viewing window.

For example, find the sixth term of the sequence

$23, 26, 29, \ldots, 23 + (n - 1)3, \ldots$

by graphing

$Y1 = 23 + (X - 1) \times 3$

by pressing

23 [+] [(] [X/θ/T] [−] **1** [)] [×] **3.**

Then set the range in the custom integer window (0, 94, 10, 0, 62, 10) described previously. Now, graph by pressing [graph key] and press [▶] to trace over to $X = 6$ and read the sequence value Y (Fig. 12.1). The sixth term of the sequence is 38.

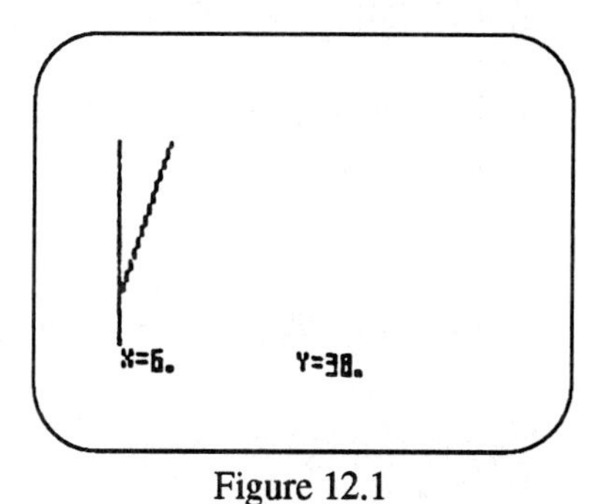

Figure 12.1

Section 12.2 Geometric Sequences and Series

Please refer to Section 12.1 of this manual to graph a geometric sequence, following the guidelines for graphing arithmetic sequences.

Section 12.3 Binomial Expansion

The n! Key

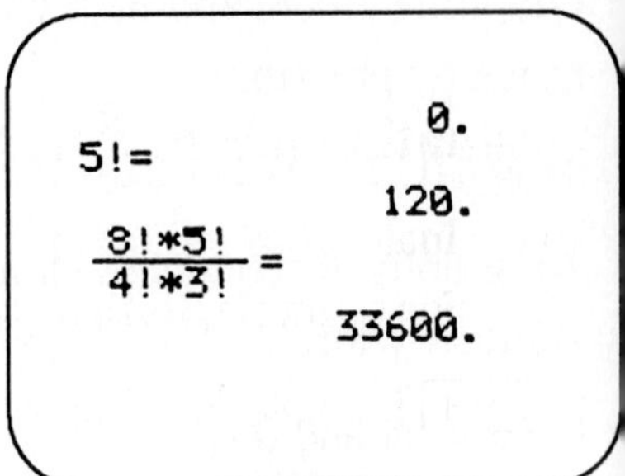

Figure 12.2

Press to enter Calculation mode. Access the factorial key n! by pressing

MATH [A:MATH] [5:n!].

To evaluate 5!, you would press

CL **5** MATH [5:n!] ENTER

(Fig. 12.2). Try evaluating

$$\frac{8! * 5!}{4! * 3!}$$

Evaluating a Binomial Coefficient on a Grapher

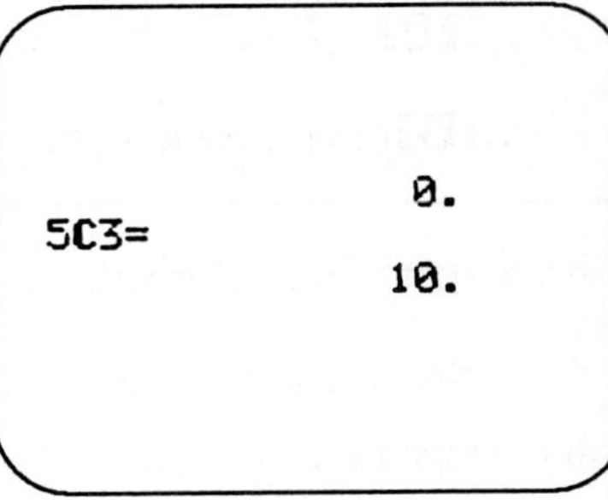

Figure 12.3

To evaluate a binomial coefficient $\binom{n}{r}$, use its equivalent notation ${}_nC_r$. For example, to find $\binom{5}{3} = {}_5C_3$ you will press

5 MATH [A:MATH] [6:nCr] **3** ENTER

(Fig. 12.3).

Section 12.4 Counting Principle, Permutations, and Combinations

Evaluating a Permutation

To evaluate a permutation ${}_nP_r$, such as ${}_{26}P_6$, you will press

2 6 MATH [A:MATH] [7:nPr] **6** ENTER

(Fig. 12.4).

Figure 12.4

Evaluating a Combination

Please refer to Section 12.3 of this manual for a discussion of the [6:nCr] key. This key is accessed through the Math menu.

(Fig. 12.2). Try evaluating

$$\frac{8! * 5!}{4! * 3!}$$

Evaluating a Binomial Coefficient on a Grapher

To evaluate a binomial coefficient $\binom{n}{r}$, use its equivalent notation ${}_nC_r$. For example, to find $\binom{5}{3} = {}_5C_3$ you will press

5 MATH [A:MATH] [6:nCr] **3** ENTER

(Fig. 12.3).

Section 12.4 Counting Principle, Permutations, and Combinations

Evaluating a Permutation

To evaluate a permutation ${}_nP_r$, such as ${}_{26}P_6$, you will press

2 6 MATH [A:MATH] [7:nPr] **6** ENTER

(Fig. 12.4).

Evaluating a Combination

Please refer to Section 12.3 of this manual for a discussion of the [6:nCr] key. This key is accessed through the Math menu.